Corrosion Inspection
and Monitoring

WILEY SERIES IN CORROSION

R.Winston Revie, Series Editor

Corrosion Inspection and Monitoring · Pierre R. Roberge

Corrosion Inspection and Monitoring

Pierre R. Roberge
Royal Military College of Canada
Ontario, Canada

WILEY-INTERSCIENCE
A John Wiley & Sons, Inc., Publication

Copyright © 2007 by John Wiley & Sons, Inc. All rights reserved.

Published by John Wiley & Sons, Inc., Hoboken, New Jersey.
Published simultaneously in Canada.

No part of this publication may be reproduced, stored in a retrieval system, or transmitted in any form or by any means, electronic, mechanical, photocopying, recording, scanning, or otherwise, except as permitted under Section 107 or 108 of the 1976 United States Copyright Act, without either the prior written permission of the Publisher, or authorization through payment of the appropriate per-copy fee to the Copyright Clearance Center, Inc., 222 Rosewood Drive, Danvers, MA 01923, (978) 750-8400, fax (978) 750-4470, or on the web at www.copyright.com. Requests to the Publisher for permission should be addressed to the Permissions Department, John Wiley & Sons, Inc., 111 River Street, Hoboken, NJ 07030, (201) 748-6011, fax (201) 748-6008, or online at http://www.wiley.com/go/permission.

Limit of Liability/Disclaimer of Warranty: While the publisher and author have used their best efforts in preparing this book, they make no representations or warranties with respect to the accuracy or completeness of the contents of this book and specifically disclaim any implied warranties of merchantability or fitness for a particular purpose. No warranty may be created or extended by sales representatives or written sales materials. The advice and strategies contained herein may not be suitable for your situation. You should consult with a professional where appropriate. Neither the publisher nor author shall be liable for any loss of profit or any other commercial damages, including but not limited to special, incidental, consequential, or other damages.

For general information on our other products and services or for technical support, please contact our Customer Care Department within the United States at (800) 762-2974, outside the United States at (317) 572-3993 or fax (317) 572-4002.

Wiley also publishes its books in a variety of electronic formats. Some content that appears in print may not be available in electronic formats. For more information about Wiley products, visit our web site at www.wiley.com.

Wiley Bicentennial Logo: Richard J. Pacifico

Library of Congress Cataloging-in-Publication Data:

Roberge, Pierre R.
  Corrosion inspection and monitoring / Pierre R. Roberge.
    p. cm.
  Includes index.
  ISBN: 978-0-471-74248-7 (cloth)
 1. Corrosion and anti-corrosives. 2. Corrosion and anti-corrosives—Testing. I. Title.
  TA462.R528 2007
  620.1′1223—dc22

                            2006023012

Printed in the United States of America.

10 9 8 7 6 5 4 3 2 1

Contents

Preface   vii

Chapter 1   Corrosion and Its Cost In a Modern World   1

Chapter 2   Corrosion Detectability   27

Chapter 3   Maintenance, Management, and Inspection Strategies   79

Chapter 4   Corrosion Monitoring   189

Chapter 5   Nondestructive Evaluation   317

Appendix A   SI Units Conversion Table   365

Appendix B   373

Index   377

Preface

Corrosion is a form of damage that has accompanied mankind since the very introduction of metals thousands of years ago. Corrosion is often insidious and hidden until striking at the worst moment of a system operation. While there are many ways to try and prevent such damage, the optimum control method relies on an early diagnosis of the problems. However, in many cases such a task is far from trivial. Consider, for example, the oil and gas systems deployed in the most remote areas, often at depths never exploited before, or the transmission pipelines traversing the harshest environments on the planet. The inspectability of these systems is very limited and extremely costly.

While there is an abundance of publications and reference documents on all aspects of corrosion science and engineering, the coverage on how the problems are prevented by inspection and monitoring is very limited. This is a poor reflection of the many advances in the tools and strategies that have been developed in many countries and organizations in recent years to improve the management of increasingly complex systems. After reviewing the principles of corrosion management as they have evolved in many industries since the introduction of risk-based thinking in the 1970's, the present book corrects this situation by bringing together descriptions of the most modern techniques developed to inspect and monitor corrosion susceptible systems.

In the first chapter the readers are introduced to the general nature of corrosion and to the complex factors that control such a universal foe. The impact of corrosion on the economy and safe operation of systems in various operational environments is also briefly introduced as a reminder that corrosion losses are a real impediment to economic growth.

The second chapter establishes the basis for assessing corrosion flaws by putting in perspective the relations between defects, faults, failures, and their consequences. The concepts of probability of failure (POF) and probability of detection (POD) are also introduced. The classic corrosion taxonomy detailing the forms of corrosion is then covered with a special focus on the detectability of the various types of corrosion damage encountered in practice.

Corrosion inspection and monitoring are maintenance tasks that should be designed to provide information for the general management and operation of systems. Chapter three reviews the maintenance strategies as they are evolving from corrective to predictive in a world increasingly focused on risk based assessments. The concepts of life cycle assessment and asset management are discussed in relation to inspection strategies and key performance indicators. The principles of risk based inspection and various risk assessment methodologies (HAZOP, FMECA, FTA, and ETA) are explained and illustrated with industrial examples. Some general guidance in carrying out failure analysis is also provided since such process can generate very valuable information for subsequent inspection tasks. The chapter concludes by describing how various industries and organizations have established very useful roadmaps for carrying out the most complex inspection and monitoring schedules.

The role of corrosion monitoring and considerations for establishing a corrosion monitoring program are discussed at the onset of chapter four. This introduction is then followed by discussions on generic aspects of probe design and selection, location of monitoring hardware, and other important points that need to be envisaged when embarking on a corrosion monitoring program. The main sections of chapter four then present a series of technical descriptions detailing important features of intrusive and non-intrusive methods that can a produce direct measurement of corrosion rates or that can provide an indirect way of monitoring corrosion by following a variable or a feature related to the corrosion degradation of a system. Chapter four also includes discussions on three additional areas of corrosion management: effects and monitoring of microbiologically induced corrosion (MIC), the monitoring of cathodic protection (CP) systems, and monitoring of atmospheric corrosion.

In many areas of modern engineering, non-destructive evaluation (NDE) techniques have provided valuable and often critical information for the safe operation of the most complex systems. Such usefulness has recently been greatly enhanced by the tremendous advances in computer and communication tools. Chapter five first provides a discussion on the purpose and other aspects specific to NDE: defect response variance, validation of inspection tools, and data representation. Following this general introduction, detailed descriptions of the main NDE inspection techniques and their many variants are presented: visual, ultrasonic, radiographic, electromagnetic, and thermographic inspection.

It is sincerely hoped that the readers will enjoy this timely reference text as much as I have enjoyed writing it.

PIERRE R. ROBERGE

Chapter 1

Corrosion and Its Cost In a Modern World

1.1. Corrosion, Why Bother?
1.2. Nature Of Corrosion
1.3. Corrosion Factors
    1.3.1. Environment Factor
        1.3.1.1. Local Cells
        1.3.1.2. Presence of Microbes
        1.3.1.3. Flow Effect
    1.3.2. Temperature Factor
    1.3.3. Stress Factor
    1.3.4. Material Factor
1.4. Strategic Impact and Cost of Corrosion Damage
References

1.1. CORROSION, WHY BOTHER?

In a modern business environment, successful enterprises cannot tolerate major corrosion failures, especially those involving personal injuries, fatalities, unscheduled shutdowns and environmental contamination. For this reason considerable efforts are generally expended in corrosion control at the design stage and in the operational phase.

Corrosion can lead to failures in plant infrastructure and machines that are usually costly to repair, costly in terms of lost or contaminated product, in terms of environmental damage, and possibly costly in terms of human safety. Decisions regarding the future integrity of a structure or its components depend on an accurate assessment of the conditions affecting its corrosion and rate of deterioration. With this information an informed decision can be made as to the type, cost and urgency of possible remedial measures.

Required levels of maintenance can vary greatly depending on the severity of the operating environments. While some of the infrastructure equipment might

---

Corrosion Inspection and Monitoring, by Pierre R. Roberge
Copyright © 2007 John Wiley & Sons, Inc.

only require regular repainting and occasional inspection of electrical and plumbing lines, some chemical processing plants, power generation plants, aircraft and marine equipment, are operated with extensive maintenance schedules.

Even the best of designs cannot be expected to anticipate all conditions that may arise during the life of a system. Corrosion inspection and monitoring are used to determine the condition of a system and to determine how well corrosion control and maintenance programs are performing. Traditional corrosion inspection practices typically require planned periodic shutdowns or service interruptions to allow the inspection process. These scheduled interruptions may be costly in terms of productivity losses, restart energy, equipment availability, and material costs. However, accidental interruptions or shutdowns are potentially much more disruptive and expensive.

1.2. NATURE OF CORROSION

The degradation of materials generally occurs via three well-known avenues: corrosion, fracture, and wear. Corrosion is traditionally related to chemical processes that break chemical bonds while fracture is related to mechanical processes that break bonds and wear to relative motion that break bonds. These are, to some extent, separate considerations, but they are also interconnected. Chemical environments accelerate fracture, chemical environments accelerate wear and vice versa as wear products produce deposits that accelerate corrosion, and fracture processes can permit one component to contaminate another (1).

Humans have most likely been trying to understand and control metallic corrosion for as long as they have been using objects made of metals. The most important periods of prerecorded history are named for the metals that were used for tools and weapons, for example, Iron Age, Bronze Age. However, these metals are unstable in ordinary aqueous environments.

The driving force that causes metals to corrode is a natural consequence of their temporary existence in metallic form. To reach this metallic state from their occurrence in Nature in the form of various chemical compounds (ores), it is necessary for them to the energy that is later returned through the corrosion process. The amount of energy required and stored varies between metals. It is relatively high for metals, such as magnesium, aluminum, and iron, and relatively low for metals, such as copper, silver, and gold. Table 1.1 lists a few metals in order of diminishing amounts of energy required to convert them from their oxides to metal.

A typical metallic cycle is illustrated by iron. The most common iron ore, hematite, is an oxide of iron (Fe_2O_3). The most common product of the corrosion of iron, rust, has a similar chemical composition. The energy required to convert iron ore to metallic iron is returned when the iron corrodes to form the original compound. Only the rate at which these energy changes occur may be different.

In contact with the environment, corrosion is a primary route by which metals deteriorate. Most metals corrode on contact with water or moisture, acids, bases, salts, oils, aggressive metal polishes, and other solid and liquid chemicals.

Table 1.1. Positions of Some Metals in the Order of Energy Required to Convert Their Oxides to Produce 1 kg of Metal

	Metal	Oxide	Energy, MJ kg^{-1}
Highest energy	Li	Li$_2$O	40.94
	Al	Al$_2$O$_3$	29.44
	Mg	MgO	23.52
	Ti	TiO$_2$	18.66
	Cr	Cr$_2$O$_3$	10.24
	Na	Na$_2$O	8.32
	Fe	Fe$_2$O$_3$	6.71
	Zn	ZnO	4.93
	K	K$_2$O	4.17
	Ni	NiO	3.65
	Cu	Cu$_2$O	1.18
	Pb	PbO	0.92
	Pt	PtO$_2$	0.44
	Ag	Ag$_2$O	0.06
Lowest energy	Au	Au$_2$O$_3$	−0.18

Metals will also corrode when exposed to gaseous materials, such as acid vapors, formaldehyde gas, ammonia gas, and sulfur-containing gases.

The fundamental nature of a corrosion process is usually electrochemical in nature, having the essential features of a battery. When metal atoms are exposed to an environment containing water molecules they can give up electrons, becoming themselves positively charged ions provided an electrical circuit can be completed.

Corrosion damage can be concentrated locally to form a pit or, sometimes, a crack, or it can extend across a wide area to produce general wastage. Localized corrosion that leads to pitting may provide sites for fatigue initiation and, additionally, corrosive agents, such as seawater, may lead to greatly enhanced growth of fatigue cracks. Pitting corrosion also occurs much faster in areas where microstructural changes have occurred due to welding operations.

1.3. CORROSION FACTORS

Considering the many complex forms and mechanisms of corrosion damage, the limitations of individual plant inspection and monitoring techniques are considerable. The large number of variables involved also implies that no single method can be expected to satisfy all possible conditions and environments. Destruction by corrosion takes many forms (see Chapter 2), and depends on the complex interaction of a multitude of factors described in the following sections.

1.3.1. Environment Factor

The environments to which materials are exposed dominate considerations in predicting and assuring their reliable performance, but environments are difficult to define and their broad and uncertain variability decreases their predictability. Environments of concern are local environments on the surface of a metal, and their characters are often quite different from bulk environments. The differences between the bulk and local chemistries are produced, for example, by heat transfer, evaporation, flow, and electrochemical cells. Environments may also vary greatly between systems.

1.3.1.1. Local Cells

Local cells can be produced by differences among small nearby areas on the metal surface. Local cells may result from differences in the metal or the environment. The differences may sometimes be simply in the thickness of a surface film at adjacent sites on a metallic surface. These differences in the environment are a driving force that can trigger the onset of a localized corrosion problem. Corrosion under insulation (CUI), for example, is a particularly severe form of localized corrosion that has been plaguing chemical process industries since the energy crisis of the 1970s forced plant designers to include much more insulation in their designs (2).

Intruding water is the key problem in CUI. Water entering an insulation material and diffusing inward will eventually reach a region of dryout at the hot pipe or equipment wall. Next to this dryout region is a zone in which the pores of the insulation are filled with a saturated salt solution. When a shutdown or process change occurs and the metal–wall temperature falls, the zone of saturated salt solution moves into the metal wall.

Upon reheating, the wall will temporarily be in contact with the saturated solution and stress-corrosion cracking (SCC) may begin. Corrosion may attack the jacketing, the insulation hardware, or the underlying piping or equipment. These drying–wetting cycles in CUI associated problems are strong accelerators of corrosion damage since they provoke the formation of an increasingly aggressive chemistry that may lead to the worst corrosion problems possible, (e.g., SCC), and premature catastrophic equipment failures.

Another important characteristic of aqueous or humid environments is the redox potential that is both a function of the chemistry of species present and the geometric conditions that can affect the potential and pH. In aqueous environments, this includes oxygen as it increases the redox potential or an oxygen scavenger, (e.g., erythorbic acid, sulfite, bisulfite, or hydrazine), as it decreases the potential. The electrochemical potential produces a powerful effect on chemical reactions and is the product not only of environmental chemistry, but of separated electrochemical cells, especially where the relative areas of anodes and cathodes are different.

1.3.1.2. Presence of Microbes

Microbes are present almost everywhere in soils, freshwater, seawater, and air. However, the mere detection of microorganisms in an environment does not necessarily indicate a corrosion problem. What is important is the number of microorganisms of the specifically corrosive types. In unaerated (anaerobic) soils, the corrosion attack is attributed to the influence of the sulfate-reducing bacteria (SRB). The mechanism is believed to involve both direct attack of the steel by hydrogen sulfide and cathodic depolarization aided by the presence of bacteria. Even in aerated (aerobic) soils, there are sufficiently large variations in aeration that the action of SRB cannot be neglected. For example, within active corrosion pits, the oxygen content becomes exceedingly low.

Microbiologically influenced corrosion (MIC) is responsible for the degradation of a wide range of materials. A useful representation of materials degradation by microbes has been made in the form of a pipe cross-section (Fig. 1.1) (3). Most metals and their alloys, (e.g. stainless steels, aluminum and copper alloys, polymers, ceramic materials, and concrete) can be attacked by microorganisms. The synergistic effect of different microbes and degradation mechanisms should be noted in Fig. 1.1.

The muddy bottom of any relatively stagnant body of water with a high biological oxygen demand often supports massive growth of SRB, and may waterlog soils. Any metallic installations buried or immersed in such environments can be expected to suffer badly from microbiological corrosion. The most serious economic problem is to pipelines, although sheet piles, ship hulls, and piers are frequently attacked. In some instances, cast iron pipes of 6.3 mm thickness have perforated within the first year of operation under such conditions, while perforation in 3 years is common.

1.3.1.3. Flow Effect

The destruction of a protective film on a metallic surface exposed to high flow rates can have a major impact on the acceleration of corrosion damage. Carbon steel pipe carrying water, for example, is usually protected by a film of rust that slows down the rate of mass transfer of dissolved oxygen to the pipe wall. The resulting corrosion rates are typically <1 mm year^{-1}. The removal of the film by flowing sand slurry has been shown to raise corrosion rates 10-fold to ~10 mm year^{-1} (4). Figure 1.2 illustrates the various states of an oxide surface film behavior as liquid velocity or surface shear stresses are increased (5, 6).

Change in the corrosion and erosion mechanisms associated with flow accelerated corrosion (FAC) are summarized in Figs. 1.3 and 1.4. In stagnant water (origin of the plot in Fig. 1.2), the corrosion rate is low and decreases parabolically with time due to the formation and growth of a corrosion protective film at

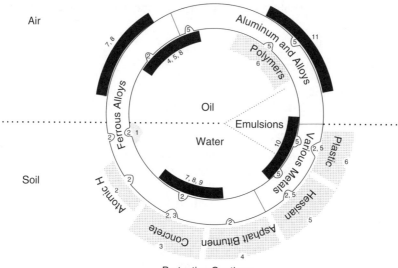

Fig. 1.1. Schematic illustration of the principal methods of microbial degradation of metallic alloys and protective coatings. 1. Tubercle leading to differential aeration corrosion cell and providing environment for "2". 2. Anaerobic sulfate reducing bacteria (SRB). 3. Sulfur-oxidizing bacteria, producing sulfates and sulfuric acid. 4. Hydrocarbon utilizers, breaking down aliphatic and bitumen coatings and allowing access of "2" to underlying metallic structure. 5. Various microbes producing organic acids as end-products of growth, attacking mainly non-ferrous metals–alloys and coatings. 6. Bacteria and molds breaking down polymers. 7. Algae-forming slimes on above-ground damp surfaces. 8. Slime-forming molds and bacteria (which may produce organic acids or utilize hydrocarbons), providing differential aeration cells and growth conditions for "2". 9. Mud on river bottoms, and so on providing matrix for heavy growth of microbes (including anaerobic conditions for "2"). 10. Sludge (inorganic debris, scale, corrosion products, etc.) providing matrix for heavy growth and differential aeration cells, and organic debris providing nutrients for growth. 11. Debris (mainly organic) on metal above ground, providing growth conditions for organic acid-producing microbes.

the surface (curve *a* in Fig. 1.3). At low flow velocities for which laminar and turbulent flow conditions coexist (parts A and B of Fig. 1.2), corrosion stems from a flow-accelerated process. The protective film that forms on the surface by corrosion is dissolved by the flowing water. It is generally accepted that the phenomenon is a steady-state process. It exhibits linear corrosion kinetics (curve *b* in Fig. 1.3), that is, the dissolved layer at the oxide–water interface is replaced by a new layer of the same thickness.

1.3.2. Temperature Factor

Metals as many other materials may be greatly affected by increasing the service temperature close to or above their stability limit. What is called high-temperature

1.3. Corrosion Factors   7

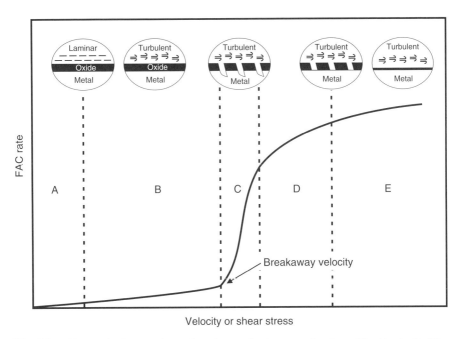

Fig. 1.2. Changes in the corrosion and erosion mechanisms as a function of liquid velocity (5).

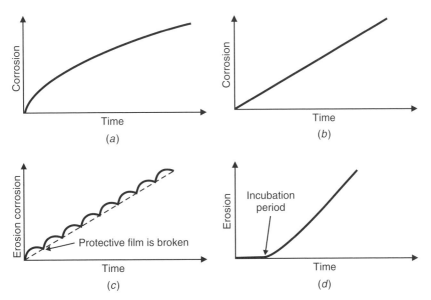

Fig. 1.3. Various time-dependent corrosion–erosion behaviors and processes: (*a*) corrosion follows a parabolic time law, (*b*) FAC follows a linear time law, (*c*) erosion and corrosion follows a quasilinear time law with repeated breaks in the protective surface film, and (*d*) erosion linear time dependency after an initial incubation period (5).

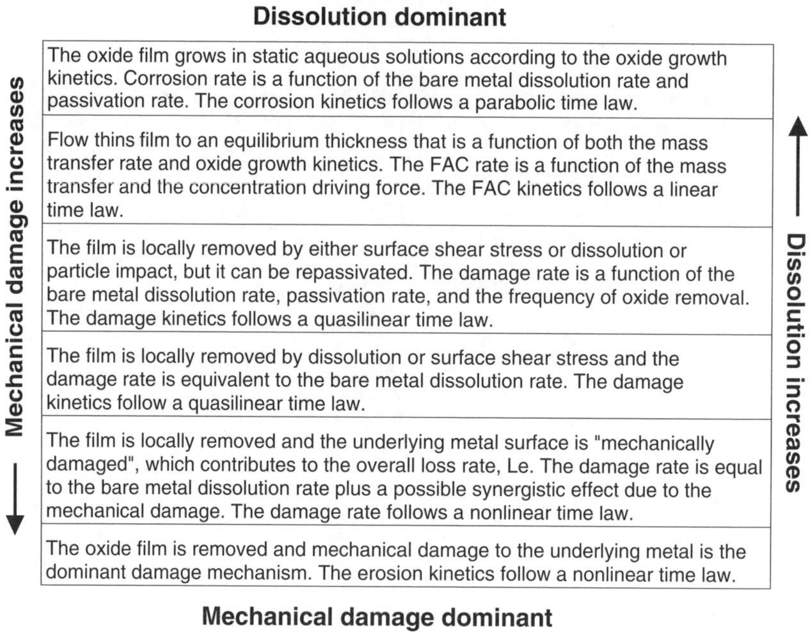

8 Chapter 1 Corrosion and Its Cost In a Modern World

Dissolution dominant

← Mechanical damage increases

The oxide film grows in static aqueous solutions according to the oxide growth kinetics. Corrosion rate is a function of the bare metal dissolution rate and passivation rate. The corrosion kinetics follows a parabolic time law.
Flow thins film to an equilibrium thickness that is a function of both the mass transfer rate and oxide growth kinetics. The FAC rate is a function of the mass transfer and the concentration driving force. The FAC kinetics follows a linear time law.
The film is locally removed by either surface shear stress or dissolution or particle impact, but it can be repassivated. The damage rate is a function of the bare metal dissolution rate, passivation rate, and the frequency of oxide removal. The damage kinetics follows a quasilinear time law.
The film is locally removed by dissolution or surface shear stress and the damage rate is equivalent to the bare metal dissolution rate. The damage kinetics follow a quasilinear time law.
The film is locally removed and the underlying metal surface is "mechanically damaged", which contributes to the overall loss rate, Le. The damage rate is equal to the bare metal dissolution rate plus a possible synergistic effect due to the mechanical damage. The damage rate follows a nonlinear time law.
The oxide film is removed and mechanical damage to the underlying metal is the dominant damage mechanism. The erosion kinetics follow a nonlinear time law.

Dissolution increases →

Mechanical damage dominant

Fig. 1.4. Summary of damage mechanisms experienced with FAC (5).

corrosion is a form of corrosion that does not require the presence of a liquid electrolyte. Sometimes, this type of damage is called "dry corrosion" or "scaling". This important and complex subject is a field of its own and is not considered here.

In aqueous environments, temperature affects reaction rates, surface temperatures, heat flux and associated surface concentrations, and temperature gradient chemical transfer. In most chemical reactions, an increase in temperature is accompanied by an increase in reaction rate. A rough rule-of-thumb suggests that the reaction rate doubles for each 10° Celsius (°C) rise in temperature. Although there are numerous exceptions to this "rule", it is important to take into consideration the influence of temperature when analyzing why materials fail, and in designing to prevent corrosion.

In chemical industries that handle very corrosive products, the choice of vessel materials depends on a multitude of factors that include service temperature. Figure 1.5 illustrates the influence of temperature and composition on the corrosion behavior of materials exposed to hydrochloric acid solutions. The iso-corrosion chart shown in Fig. 1.5 is drawn for a corrosion rate of 0.5 mm/year or less. The upper (high-temperature) boundary of each area represents a corrosion rate of 0.5 mm year^{-1}. The materials listed in each area are metals having a corrosion rate of 0.5 mm/year or less under the indicated conditions of temperature and acid concentration. The boiling point curve for atmospheric pressure

1.3. Corrosion Factors

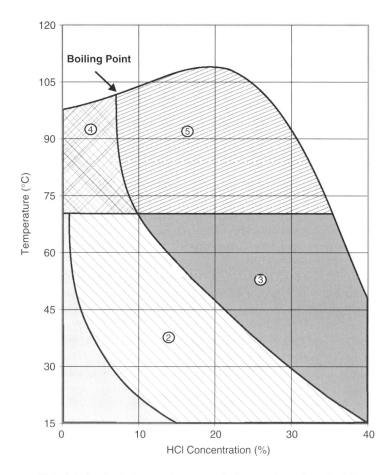

Materials in shaded zones have reported corrosion rates of < 0.5mm year^{-1}

Zone 1	Zone 2	Zone 3	Zone 4	Zone 5
20Cr 30Ni[a]	Zirconium	62Ni 28Mo[e]	66Ni 32Cu[b, f]	62Ni 28Mo[e]
66Ni 32Cu[b]	62Ni 32Cu	Molybdenum	62Ni 28Mo[e]	Platinum
62Ni 28Mo	Molybdenum	Platinum	Platinum	Silver
Copper[b]	Platinum	Silver	Silver	Tantalum
Nickel	Silicon bronze[b]	Tantalum		Zirconium
Platinum	Silicon cast iron[c]	Zirconium		
Silicon bronze[b]	Silver			
Silicon cast iron[c]	Tantalum			
Silver	Zirconium			
Tantalum				
Titanium[d]				
Tungsten				
Zirconium				
Tantalum				
Tungsten				

[a] < 2% at 25°C
[b] No air
[c] No FeCl$_3$
[d] <10% at 25°C
[e] No chlorine
[f] < 0.05% concentration

Fig. 1.5. Corrosion rate map of metallic materials in hydrochloric acid (7).

is shown because at temperatures above this curve, the metal is no longer in a liquid environment at 1-atm pressure (7).

An important corrosion contributor to most aqueous corrosion situations is the quantity of dissolved oxygen (DO) from ambient air that is present in the local environment shown as a function of temperature in Fig. 1.6. Oxygen is a cathodic depolarizer that reacts with and removes the hydrogen atoms produced by the cathodic branch of a corrosion reaction, thereby permitting corrosion attack to continue and progress at an accelerated rate. The combined effects of the DO level and temperature on the corrosion rate of low-carbon steel in tap water at different temperatures is illustrated in Fig. 1.7 (8).

The effect of oxygen and pH on the corrosion rate of steel at two temperatures is also shown in Fig. 1.8 (9). In a broad range of ~pH 5–9, the corrosion rate can be expressed simply in terms of the amount of DO present (e.g., $\mu m\ year^{-1}\ mL^{-1}$. DO per liter of water or ppm). At ~pH 4.5, acid corrosion is initiated, overwhelming the oxygen control. At ~pH 9.5 and above, the deposition of insoluble ferric hydroxide tends to stifle the corrosion attack.

Changing the temperature of a solution can influence the corrosion tendency. Many household hot water heater tanks, for example, have historically been made of galvanized steel. The zinc coating on the mild steel base offered a certain amount of cathodic protection to the underlying steel, and the service life (usually judged by how long it took to produce red water, i.e., rusty water) was considered adequate. Water tanks seldom were operated >60°C (10).

With the development of automatic dishwashers and automatic laundry equipment, the average water temperature was increased so that temperatures of ~80°C are not unusual in household hot water tanks. Coinciding with the widespread use of automatic dishwashers and laundry equipment was a sudden upsurge of

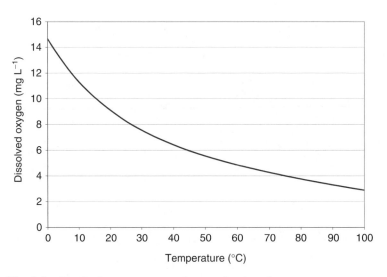

Fig. 1.6. Dissolved oxygen concentration as a function of temperature.

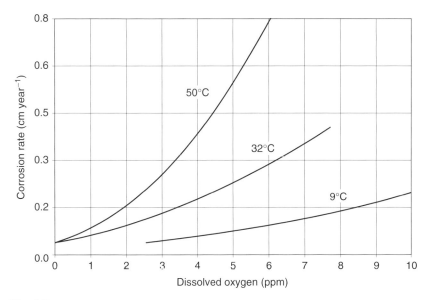

Fig. 1.7. Effect of oxygen concentration on the corrosion of low-carbon steel in tap water at different temperatures.

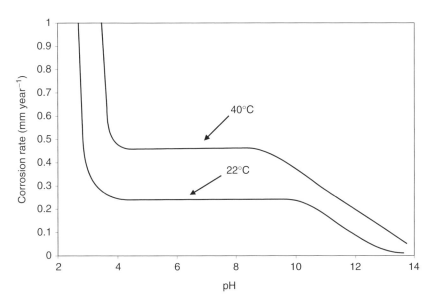

Fig. 1.8. Corrosion of steel in water containing 5 ppm of DO at two different temperatures as a function of the water pH (9).

complaints of short-life of galvanized steel water heater tanks. Electrochemical measurements showed that in many cases, iron was anodic to zinc at 77°C, whereas zinc was anodic to iron at temperatures <60°C. This explained why zinc offered no cathodic protection at 77°C, and why red water and premature perforation of galvanized water tanks occurred so readily at higher temperatures. The problem was partly solved by using either magnesium sacrificial anodes, protective coatings, or by the replacement of galvanized steel with new alloys.

1.3.3. Stress Factor

Environmental cracking refers to corrosion cracking caused by a combination of conditions that can specifically result in one of three forms of corrosion damage, that is, SCC, corrosion fatigue, and hydrogen embrittlement. These are described in more details in Chapter 2. Stresses that cause environmental cracking arise from residual cold work, welding, grinding, thermal treatment, or may be externally applied during service and must be tensile (as opposed to compressive) to be damaging:

- Stress variables
    - Mean stress
    - Maximum stress
    - Minimum stress
    - Constant load/constant strain
    - Strain rate
    - Plane stress/plane strain
    - Modes I, II, or III
    - Biaxial
    - Cyclic frequency
    - Wave shape
- Stress origin
    - Intentional
    - Applied
        - Quenching
        - Thermal cycling
        - Thermal expansion
        - Vibration
        - Rotation
        - Bolting
        - Dead load
        - Pressure
    - Residual
        - Shearing, punching, cutting
        - Bending, crimping, riveting
        - Welding

- Machining
- Grinding
  ○ Products of corrosion reaction

Stress cells can exist in a single piece of metal where a portion of the metal's microstructure possesses more stored strain energy than the rest of the metal. Metal atoms are at their lowest strain energy state when situated in a regular crystal array. Applied stresses include those due to cyclic stressing including cyclic frequencies and wave shapes. Less well defined, but often the critical contribution to modes of corrosion, such as SCC, are residual stresses. Whereas, the applied stresses are usually required to be something less than half the yield stress or lower, residual stresses are usually in the range of the yield stress (1). Quantifying such residual stresses is often omitted in design with the erroneous conclusion that such stresses are irrelevant to design and performance.

The stresses produced by an accumulation of corrosion products are typically formed in restricted geometries where the specific volume of the corrosion product is greater than that of the metal that is corroded. Such stresses can cause cracks to initiate and grow. Stresses from expanding corrosion products can readily cause adjacent metals to flow plastically, as occurs in nuclear steam generators in a process called "denting". Denting results from the corrosion of the carbon steel support plates and the buildup of corrosion product in the crevices between tubes and the tube support plates. This process is called "denting" since, when seen from the inside of the tubes, these deformations seem to produce dents at the tubesheet locations. Similar stresses from the build up of steel corrosion products cause the degradation of reinforced concrete (Fig. 1.9). An equivalent expansion

Fig. 1.9. Growth of the corrosion oxide products can lead to cracking and spalling of the concrete cover.

ratio of 3.0–3.2 has been measured due to the formation of corrosion products on steel bars embedded in concrete (11).

"Pack rust"[1] is another example of the tremendous forces created by expanding steel corrosion products. Figure 1.10 illustrates the effect of pack rust that has developed on an important steel bridge under repair. The force of expansion in this particular example was sufficient to break three of the bridge rivets shown in Figure 1.10.This specific type of localized corrosion is known to be a serious derating factor when the load bearing capacity of a bridge or of any other infrastructure component is evaluated during inspection.

In the following example, the deformation due to the corrosion of aluminum in lap joints of commercial airlines is accompanied by a bulging (pillowing) between rivets, due to the increased volume of the corrosion products over the original material. This problem was said to be the primary cause of the Aloha incident in which a 19-year old Boeing 737, operated by Aloha airlines, lost a major portion of the upper fuselage (Fig. 1.11) in full flight at 24,000 f (12). The

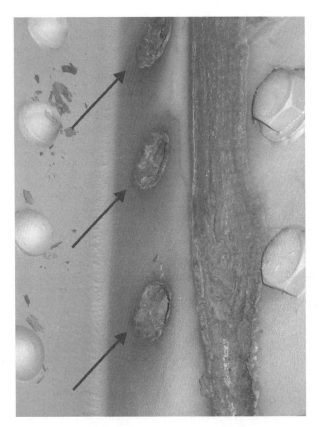

Fig. 1.10 The effect of "pack rust" that has developed on an important steel bridge under repair. (Courtesy Wayne Senick Termarust Technologies www.termarust.com.)

---

[1]The expression "pack rust" is often used in the context of bridge inspection to describe built-up members of steel bridges that are showing signs of rust packing between steel plates.

Fig. 1.11. Photograph of the Boeing 737 operated by Aloha airlines that lost a major portion of the upper fuselage in 1988.

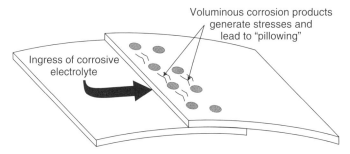

Fig. 1.12. Schematic description of the "pillowing" of lap splices.

"pillowing" process in which the faying surfaces are forced apart is schematically illustrated in Fig. 1.12.

The prevalent corrosion product identified in corroded fuselage joints is hydrated alumina, $Al(OH)_3$, with a particularly high volume expansion relative to aluminum as shown in Fig. 1.13 (13). This build-up of voluminous corrosion products can lead to an undesirable increase in stress levels near critical fastener holes (Fig. 1.14) and subsequent fracture due to the high tensile stresses resulting from the "pillowing".

1.3.4. Material Factor

In environments that can be reasonably well defined, unexpected corrosion may result from lack of definition of the material itself. Defining a material by its specification is often, and especially for new designs, quite inadequate. A thorough

16 Chapter 1 Corrosion and Its Cost In a Modern World

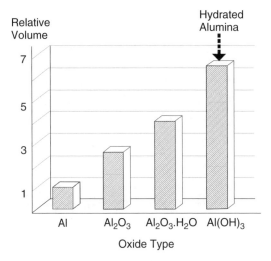

Fig. 1.13. Relative volume of aluminum corrosion products.

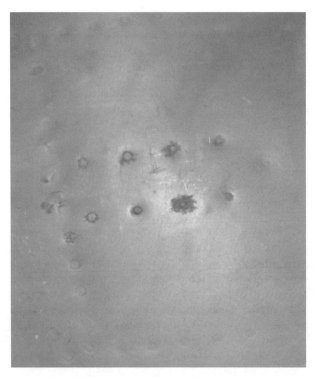

Fig. 1.14 Advanced stage on the belly of a Boeing 737, where the corrosion products have expanded to the point where a number of rivets have actually popped their heads. (Courtesy Mike Dahlager Pacific Corrosion Control Corp.)

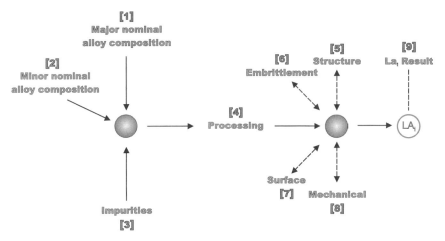

Fig. 1.15. Analysis sequence for determining materials at a location for analysis (LA) matrix.

material definition process has been proposed to overcome this problem. A brief explanation of the individual elements of the process illustrated in Fig. 1.15 follows (1):

1. "Major Nominal Alloy Composition" includes those species added in relatively large concentrations, such as the iron, chromium, nickel, and molybdenum in S31600 stainless steel.
2. "Minor Nominal Alloy Composition" includes species, such as sulfur, nitrogen, phosphorous, silicon, and manganese usually found in steels or the oxygen and nitrogen usually found in titanium.
3. "Impurities" include species, such as arsenic, selenium, and similar elements that contribute to temper brittleness or to excess oxygen and nitrogen picked up during welding of titanium.
4. "Processing" includes fabrication to shape with associated times and temperatures of hot deformation and heat treatments, cold deformation to size, welding, machining, and grinding.
5. "Structure" includes single and multiple phases, anisotropy in the grain structure, second phases, and grain size.
6. "Embrittlement" includes grain boundary composition and the formation of phases that may embrittle, such as sigma phases in iron–chromium base alloys.
7. "Surface" includes surface machining, surface stresses that result from thermal gradients, grinding, shot peening, and impurities adsorbed at the surface.
8. "Mechanical" includes the mechanical properties of tensile strength, ductility and toughness.

18 Chapter 1 Corrosion and Its Cost In a Modern World

The major nominal composition is the backbone of an alloy and controls its most obvious functions. Figure 1.16 illustrates the behavior of different compositions of a single-phase copper–nickel alloy system when exposed to seawater (14). The overall region where maximum pitting and biofouling occur is evidence of the passive state of the alloy. The electronic d band of the alloy, as indicated by physical data, is filled at 60 at % copper or unfilled at 40 at % nickel and above. Predicted minimum nickel content for passivity is at 40 at % or 38 wt %.

However, a less obvious, but most important, consideration in defining materials is defining the compositions of their grain boundaries, since many degradation reactions either follow grain boundaries or are initiated at grain boundaries. In addition to corrosion reactions that follow compositionally altered grain boundaries, embrittlement processes often follow the same compositionally altered boundaries, producing, for example, the phenomenon of "temper embrittlement".

A good example of such alteration is the depletion of chromium associated with the formation of chromium carbides at grain boundaries in stainless steels when they are exposed to a particular range of temperatures that produces "sensitization", for example, welding, because the resulting low chromium concentration adjacent to the grain boundaries sustains rapid corrosion in aggressive environments. The deficient alloy in this area can then be attacked so severely in certain environments that the entire grain is surrounded and will fall from the structure (Fig. 1.17). Welding of the unstabilized stainless steels can be a major source of this problem (Fig. 1.18).

The austenitic stainless steels may be "sensitized" to intergranular attack by heating in the temperature range 400–900°C (Fig. 1.19). A short time (minutes)

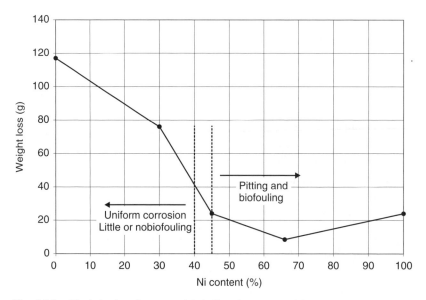

Fig. 1.16. The behavior of copper–nickel alloys in seawater (14).

1.3. Corrosion Factors   19

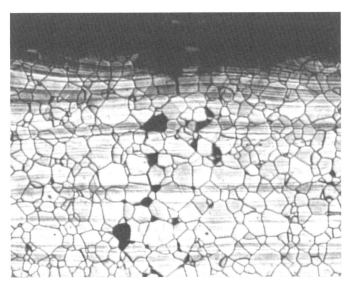

Fig. 1.17 Microstructure typical of sensitized austenitic stainless steel (X67). This particular problem occurred on Type 316H stainless steel after 7 years of service. (Courtesy of Corrosion Testing Laboratories, Newark, Delaware.)

Fig. 1.18 End fitting of a welded support rod (S30300) after service in seawater. This particular stainless steel is not recommended for welding. (Courtesy Kingston Technical Software.)

>750°C will produce sensitization and longer times (hours) are required to produce sensitization at lower temperatures. The commonly accepted mechanism attributes intergranular corrosion to a difference in potential between a chromium-depleted zone (at the grain boundaries) and the chromium-rich central zones of the grains (Fig. 1.20).

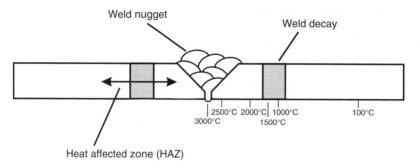

Fig. 1.19. Weld decay zone as a function of the welding temperature of stainless steel.

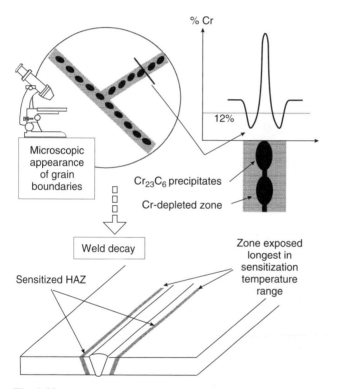

Fig. 1.20. Sensitization of stainless steel in the heat adjacent zone.

Degradation reactions are also affected by the presence and distribution of second phases whether at grain boundaries or distributed through the grains. Such second phases are often loci of pitting or, when lined up, may provide preferential paths for chemical degradation. Second phases also produce certain embrittlement, such as the formation of sigma phase in Fe–Cr base alloys.

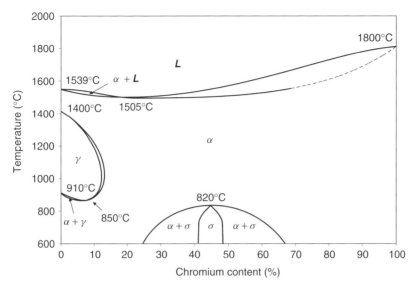

Fig. 1.21. Iron–chromium phase diagram.

Figure 1.21 shows a phase diagram that is extremely important to stainless steel technology, the iron–chromium system. Although this diagram is for the Fe–Cr binary system, the influence of a third important alloying element, carbon, will cause an enlargement of the gamma (γ, face-centered cubic, fcc) region with increasing carbon content. The sigma (σ) phase is also noted in this diagram. This phase is important because its formation usually results in a reduction in the mechanical properties and corrosion resistance of the alloy.

Many wrought aluminum alloy products leave highly directional grain structures. Figure 1.22 shows the three directional longitudinal (principal working direction), long transverse, and short transverse grain structure typically present in rolled plate (7). Almost all forms of corrosion, even pitting, are affected to some degree by this grain directionality. However, highly localized forms of corrosion, such as exfoliation and SCC that proceed along grain boundaries, are highly affected by grain structure. Long, wide, and very thin pancake-shaped grains are virtually a prerequisite for a high degree of susceptibility to exfoliation. Such products are highly anisotropic with respect to resistance to SCC (Fig. 1.23).

Resistance, which is measured by magnitude of tensile stress required to cause cracking, is highest for aluminum alloys when the stress is applied in the longitudinal direction, lowest in the short-transverse direction, and intermediate in other directions. These differences are most noticeable in the more susceptible tempers, but are usually much lower in tempers produced by extended precipitation treatments, such as T6 and T8 tempers for 2xxx alloys and T73, T736, and T76 tempers for 7xxx alloys (15).

22 Chapter 1 Corrosion and Its Cost In a Modern World

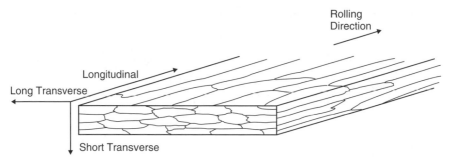

Fig. 1.22. Schematic representation of the three-dimensional (3D) grain structure typically present in rolled aluminum plates.

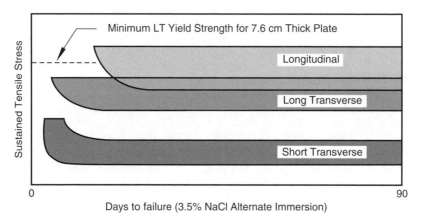

Fig. 1.23. SCC of alloy 7075–7651 plate. Shaded bands indicate combinations of stress and time known to produce SCC in specimens intermittently immersed in 3.5% NaCl solution. Point A is minimum yield strength in the long-transverse direction for a 75-mm thick plate (15).

During the design and prototyping processes, it is not uncommon for materials to be changed for reasons of performance, compatibility, or cost. When such changes occur, the materials should be reevaluated according to the steps defined in this section.

1.4. STRATEGIC IMPACT AND COST OF CORROSION DAMAGE

It is the belief of many that corrosion is an inevitable foe that should be accepted as an inevitable process. Actually, something can and should be done to prolong the life of metallic structures and components exposed to the environments. As products and manufacturing processes have become more complex and the penalties of failures from corrosion, including safety hazards and interruptions in plant operations, have become more costly and more specifically recognized, the attention that is being given to the control and prevention of corrosion has increased.

Since the first significant report by Uhlig in 1949 that the cost of corrosion to nations is indeed great (16), the conclusion of all subsequent studies has been that corrosion represents a constant charge to a nation's gross national product (GNP). The annual cost of corrosion to the United States was estimated in Uhlig's report to be $5.5 billion or 2.1% of the 1949 GNP. This study attempted to measure the total costs associated to corroding components by summing up the cost for both the owner and operator (direct cost) and for the users (indirect cost).

Corrosion costs studies of various forms and importance have since then been undertaken by several countries including, the United States, United Kingdom, Japan, Australia, Kuwait, Germany, Finland, Sweden, India, and China (17). A common finding of these studies has been that the annual corrosion costs range from ∼1 to 5% of the GNP of each nation. Several studies separated the total corrosion costs into two parts:

1. The portion of the total corrosion cost that could be avoided if better corrosion control practices were used.
2. Costs where savings required new and advanced technology (currently unavoidable costs).

Estimates of avoidable corrosion costs in these studies have varied widely with a range from 10 to 40% of the total cost. Most studies have categorized corrosion costs according to industrial sectors or to types of corrosion control products and services. All studies have focused on direct costs even if it has been estimated that indirect costs due to corrosion damage were often significantly greater than direct costs. Indirect costs have been typically excluded from these studies simply because they are more difficult to estimate.

Potential savings and recommendations in terms of ways to realize savings from corrosion damage were included in most of the reports as formal results or as informal directions and discussion. Two of the most important and common findings were

1. Better dissemination of the existing information through education and training, technical advisory and consulting services, and research and development activities.
2. The opportunity for large savings through more cost-effective use of currently available means to reduce corrosion.

The most recent study resulted from discussions between NACE International representatives, members of the U.S. Congress, and the Department of Transportation (DOT). An amendment for the cost of corrosion was included in the Transportation Equity Act for the twenty-first century, which was passed by the U.S. legislature in 1998. The amendment requested that a study be conducted in conjunction with an interdisciplinary team of experts from the fields of metallurgy, chemistry, economics, and others, as appropriate.

24　Chapter 1　Corrosion and Its Cost In a Modern World

Two different approaches were taken in the ensuing study to estimate the cost of corrosion. The first approach followed a method where the cost was determined by summing the costs for corrosion control methods and contract services. The costs of materials were obtained from various sources, such as the U.S. Department of Commerce Census Bureau, existing industrial surveys, trade organizations, industry groups, and individual companies. Data on corrosion control services, such as engineering services, research and testing, and education and training, were obtained primarily from trade organizations, educational institutions, and individual experts. These services included only contract services and not service personnel within the owner–operator companies.

The second approach followed a method where the cost of corrosion was first determined for specific industry sectors, and then extrapolated to calculate a national total corrosion cost. Data collection for the sector-specific analyses differed significantly from sector to sector, depending on the availability of data and the form in which data were available. In order to determine the annual corrosion costs for the reference year of 1998, data were obtained for various years in the surrounding decade, but mainly for the years 1996–1999.

The total cost due to the impact of corrosion for the individual economic sectors was \$137.9 billion year^{-1} (Table 1.2). A breakdown of these costs by individual sectors is shown in Fig. 1.24. Since not all economic sectors were examined, the sum of the estimated costs for the analyzed sectors did not represent the total cost of corrosion for the entire U.S. economy.

By estimating the percentage of U.S. GNP for the sectors for which corrosion costs were determined and by extrapolating the figures to the entire U.S. economy, a total cost of corrosion of \$276 billion was estimated. This value shows that the impact of corrosion is $\sim$3.1% of United States' GNP. This cost is considered to be a conservative estimate since only well-documented costs were used in the

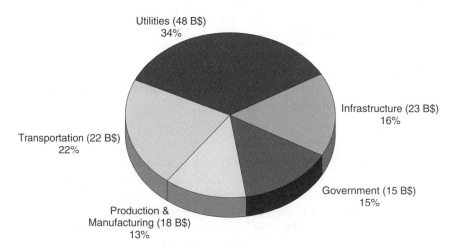

Fig. 1.24. Corrosion costs breakdown across industrial sectors.

Table 1.2. Summary of Estimated Direct Cost of Corrosion for Industry Sectors Analyzed in the 2001 Study

Category	Industry sectors	Appendix	Estimated direct cost of corrosion per sector	
			$ x billion	percent
Infrastructure (16.4% of total)	Highway bridges	D	8.3	37
	Gas and liquid transmission pipelines	E	7.0	27
	Waterways and ports	F	0.3	1
	Hazardous materials storage	G	7.0	31
	Airports	H	**	**
	Railroads	I	**	**
	Subtotal		$22.6	100%
Utilities (34.7% of total)	Gas distribution	J	5.0	10
	Drinking water and sewer systems	K	36.0	75
	Electrical utilities	L	6.9	14
	Telecommunications	M	**	**
	Subtotal		$47.9	100%
Transportation (21.5% of total)	Motor vehicles	N	23.4	79
	Ships	O	2.7	9
	Aircraft	P	2.2	7
	Railroad cars	Q	0.5	2
	Hazardous materials transport	R	0.9	3
	Subtotal		$29.7	100%
Production and Manufacturing (12.8% of total)	Oil and gas exploration and production	S	1.4	8
	Mining	T	0.1	1
	Petroleum refining	U	3.7	21
	Chemical, petrochemical, and pharmaceutical	V	1.7	10
	Pulp and paper	W	6.0	34
	Agricultural	X	1.1	6
	Food processing	Y	2.1	12
	Electronics	Z	**	**
	Home appliances	AA	1.5	9
	Subtotal		$17.6	100%
Government (14.6% of total)	Defense	BB	20.0	99.5
	Nuclear waste storage	CC	0.1	0.5
	Subtotal		$20.1	100%
	Total		$137.9	

study. The indirect cost of corrosion was conservatively estimated to be equal to the direct cost, giving a total direct plus indirect cost of $552 billion or 6% of the GNP.

REFERENCES

1. STAEHLE RW. Lifetime Prediction of Materials in Environments. In: Revie RW, ed. *Uhlig's Corrosion Handbook*. New York: Wiley-Interscience, 2000; pp. 27–84.
2. LISS VM. Preventing Corrosion under Insulation. *Chemical Engineering* 1987; 97–100.
3. HILL EC. *Microbial Aspects of Metallurgy*. New York: American Elsevier, 1970.
4. POSTLETHWAITE J, DOBBIN MH, and BERGEVIN K. The Role of Oxygen Mass Transfer in the Erosion-Corrosion of Slurry Pipelines. *Corrosion* 1986; **42**: 514–521.
5. CHEXAL B, HOROWITZ J, DOOLEY B, MILLETT P, WOOD C, and JONES R. *Flow-Accelerated Corrosion in Power Plants- Revision 1. EPRI TR-106611-R1*. 1998. Palo Alto, CA, Electric Power Research Institute.
6. ROBERGE PR. *Corrosion Testing Made Easy: Erosion-Corrosion Testing*. Houston, TX: NACE International, 2004.
7. ROBERGE PR. *Handbook of Corrosion Engineering*. New York: McGraw-Hill, 2000.
8. BERRY WE. Water Corrosion. In: Van Delinder LS and Brasunas Ad, eds. *Corrosion Basics*. Houston, TX: NACE International, 1984; 149–176.
9. UHLIG HH. Iron and Steel. In: Uhlig HH, ed. *The Corrosion Handbook*. New York: John Wiley & Sons, 1948; 125–143.
10. ROBERGE PR. *Corrosion Basics—An Introduction*. 2nd edn. Houston, TX: NACE International, 2005.
11. SUDA K, MISRA S, and MOTOHASHI K. Corrosion Products of Reinforcing Bars Embedded in Concrete. *Corrosion Science* 1993; **35**: 1543–1549.
12. MILLER D. Corrosion control on aging aircraft: What is being done ? *Materials Performance* 1990; **29**: 10–11.
13. KOMOROWSKI JP, KRISHNAKUMAR S, GOULD RW, BELLINGER NC, KARPALA F, and HAGENIERS OL. Double pass Retroreflection for Corrosion Detection in Aircraft Structures. *Materials Evaluation* 1996; **54**: 80–86.
14. UHLIG HH and MEARS RB. Passivity. In: Uhlig HH, ed. *The Corrosion Handbook*. New York: John Wiley & Sons, 1948; 20–33.
15. HOLLINGSWORTH EH and HUNSICKER HY. Corrosion of Aluminum and Aluminum Alloys. In: *Metals Handbook: Corrosion Vol. 13*. Metals Park, OH: American Society for Metals, 1987; pp. 583–609.
16. UHLIG HH. The Cost of Corrosion in the United States. *Chemical and Engineering News* 1949; **27**: 2764.
17. KOCH GH, BRONGERS MPH, THOMPSON NG, VIRMANI YP, and PAYER JH Corrosion Costs and Preventive Strategies in the United States. FHWA-RD-01-156. 2001. Springfield, VA, National Technical Information Service.

Chapter 2

Corrosion Detectability

2.1. Defects And Failures
    2.1.1. What Is a Defect?
    2.1.2. What Is a Fault?
    2.1.3. What Is a Failure?
        2.1.3.1. Functional Failure
        2.1.3.2. Potential Failure
    2.1.4. What Are the Consequences of a Failure?
        2.1.4.1. Safety Consequences
        2.1.4.2. Operational Consequences
        2.1.4.3. Nonoperational Consequences
        2.1.4.4. Hidden Failure Consequences
    2.1.5. Corrosion Failure Examples
        2.1.5.1. Bhopal Accident
        2.1.5.2. Carlsbad Pipeline Explosion
        2.1.5.3. Guadalajara Sewer Explosion
        2.1.5.4. El Al Boeing 747 Crash
        2.1.5.5. Nuclear Reactor With a Hole in the Head
        2.1.5.6. Piping Rupture Caused by Flow Accelerated Corrosion
        2.1.5.7. Sinking Ships
        2.1.5.8. SCC of Chemical Reactor: The Flixborough Explosion
        2.1.5.9. Swimming Pool Roof Collapse
    2.1.6. Age-Reliability Characteristics
    2.1.7. Probability of Failure
    2.1.8. Probability of Detection
2.2. Forms of Corrosion
    2.2.1. Uniform Attack
    2.2.2. Pitting Corrosion
    2.2.3. Crevice Corrosion
    2.2.4. Galvanic Corrosion
    2.2.5. Flow Influenced Corrosion
    2.2.6. Fretting Corrosion
    2.2.7. Intergranular Corrosion
    2.2.8. Dealloying
    2.2.9. Environmental Cracking
        2.2.9.1. Stress Corrosion Cracking

Corrosion Inspection and Monitoring, by Pierre R. Roberge
Copyright © 2007 John Wiley & Sons, Inc.

2.2.9.2. Corrosion Fatigue
2.2.9.3. Hydrogen-Induced Cracking (HIC) and Hydrogen Embrittlement (HE)
2.2.9.3.1. Loss of Ductility and Strength
2.2.9.3.2. Hydrogen Blistering
2.2.9.3.3. High-Temperature Effects
References

2.1. DEFECTS AND FAILURES

Corrosion damage, defects, and failures can have all sorts of consequences on the operation of a system. Whereas Chapter 3 introduces management methodologies that have been created to reduce these consequences to manageable levels, this chapter discusses the possible types of corrosion damage that exist, their initiation and propagation rates, and their detectability.

A universal representation describing the interactions between defects, faults, and failures of a system is shown in Fig. 2.1. The arrows in this figure imply that quantifiable relations possibly exist between a defect, a fault, and a failure.

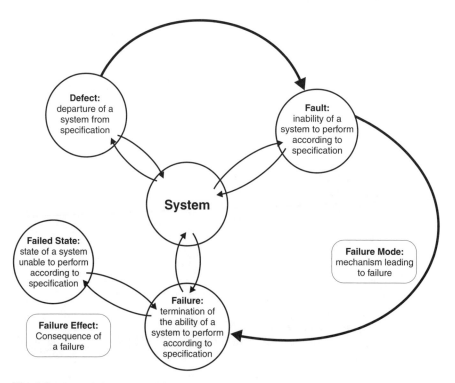

Fig. 2.1. Interrelation among defects, failures, and faults.

2.1.1. What Is a Defect?

In materials science, a defect is any microstructural feature representing a disruption in the perfect periodic arrangement of atoms in a crystalline material. These fundamental defects and their distribution in a given material can have a great importance on the overall properties of that material. While such defects do not constitute flaws in the normal sense of the word they can nonetheless serve as anchors for the initiation of actual faults and subsequent failures. There are four fundamental defect types:

1. **Point defects:** Point defects or sites are vacancies that are usually occupied by an atom, but are presently unoccupied. If a neighboring atom moves to occupy the vacant site, the vacancy moves in the opposite direction to the site that used to be occupied by the moving atom. The stability of the surrounding crystal structure guarantees that the neighboring atoms will not simply collapse around the vacancy. In some materials, neighboring atoms actually move away from a vacancy, because they can better form bonds with atoms in the other directions.

2. **Line defects:** Line defects are dislocations around which some of the atoms of the crystal lattice are misaligned. There are two basic types of dislocations: the edge dislocation and the screw dislocation. Dislocations are caused by the termination of a plane of atoms in the middle of a crystal. In such a case, the surrounding planes are not straight, but instead bend around the edge of the terminating plane so that the crystal structure is perfectly ordered on either side.

3. **Planar and surface defects:** An important planar defect example is the grain boundaries that occur where the crystallographic direction of the lattice abruptly changes. This commonly occurs when two crystals begin growing separately and then meet. Many types of corrosion discussed later are directly related to the nature and geometry of grain boundary structures.

4. **Bulk defects:** An example of bulk defects are voids where there are simply no structural atoms. Another example are impurities that can cluster together to form small regions of a different phase, (e.g., precipitates). Impurities and precipitates also play an important role in the corrosion resistance of metallic materials.

2.1.2. What Is a Fault?

The growth of a defect into what becomes a fault or a faulty component really depends on a multitude of factors that include predominantly the type of corrosion that is indeed progressing, as will be described later in the chapter. In the fault-tree analysis context, the fault event of a component is defined as a state transition from the normal state to a faulty state of that component. These state transitions

30 Chapter 2 Corrosion Detectability

are irreversible, which means that a faulty state does not return to the intended state even if the influences that caused the fault event in the first place disappear.

Corrosion processes are irreversible by nature since they change a metal into more stable oxidized states. In fact, the corrosion products can only be returned to metals by complicated and energetically expensive processes that eventually end up with molten metals. However, not all corrosion processes lead to faulty systems if, for example, a corrosion allowance has been included in the system at the design stage. Some clear examples of corrosion faults can be found in electronic components where even very small amounts of surface corrosion can drastically alter the intended behavior of components.

Connector corrosion is well understood as an age-related problem that contributes greatly to electrical wiring failures. Connector corrosion is also the prime suspect in several military and commercial aircraft incidents and accidents. Fretting corrosion in electronic components, for example, is the result of flaking of tin oxide from a mated surface on tin-containing contacts. The problem becomes more frequent as tin is used to replace gold as a cheaper plating route. The only solution for this hard-to-diagnose, and often intermittent, problem is to replace the faulty part.

One problem of this type discovered by an air force corrosion engineer was the corrosion of tin-plated electrical connector pins mated with gold-plated sockets. Fretting corrosion between these very small contacts appears to have been implicated in at least six F-16 fighter aircraft crashes when their main fuel shutoff valves closed uncommanded (1). Figure 2.2 shows an example of a failed tin-plated pin corrosion of an F-16 actuator.

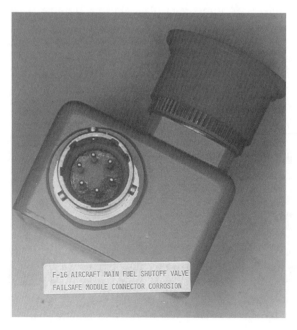

Fig. 2.2 Tin-plated pin corrosion in failsafe module connector on F-16 actuator found in the closed position with master fuel switch to master. (Courtesy David H. Horne, Hill AFB, Utah.)

Fig. 2.3 Dendrite on a circuit board after its exposure in a test chamber. (Courtesy David A. Douthit LoCanLLC.)

Another problem in which a microscopic quantity of corrosion products can create havoc in a complex electronic system is the formation of dendrites across circuit channels. In the presence of moisture and an electric field, metals in their ionic state can migrate to a cathodically (negatively) charged surface and plate out, forming dendrites. Dendrites can be silver, copper, tin, lead, or a combination of metals. The dendrites grow and eventually bridge the gap between the contacts, causing an electric short and possibly arcing and fire. Even a small volume of dissolved metal can result in formation of a relatively large dendrite (Fig. 2.3).

Dendrite growth can be very rapid. Failures have been known to occur in <30 min, but can take several months or more. The rate of growth depends on the applied voltage, the quantity of contamination, and surface moisture. The amount of contamination required for silver dendrites can be extremely small.

2.1.3. What Is a Failure?

A failure is an unsatisfactory condition or a deviation from the original condition which is unsatisfactory to a particular user. The determination that a condition is unsatisfactory, however, depends on the failure consequences in a given operating context (2).

The exact dividing line between satisfactory and unsatisfactory conditions will depend not only on the function of the item in question, but on the nature of the equipment in which it is installed and the operating context in which that equipment is used. The determination will therefore vary from one operating organization to another. Within a given organization, however, it is essential that the boundaries between satisfactory and unsatisfactory conditions be defined for each item in clear and unmistakable terms.

The judgment that a condition is unsatisfactory implies that there must be some condition or performance standard on which this judgment can be based. However, an unsatisfactory condition can range from the complete inability of an item to perform its intended function to some physical evidence that it will soon be unable to do so. For maintenance purposes, failures must therefore be further classified as either functional failures or potential failures.

2.1.3.1. Functional Failure

A functional failure is the inability of a system to meet a specified performance standard. A complete loss of function is clearly a functional failure. However, a functional failure also includes the inability of an item to function at the level of performance that has been specified as satisfactory.

To define a functional failure for any component or system, a clear understanding of its functions is required. It is extremely important to determine all the functions of an item that are significant in a given operational context, since it is only in these terms that its functional failures can be defined.

2.1.3.2. Potential Failure

Once a particular functional failure has been defined, some physical condition may often be identified to indicate that this failure is imminent. Under these circumstances it might be possible to remove the component or system from service before the occurrence of a functional failure. When such conditions can be identified, they are defused as potential failures.

A potential failure is an identifiable physical condition that indicates when a functional failure is imminent. The fact that potential failures can be identified is an important aspect of modern maintenance theory, because it permits maximum use of each system without the consequences associated with a functional failure. Units are removed or repaired at the potential failure stage, so that potential failures preempt functional failures.

For some items, the identifiable condition that indicates imminent failure is directly related to the performance criterion that defines the functional failure. The ability to identify either a functional or a potential failure thus depends on three factors:

1. Clear definitions of the functions of a component or system as they relate to the equipment or operating context in which the item is to be used.
2. A clear definition of the conditions that constitute a functional failure.

3. A clear definition of the conditions that indicate the imminence of this failure.

In other words, it is important not only to define the failure, but also to specify the precise evidence by which it can be recognized.

2.1.4. What Are the Consequences of a Failure?

Failure analysis is the traditional method of relating a failure to its consequences. These may range from the modest cost of replacing a failed component to the possible destruction of a piece of equipment and the loss of lives. The consequences of a failure determine the priority of the maintenance activities or design improvement required to prevent its occurrence. The more complex a piece of equipment is, the more ways there is by which it can fail. Failure consequences can be grouped into the four categories described in the following sections (2).

2.1.4.1. Safety Consequences

The first consideration in evaluating a failure possibility is safety, that is, does the failure cause a loss of function or secondary damage that could have a direct adverse effect on operating safety?

A critical failure is any failure that could have a direct effect on safety. Note, however, that the term direct implies certain limitations. The impact of the failure must be immediate if it is to be considered direct. In addition, these consequences must result from a single failure and not from a combination of this failure with one that has not yet occurred. If a failure has no evident results, it cannot, by definition, have a direct effect on safety.

Not every critical failure results in an accident. However, the issue is not whether such consequences are inevitable, but whether they are possible. Safety consequences are always assessed at the most conservative level, and in the absence of proof that a failure cannot affect safety, it is classified, by default, as critical.

In the presence of a possible critical failure, every attempt should be made to prevent its recurrence. Often redesign of one or more vulnerable items is necessary. However, the design and manufacture of new parts and their subsequent incorporation in in-service equipment can take months, and sometimes years. Hence, temporary measures are often required in the meantime.

2.1.4.2. Operational Consequences

Once safety consequences have been ruled out, a second set of consequences must be considered, that is, has the failure any direct adverse effect on operational capability?

A failure has operational consequences whenever the need to correct a failure disrupts planned operations. Thus operational consequences include the need

34 Chapter 2 Corrosion Detectability

to abort an operation after a failure occurs, the delay or cancellation of other operations to make unanticipated repairs, or the need for operating restrictions until repairs can be made. A critical failure can, of course, be viewed as a special case of a failure with operational consequences. In this case, the consequences are economic and consist of imputed cost of lost operational capability.

2.1.4.3. Nonoperational Consequences

There are many kinds of functional failures that have no direct adverse effect on operational capability. One common example is the failure of a navigation unit in a plane equipped with a highly redundant navigation system. Since other units ensure availability of the required function the failed unit can be replaced at some convenient time. Thus the costs generated by such a failure are limited to the cost of corrective maintenance.

2.1.4.4. Hidden Failure Consequences

Another important class of failures with no immediate consequences consists of failures of hidden-function items. By definition, hidden failures have no direct adverse effects, that is, if they did these failures would not be hidden. However, the ultimate consequences can be major if a hidden failure is not detected and corrected. In other words, the consequence of any hidden function failure is increased exposure to the consequences of a multiple failure.

2.1.5. Corrosion Failure Examples

As mentioned in Chapter 1, corrosion failures are very environmentally context specific. As it is becoming obvious here, corrosion failures are also very consequentially context specific. In much industrial equipment, a corrosion rate of 25 μm year^{-1} for steel is quite acceptable, whereas, such amounts of rust from the point of view of the food industry would be unacceptable.

As another example, in the development of a burial site for storing radioactive waste a corrosion failure would occur if a minimum amount of radioactivity would leach in the groundwater after somewhere in the range of 10 to 1 million years. Thus, the overall concept of failure in this context may have nothing to do with the integrity of the storage container, but everything to do with the transport of radioactive species in the surrounding environment. This particular problem is challenging since it is not so easy to monitor the performance of containers of the radioactive waste owing to the long times and their relative inaccessibility associated with radioactivity (3).

The F-16 fighter aircraft crashes mentioned earlier in this chapter and the Aloha incident described in Chapter 1 are good examples of documented corrosion related failures. In addition to these, the following recent failures have been selected for their importance and consequences. Note at the onset that all these accidents could have been prevented if proper maintenance and inspection had been carried out.

2.1.5.1. Bhopal Accident

Bhopal is probably the site of the greatest industrial disaster in history. Between 1977 and 1984, Union Carbide India Limited, located within a crowded working class neighborhood in Bhopal, was licensed by the Madhya Pradesh Government to manufacture phosgene, monomethylamine, methylisocyanate, and the pesticide carbaryl, also known as Sevin.

In the early morning of December 3, 1984, water inadvertently entered the methylisocyanate storage tank, where >40 metric tons of methylisocyanate were being stored. The addition of water to the tank caused a runaway chemical reaction, resulting in a rapid rise in pressure and temperature. The heat generated by the reaction, the presence of higher than normal concentrations of chloroform, and the presence of an iron catalyst, produced by the corrosion of the stainless steel tank wall, resulted in a reaction of such momentum that gases formed could not be contained by safety systems (4).

Consequently, methylisocyanate and other reaction products, in liquid and vapor form, escaped from the plant into the surrounding areas. There was no warning for people surrounding the plant since the emergency sirens had been switched off. The effect on the people living in the shanty settlements just over the fence was immediate and devastating. Many died in their beds, others staggered from their homes, blinded and choking to die in the street. It has been estimated that at least 3000 people died as a result of this accident, while figures for the number of people injured currently range from 200,000 to 600,000, with an estimated 500,000 typically quoted. The processing plant was closed down after the accident.

The Bhopal disaster was the result of a combination of legal, technological, organizational, and human errors. The immediate cause of the chemical reaction was the seepage of 500 L of water into the methylisocyanate storage tank. The results of this reaction were exacerbated by the failure of containment and safety measures and by a complete absence of community information and emergency procedures.

The long-term effects were made worse by the absence of systems to care for and compensate the victims. Furthermore, safety standards and maintenance procedures at the plant had been deteriorating and ignored for months.

2.1.5.2. Carlsbad Pipeline Explosion

At 5:26 a.m. on August 19, 2000, a 75-cm diameter natural gas transmission pipeline operated by El Paso Natural Gas Company (EPNG) ruptured adjacent to the Pecos River near Carlsbad, New Mexico. The released gas ignited and burned for 55 min. Twelve persons who were camping under a concrete-decked steel bridge that supported the pipeline across the river were killed and their three vehicles destroyed. Two nearby steel suspension bridges for gas pipelines crossing the river were extensively damaged with ~$1 million in property and other damages or losses (5).

The force of the rupture and the violent ignition of the escaping gas created a 16-m wide crater ~34 m along the pipe. A 15-m section of the pipe was ejected from the crater in three pieces measuring ~1, 6, and 8 m in length. The largest piece was found ~90 m northwest of the crater in the direction of the suspension bridges. Investigators visually examined the pipeline that remained in the crater as well as the three ejected pieces. All three ejected pieces showed evidence of internal corrosion damage, but one of the pieces showed significantly more corrosion damage than the other two. Pits were visible on the inside surface of this piece, and at various locations, the pipe wall evidenced significant thinning. At one location, a through-wall perforation was visible. No significant corrosion damage was visible on the outside surfaces of the three pieces or on the two ends of the pipeline remaining in the crater. Pieces were cut from the ruptured pipeline segments and shipped to the Safety Board's Materials Laboratory in Washington, D.C., for further evaluation.

The drip between the closest block valve and the rupture site was removed from the pipeline and visually examined. The drip was found to contain a blackish oily powdery grainy material. At the area of its heaviest concentration, ~4 m from the drip opening, this material filled ~70% of the cross-sectional area of the drip. No significant material was observed in the area just underneath and several centimeters away from the siphon drain at the closed end of the drip. No significant internal corrosion was observed in the drip.

Interconnecting pits were observed on the inside of the pipe in the ruptured area. Typically, these pits showed the striations and undercutting features that are often associated with microbial corrosion. A pit profile showed that chloride concentration in the pits increased steadily from top to bottom. Increased chloride concentration can result from certain types of microbial activity. All four types of microbes (sulfate reducing, acid-producing, general aerobic, and anaerobic) were observed in samples collected from two pit areas in the piece of line where internal corrosion was discovered after the accident ~630-m downstream of the rupture site.

This accident created a sense of urgency in the corrosion engineering community that NACE International took seriously by forming a special training task group. The Internal Corrosion for pipelines course that resulted from this effort was soon followed by the production of a field guide to help inspectors recognize this serious problem (6).

2.1.5.3. Guadalajara Sewer Explosion

The following corrosion failure was also due to a combination of human errors and shared responsibilities. The sewer explosion that killed 215 people in Guadalajara, Mexico, in April 1992, also caused a series of blasts that damaged 1600 buildings and injured 1500 people. At least nine separate explosions were

Fig. 2.4 Pits in the galvanized water pipe that contributed to the erosion corrosion of the gas line and the subsequent leak in a sewer main that caused the Guadalajara 1992 explosion (7). (Courtesy of Dr. Jose M. Malo, Electric Research Institute, Mexico.)

heard, starting at ~10:30 a.m., ripping a jagged trench that ran almost 2 km. The trench was contiguous with the city sewer system, and the open holes at least 6 m deep and 3 m across. In several locations, much larger craters of 50 m in diameter were evident with numerous vehicles buried or toppled into them. An eyewitness said that a bus was "swallowed up by the hole".

Damage costs were estimated at $75 million. The sewer explosion was traced to the installation of a water pipe by a contractor several years before the explosion that leaked water on a gasoline line lying underneath. The cathodically protected gasoline pipeline had a hole within a cavity and an eroded area, all in a longitudinal direction. A second hole did not perforate the internal wall. The galvanized water pipe obviously had suffered stray current corrosion effects that were visible in pits of different sizes (Fig. 2.4) (7). The subsequent corrosion of the gasoline pipeline, in turn, caused leakage of gasoline into a main sewer line. The Mexican attorney general sought negligent homicide charges against four officials of Pemex, the government-owned oil company. Also cited were three representatives of the regional sewer system and the city's mayor.

2.1.5.4. El Al Boeing 747 Crash

On October 4, 1992 an EL AL 747 freighter crashed in Amsterdam, killing all four people on board and >50 people on the ground. The cause of the crash was attributed to the loss of the number 3 and 4 engines from the wing that in turn caused a complete loss of control of the airplane. The reason for the number 3 engine separation was a breakage of the fuse pin. The pin was designed to break when an engine seizes in flight, producing a large amount of torque. Both of the engines were stripped off the right wing causing the Boeing 747-200 Freighter to crash as it maneuvered toward the airport (8).

Unfortunately, this was not the first Boeing 747 to crash in this way. In December 1991 a China Airlines Boeing 747-200F freighter crashed shortly after takeoff. A possible reason for the shearing away of the two right engines was that corrosion pits and fatigue weakened the fuse pins that hold the strut to the wings. Constant pressure variance coupled with corrosion is a terrible force that can cause corrosion pits to expand into cracks, such as the 4.3-cm crack found in one of the fuse pins of the El Al 747.

In both the El Al crash and the China Airlines crash, the No.3 and No.4 engines on the right side of the plane ripped away from the fuselage. It is believed that in the El Al crash the inboard fuse pin failed due to corrosion cracking and fatigue, which caused the outboard fuse pin, already weakened by a crack, to fail. With these two pins malfunctioning the No.3 engine tore off the plane in such a way that it may have taken the No.4 engine with it. Boeing had just begun distributing a safety bulletin pertaining to the inspection of all fuse pins on their 747- 100/200/300 that used Pratt & Whitney and Rolls-Royce engines. Both the El Al and China Airlines planes were Boeing 747-200's with Pratt & Whitney engines.

This design of the fuse pin has been used since 1982 and in a 7-year period there have been 15 reports of cracked pins. It was discovered that these pin failures resulted from the absence of primer, cadmium plating, and a corrosion prevention compound. Since the El Al 747 crash, Boeing has also been trying to upgrade the 747. Specific targets on this model included fabricating new parts for the pylon-to-wing attachment for the Pratt & Whitney engines and improve the cost and time efficiency of inspection.

2.1.5.5. Nuclear Reactor With a Hole in the Head

On March 6, 2002, Workers repairing one of five cracked control rod drive mechanism (CRDM) nozzles at Davis–Besse nuclear plant discovered extensive damage to the reactor vessel head. The reactor vessel head is the dome-shaped upper portion of the carbon steel vessel housing the reactor core. It can be removed when the plant is shut down to allow spent nuclear fuel to be replaced with fresh fuel. The CRDM nozzles connect motors mounted on a platform above the reactor vessel head to control rods within the reactor vessel. Operators withdraw control rods from the reactor core to startup the plant and insert them to shut down the reactor.

The reactor core at the Davis–Besse nuclear plant sits within a metal pot designed to withstand pressures up to 17 MPa. The reactor vessel has carbon steel walls nearly 15 cm thick to provide the necessary strength. Because the water cooling the reactor contains boric acid, which is highly corrosive to carbon steel, the entire inner surface of the reactor vessel is covered with 0.6-mm thick stainless steel. But water routinely leaked onto the reactor vessel's outer surface.

Because the outer surface lacked a protective stainless steel coating, boric acid ate its way through the carbon steel wall until it reached the backside of the inner liner. High pressure inside the reactor vessel pushed the stainless steel outward into the cavity formed by the boric acid. The stainless steel bent, but did not break. Cooling water remained inside the reactor vessel not because of thick carbon steel, but due to the thin stainless steel layer. The plant's owner ignored numerous warning signs spanning many years, which lead to this highly visible incident.

The corrosion incident also exposed problems within the staff of the regulatory commission, which initially wanted prompt inspections of all 68 plants that could be vulnerable to the problem, but relented and gave the owners permission to delay, leaving time for the hole in the lid to grow. Plants are designed with emergency equipment to cope with leaks, but the designs do not contemplate failure of the thick steel in that location.

A subsequent investigation by the commission's inspector general found poor communications within the agency itself. The commission had a photo taken during a refueling shutdown in 2000 that showed evidence of the corrosion damage, but, according to the inspector general, officials failed to act on it.

2.1.5.6. Piping Rupture Caused by Flow Accelerated Corrosion

Piping rupture caused by flow accelerated corrosion occurred at Mihama-3 at 3:28 pm on August 9, 2004, killing four and injuring seven. One of the injured men later died, bringing the total to five fatalities. The rupture was in the condensate system, upstream of the feedwater pumps. According to Japan's Nuclear and Industrial Safety Agency (NISA), the rupture was 60 cm in size. The pipe wall at the rupture location had thinned from 10 to 1.5 mm.

Mihama-3 is a 826 MW Mitsubishi-built pressurized water reactor (PWR) plant located 320 km west of Tokyo. Although the carbon steel pipe carried the high temperature steam at high pressure, it had not been inspected since the power plant opened in 1976. In April 2003, Nihon Arm, a maintenance subcontractor, informed Kansai Electric Power Co., the plant owner, that there could be a problem, following which the power company had scheduled an ultrasonic inspection for August 2004. Four days before the scheduled shutdown and inspection, superheated steam blew the 60-cm wide hole in the pipe. The steam that escaped had not been in contact with the nuclear reactor, and no nuclear contamination has been reported.

In response to the accident, Japan's Nuclear and Industry Safety Agency ordered four other power companies that owned nuclear plants with the same type of pressurized water reactors to conduct ultrasonic inspections of their pipes. The inspections were to involve nearly half of the country's 52 nuclear power plants.

Japan has the world's third-largest nuclear power industry, after the United States and France. The government was planning to build 11 more reactors in the decade, increasing Japan's reliance on home-based nuclear power to 40% of electricity needs. Already slowed by local opposition, this program was in danger of being stalled by the accident, the most deadly in the history of nuclear power in Japan.

2.1.5.7. Sinking Ships

On December 12, 1999, the Maltese registered tanker Erika broke in two some 70 km from the French coast off of Brittany, while carrying ~30,000 tons of heavy fuel oil. Some 19,000 tons were spilled. This is equal to the total amount of oil spilled worldwide in 1998. The sunken bow section still contained 6400 tons of cargo and the stern had a further 4700 tons. The bow section sank within 24 h. The stern section sank on December 13 while under tow.

The economic consequences of the incident have been felt across the region; a drop in the income from tourism, loss of income from fishing, and a ban on the trade of sea products, including oysters and crabs, have added to the discomfort of local populations.

Corrosion problems had been apparent on the Erika since at least 1994, with details readily available to port state control authorities and potential charterers. In addition, there were numerous deficiencies in her firefighting and inert gas systems, pointing to a potential explosion risk on the tanker. Lloyd's list reported that severe corrosion had been discovered by class just weeks before the incident. However, no immediate remedial action had been taken.

According to publicly available, U.S. Coast Guard records obtained by Lloyd's List, Erika had been inspected in a variety of U.S. ports on several occasions since 1994. Her certificate of financial responsibility, a document legally required by tankers wishing to trade in U.S. waters, had expired in March 1999, and had not been renewed as of November 30. In an inspection in Portland in July 1994, holes had been discovered in the main deck coaming, indicating that signs of corrosion were already in place >5 years prior to the catastrophe. It was also found that there were holes in both the portside and starboard inert gas system risers, which are critical items of safety equipment. Malfunctions would tend to increase vulnerability to explosions.

Many of Erika's problems had simply been patched up, rather than properly repaired. In an inspection in 1997 in New Orleans, the U.S. Coast Guard ordered that no cargo operations requiring the use of inert gas systems should be conducted until permanent repairs had been effected. Pinhole leaks remained in the firemain, contrary to Safety of Life at Sea convention regulations. There was yet more evidence of corrosion, with the ship's watertight doors not sealing properly and wasting on the door coamings. Erika switched from Bureau Veritas to Registro Italiano Navale in 1998, which authorized her to continue operations despite the French society's order for a full inspection.

Another example of major losses to corrosion that could have been prevented and that was brought to public attention on numerous occasions since the 1960s, is related to the design, construction, and operating practices of bulk carriers. In 1991, 44 large bulk carriers were either lost or critically damaged and >120 seamen lost their lives (9).

A highly visible case was the MV KIRKI, built in Spain in 1969 to Danish designs. In 1990, while operating off the coast of Australia, the complete bow section became detached from the vessel. Miraculously, no lives were lost, there was little pollution, and the vessel was salvaged. Throughout this period it seems to have been common practice to neither use coatings nor cathodic protection inside ballast tanks. Not surprisingly therefore, evidence was produced that serious corrosion had greatly reduced the thickness of the plate and that this, combined with poor design to fatigue loading, were the primary cause of the failure. The case led to an Australian Government report entitled "Ships of Shame". The MV KIRKI was not an isolated case. There have been many others involving large catastrophic failures, although in many of these cases there was little or no hard evidence on what actually caused the ships to go to the bottom.

2.1.5.8. SCC of Chemical Reactor: The Flixborough Explosion

The June 1974, Flixborough explosion was the largest ever peacetime explosion in the United Kingdom. There were 28 fatalities, as well as the near complete destruction of the NYPRO plant in North Lincolnshire by blast and then fire. This catastrophic explosion has been traced to the failure of a bypass assembly introduced into a train of six cyclohexane oxidation reactors after one of the reactors was removed owing to the development of a leak. The leaking reactor, like the others, was constructed of 12.3-mm mild steel plate with 3-mm stainless steel bonded to it, and it developed a vertical crack in the mild steel outer layer of the reactor from which cyclohexane leaked leading to the removal of the reactor. One of the factors contributing to the crack was SCC, resulting from the presence of nitrates from the contaminated river water being used to cool a leaking flange.

2.1.5.9. Swimming Pool Roof Collapse

In 1985, 12 people were killed in Uster, Switzerland when the concrete roof of a swimming pool collapsed after only 13 years of use. The roof was supported by stainless steel rods in tension, which failed due to SCC. There have been other incidents associated with the use of stainless steel in safety critical load bearing applications in the environment created by modern indoor swimming pools and leisure centers. The collapse of this ceiling above a swimming pool showed how a simple structural concept could be sensitive to the loss, through corrosion, of support from one of many hangers.

The Federal Materials Testing Institute, based in Duebendorf, Switzerland, and the Federal Materials Research and Testing Institute of Berlin concluded that the collapse was the result of chloride induced SCC. The steel rods had been pitted, causing the roof to cave in. The roof collapsed in a zipper-like fashion, starting with the corroded rods. The collapse continued as the remaining rods were unable to bear the increased load. The chloride was either already present in the concrete or came from the pool via water vapor. Chloride can overcome the passivity of the natural oxide film on the surface of the steel. The inspection of safety-critical stainless steel components for SCC and loss of section by pitting should be viewed as a priority. The following inspection procedures have since that accident been recommended:

- Compile an inventory of all stainless steel components within the pool building, identifying their grade, location, and function.

- SCC may be difficult to detect in its early stages and, although normally accompanied by visual signs of other types of corrosion, such as pitting corrosion, reliance should not be placed on visual means alone. Although not definitive, a normal telltale sign is brown staining, varying from a pale, dry discoloration to wet pustules.
- Specifically inspect all stainless steel components at least twice a year for any evidence of staining or corrosion. During inspection special attention should be given to components that are load bearing and/or safety critical;
- Where staining or corrosion is found, the corrosion products should be removed and the loss of cross-section and integrity assessed.
- Where the component is load bearing and/or safety critical, samples of the component should be tested for SCC. Where tests reveal the presence of SCC a full risk assessment of the affected components should be carried out by qualified personnel.
- If necessary, components should be replaced with a more corrosion resistant stainless steel grade

2.1.6. Age-Reliability Characteristics

The failure patterns of complex systems have been extensively studied to improve maintenance strategies and operational procedures. Six basic patterns shown in Fig. 2.5 seem to describe all failures (2):

Pattern A is often referred to in reliability literature as the bathtub curve. This type of curve has three identifiable regions:

1. An infant-mortality region, the period immediately after manufacture or overhaul in which there is a relatively high probability of failure.
2. A region of constant and relatively low failure probability.
3. A wearout region, in which the probability of failure begins to increase rapidly with age.

Pattern B is characterized by a constant or gradually increasing failure probability, followed by a pronounced wearout region. Once again, an age limit may be desirable.

Pattern C shows a gradually increasing failure probability, but with no identifiable wearout age. It is usually not desirable to impose an age limit in such cases.

Pattern D starts with a low failure probability when the item is new or just out of the shop, followed by a quick increase to a constant level.

Pattern E, which is characterized by a constant probability of failure at all ages.

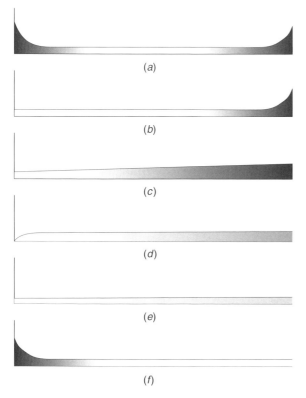

Fig. 2.5. Six patterns of failure (2).

Pattern F in which infant mortality is followed by a constant or very slowly increasing failure probability. This behavior is particularly applicable to electronic equipment.

In the 1960s United Airlines carried out a systematic study of component failures to reveal their failure patterns. In this study, it was found that the relative frequency of each type of conditional-probability curve proved especially interesting. Some 89% of the items analyzed had no wearout zone; therefore their performance could not he improved by the imposition of an age limit. In fact, after a certain age the conditional probability of failure continued on at a constant rate (curves D, F). Another 5% had no well-defined wearout zone (curve C), but did become steadily more likely to fail as age increased (2).

However, it is obvious that if the failure pattern of a system has a wearout behavior, that is, curves A and B, it would be justified to recommend that a corrective action be taken before this system enters the wearout zone in order to reduce the overall failure rate. In such cases, allowing the system to age well into the wearout region would cause an appreciable increase in the failure rate. While the presence of a well-defined wearout region exists in only two of the six curves in Fig. 2.5 it happens that these two curves are associated with a great many single components or simple systems. Since corrosion is definitively an aging process, its influence typically follows pattern B in Fig. 2.5.

2.1.7. Probability of Failure

To determine the probability of a failure (POF), two fundamental issues must be considered:

1. What are the specific forms of corrosion and their rates?
2. What is the possible effectiveness of corrosion inspection or monitoring?

The input of corrosion experts is required to identify the relevant forms of corrosion in a given situation and to determine the key variables affecting the propagation rate. It is also important to realize that full consensus and supporting data on the variables involved is highly unlikely in real-life complex systems and that simplification will often be necessary.

One semiquantitative approach for ranking process equipment is based on internal POF. The procedure consists in carrying out an analysis of equipment process and inspection parameters, and then ranking equipment on a scale of one to three, with "one" being the highest priority. The procedure requires a fair degree of engineering judgment and experience and, as such, is dependent on the background and expertise of the analyst. This procedure can facilitate an efficient use of finite inspection resources when 100% inspection is not practical.

The POF approach is based on a set of rules heavily dependent on detailed inspection histories, knowledge of corrosion processes, and knowledge of normal and upset conditions. The equipment rankings may have to be changed and could require updating as additional knowledge is gained, process conditions change, and equipment ages. Maximum benefits of the procedure depend on fixed equipment inspection programs that permit the capture, documentation, and retrieval of inspection, maintenance, and corrosion–failure mechanism information.

2.1.8. Probability of Detection

The probability of detection (POD) concept was introduced in 1973 and was incorporated into design requirements for the National Aeronautics and Space Administration's (NASA), Space Shuttle program (10). The POD concept and methodology have since then gained widespread acceptance and continuing improvements have enhanced its acceptance as a useful metric for quantifying and assessing nondestructive evaluation (NDE) capabilities. Since a wide range of NDE methods and procedures are used in "fracture control" of engineering hardware and systems, a large volume of POD data has been generated to validate the capabilities of specific NDE procedures in a multitude of applications (see Chapter 5 for more details).

2.2. FORMS OF CORROSION

Corrosion can take many forms (Fig. 2.6), and the statistics associated with each will be different. The actual importance of each corrosion type will also differ between systems, environments, and other operational variables. However, there are surprising similarities in the corrosion failure distributions within the same industries as can be seen by comparing Fig. 2.7*a* and *b*. Both corrosion failure distribution charts represent data for two large chemical plants, but they are from two different continents. These corrosion failure statistics are for a chemical plant located in Germany (Fig. 2.7*a*) and in the United States (Fig. 2.7*b*). This surprising similarity indicates that corrosion problems are very similar worldwide in the chemical industry and thus, ways to prevent these corrosion service failures should be similar (12).

The multifaceted nature of corrosion damage has long been recognized. The classification that seems to have gained the widest acceptance was first presented

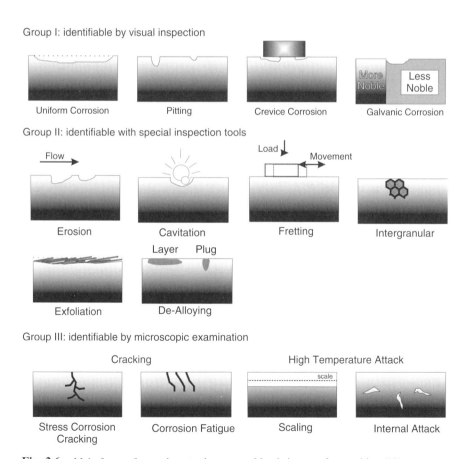

Fig. 2.6. Main forms of corrosion attack regrouped by their ease of recognition (11).

46 Chapter 2 Corrosion Detectability

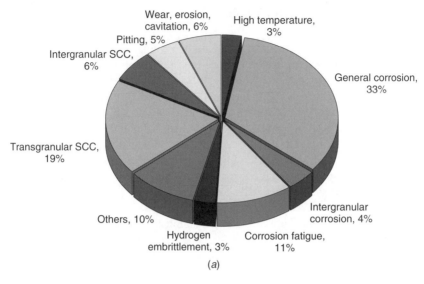

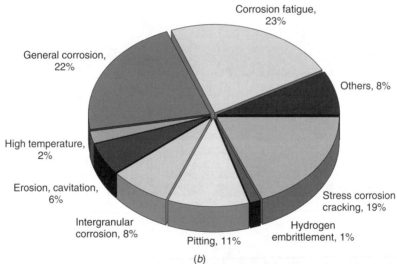

Fig. 2.7. Failure statistics of large chemical process plant in Germany (*a*), and in the United States (*b*).

in a landmark textbook on corrosion engineering published in the late 1960s (13). The forms of corrosion shown in Fig. 2.6 have been further divided into three groups by a veteran corrosion engineer (11):

1. Those recognizable with the unaided eye.
2. Those that are more easily discerned with specific aids (e.g., dye penetrants, magnetic particles, or low-power microscopy).

3. Those that can only be identified definitely by optical or electronic microscopy.

These types of corrosion could also be organized on the basis of other factors (e.g., degree of localization and metallurgical features). The degree of localization refers to the selective attack by corrosion at small special areas or zones on a metal surface in contact with an environment. It usually occurs under conditions where the largest part of the exposed surface is either not attacked or is attacked to a much smaller degree than at local sites. The degree of localization is an important aspect of any type of corrosion for many reasons. One reason is that the degree of corrosion severity usually increases with the degree of localization, often leading to a more serious problem, such as stress corrosion cracking (SCC). Another reason is that the detectability of a corrosion defect decreases with its degree of localization.

The most common type of localized corrosion is pitting, in which small volumes of metal are removed by corrosion from certain areas on the surface to produce craters or pits. Pitting corrosion may occur on a metal surface in a stagnant or slow moving liquid. It also may be caused by crevice corrosion, poultice corrosion, deposition corrosion, cavitation, impingement, and fretting corrosion.

Among the metallurgical features of importance, the grain structure of a metallic material is probably the most determining in relation to corrosion damage. A common type of corrosion attack for which the grain structure is important is intergranular or intercrystalline corrosion, during which a small volume of metal is preferentially removed along paths that follow the grain boundaries to produce what might appear to be fissures or cracks. The same kind of subsurface fissures can be produced by transgranular or transcrystalline corrosion. In this, a small volume of metal is removed in preferential paths that proceed across or through the grains.

Intergranular and transgranular corrosion sometimes are accelerated by tensile stress. In extreme cases, the cracks proceed entirely through the metal, causing rupture or perforation. This condition is known as SCC.

In a completely different type of corrosion, which is highly dependent on the metallurgical make-up, one of the metals in an alloy may be selectively leached out without producing visible pits or cracks and without changing the dimensions of the metal. At a casual glance, the metal may appear to be intact. Under a microscope, however, it can be seen to be porous. The mechanical properties of the alloy are greatly reduced by the selective attack.

The most common example of this type is dezincification of brass in which the zinc is selectively dissolved out of the alloy. Another case is graphitic corrosion of cast iron, in which the iron that constitutes 95% of the mass is selectively dissolved or leached away leaving a porous mass that appears intact while it only consists of graphite.

While the types of corrosion identified in Fig. 2.6 are described individually in the following sections it should be recognized at the onset that during any damaging corrosion process these types often act in synergy. The unfolding of a crevice situation, for example, will typically create an environment favorable for pitting, intergranular attack and even cracking.

2.2.1. Uniform Attack

General corrosion is usually the least threatening type of attack, allowing one to forecast with some accuracy the probable life of equipment. Except in rare cases of a grossly improper choice of material for a particular service, or an unanticipated drastic change in the corrosive nature of the environment or complete misunderstanding of its nature, failures of metals by rapid general attack (wasting away) are not often encountered.

From a corrosion inspection point of view, uniform attack is relatively detectable and its effects predictable, hence it may be less troublesome than other forms of corrosion unless the corroding material is hidden from sight. The internal corrosion of pipeline and that of any other buried or immerged structures or the corrosion of hidden components are good examples that even the simplest corrosion process needs to be monitored.

2.2.2. Pitting Corrosion

Pitting corrosion is a localized form of corrosion by which cavities or "holes" are produced in the material. Pitting is considered to be more dangerous than uniform corrosion damage because it is more difficult to detect, predict, and design against. A small, narrow pit with minimal overall metal loss can lead to the failure of an entire engineering system. Only a small amount of metal is corroded, but perforations can lead to costly repair of expensive equipment.

Pitting cavities may or may not become filled with corrosion products. Corrosion products may form caps over pit cavities that are described as nodules or tubercles. While the shapes of pits vary widely (Fig. 2.8), they usually are roughly saucer-shaped, conical, or hemispherical for iron-based alloys, such as steels and stainless steels. Pit walls usually are irregular when viewed under a microscope. The following are some factors contributing to the onset and propagation of pitting corrosion:

- Localized chemical or mechanical damage to a protective oxide film that is almost, but not completely, resistant to corrosion.
- Water chemistry factors that can cause breakdown of a passive film are acidity, low, dissolved oxygen concentrations (which tend to render a protective oxide film less stable), and high concentrations of chloride (as in seawater).
- Localized damage to or poor application of a protective coating.
- The presence of nonuniformities in the metal structure of the component (e.g., nonmetallic inclusions).

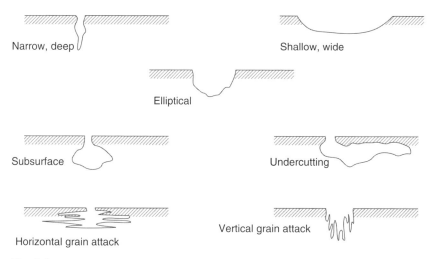

Fig. 2.8. Typical cross-sectional shapes of corrosion pits.

The practical importance of pitting corrosion depends on the thickness of the metal and on the penetration rate, which could be small. The rate usually decreases with time. Thus, on thin sections pitting may be serious, while on a thick section it might be unimportant. In general, the rate of penetration decreases if the number of pits increases. This is because adjacent pits have to share the available adjacent cathodic area, which controls the corrosion current that can flow. Movement of the solution over a metal surface often reduces and may even prevent pitting that otherwise would occur if the liquid was stagnant.

Pitting is often associated with other forms of corrosion. Intergranular corrosion and cracks, for example, may progress from the main pit cavity further into the metal. In the example shown in Fig. 2.9, pitting at the edge of an aluminum/lithium sheet has progressed as intergranular corrosion at the root of the pits. And crevice corrosion described later can be considered to be an aggravated case of pitting corrosion. Stray current corrosion that occurs when an electric current leaves a metal surface and flows into the environment can cause a very characteristic form of macroscopic pits as illustrated in Fig. 2.10.

Pitting corrosion can be assessed by various methods, including simple visual examination of a corroded specimen or monitoring coupon. When the sites of attack are numerous, the tedious job of counting can be eliminated by using rating charts, as shown in Fig. 2.11 (14). Efforts have also been made to treat the rate of penetration statistically. For example, equations have been developed to predict the perforation of buried steel pipelines. The depth of the deepest pit on a specimen is the upper tail of the distribution of depths of all pits on that specimen (15).

Engineers concerned with soil corrosion and underground steel piping are aware that the maximum pit depth found on a buried structure is somehow related to the percentage of the structure inspected. Finding the deepest actual

50   Chapter 2   Corrosion Detectability

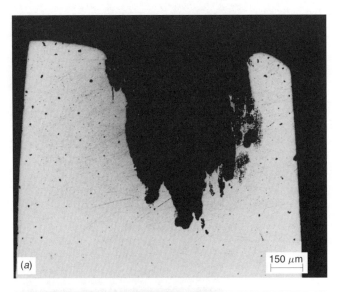

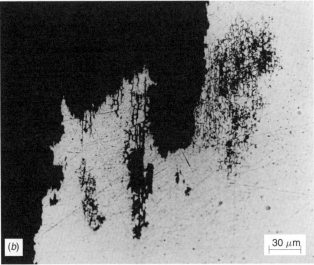

Fig. 2.9. Photomicrograph of a section through an edge of the 8090-T851 panel immersed in seawater during 4 months; (*a*) at 64X and (*b*) at 320X to illustrate the intergranular nature of the corrosion attack.

pit requires a detailed inspection of the whole structure. As the area of the structure inspected decreases, so does the probability of finding the deepest actual pit. A number of statistical transformations have been proposed to quantify the distributions in pitting variables. Gumbel is given the credit for the original development of extreme value statistics (EVS) for the characterization of pit depth distribution (16).

Fig. 2.10 Pits in 16 in. Sheet Iron Water Main in San Francisco's East Bay that failed in 1 month due to leakage of current from electric streetcars. (Photo dated August 10, 1926, Courtesy of East Bay Municipal Utility District.)

The EVS procedure is to measure maximum pit depths on several replicate specimens that have pitted, then arrange the pit depth values in order of increasing rank. The Gumbel or extreme value cumulative probability function $F(x)$, is shown in Eq. (2.1), where λ and α are the location and scale parameters, respectively. This probability function can be used to characterize the data set and estimate the extreme pit depth that possibly can affect the system from which the data was initially produced.

$$F(x) = \exp\left(-\exp\left[-\frac{x-\lambda}{\alpha}\right]\right) \qquad (2.1)$$

2.2.3. Crevice Corrosion

Crevice corrosion occurs in cracks or crevices formed between mating surfaces of metal assemblies (e.g., the "pillowing" corrosion of fayed surfaces described in Chapter 1), and usually takes the form of pitting or etched patches. Both surfaces may be of the same metal or of dissimilar metals, or one surface may be a nonmetal (e.g., corrosion under insulation also described in Chapter 1 or under rocks as shown in Fig. 2.12. It can also occur under scale and surface deposits and under loose-fitting washers and gaskets that do not prevent the entry of liquid between them and the metal surface. Crevices may proceed inward from a surface exposed to air, or may exist in an immersed structure.

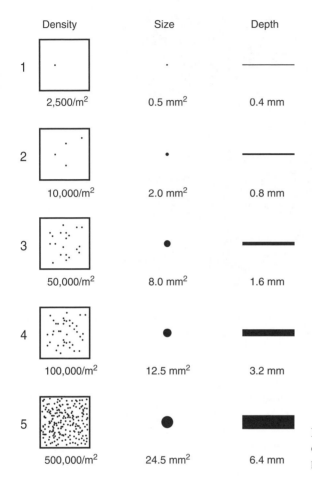

Fig. 2.11. Rating charts that eliminate tedious counting of pitting sites.

The series of events leading to the formation of a severe crevice can be summarized in the following three stages. First, crevice corrosion initiates as the result of a differential aeration process. Dissolved oxygen in the liquid that is deep in the crevice is consumed by reaction with the metal (Fig. 2.13a). Second, as oxygen diffusion into the crevice is restricted, a differential aeration cell develops between the crevice microenvironment and the external surface exposed to the bulk environment. The corrosion reactions then specialize in the crevice (anodic) and on the open surface (cathodic). The large cathodic surface (S_c) versus anodic surface (S_a) ratio (S_c/S_a), which forms in these conditions, is a definitive aggravating factor of the anodic (corrosion) reaction (Fig. 2.13b).

The cathodic oxygen reduction reaction cannot be sustained in the crevice area, giving it an anodic character in the concentration cell. This anodic imbalance can lead to the creation of highly corrosive micro-environmental conditions in the crevice, conducive to further metal dissolution. It is also thought that subsequent pH changes at anodic and cathodic sites further stimulate local cell action

Fig. 2.12 Large holes created in a bridge structural steel by the presence of rocks. (Courtesy Kingston Technical Software.)

(Fig. 2.13c). The aggravating factors present in a fully developed crevice can be summarized in the following points:

- Metal ions produced by the anodic corrosion reaction readily hydrolyze giving off protons (acid) and forming corrosion products. The pH in a crevice can reach very acidic values, sometimes equivalent to pure acids.
- The acidification of the local environment can produce a serious increase in the corrosion rate of most metals.
- The corrosion products even further seal the crevice environment from oxygen penetration also preventing the dilution of the acidic build-up.
- The accumulation of positive charge in the crevice becomes a strong attractor of negative ions in the environment, such as chlorides and sulfates that can be corrosive in their own right.

2.2.4. Galvanic Corrosion

Galvanic corrosion (also called "dissimilar metal corrosion" or wrongly "electrolysis") refers to corrosion damage induced when two dissimilar materials are coupled in a corrosive electrolyte. In a bimetallic couple, the less noble material becomes the anode and tends to corrode at an accelerated rate, compared with the uncoupled condition and the more noble material will act as the cathode in the corrosion cell.

Since the corrosion potential of a metal in an environment is related to the energy that is released when the metal corrodes, differences in corrosion

54 Chapter 2 Corrosion Detectability

potentials of dissimilar metals may be used to estimate the direction of the current that is generated by the galvanic action of these metals when immersed in a given environment. If such measurements were repeated with all the possible combinations of metals, the results would make it possible to arrange the metals in what is called a galvanic series.

The relative tendencies of metals to corrode remain about the same in many of the environments in which they are likely to be used. Consequently, their relative positions in a galvanic series may be about the same in many environments. Since more observations of potentials and galvanic behavior have been made in seawater than in any other single environment, an arrangement of metals in a galvanic series based on observations in seawater, as shown in Table 2.1 (17), is frequently used as a first approximation of the probable direction of the galvanic effects in other environments.

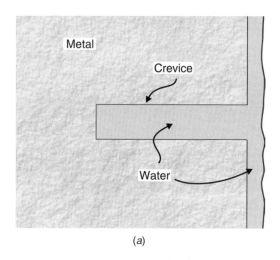

(a)

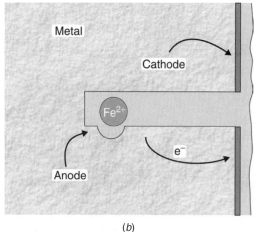

(b)

Fig. 2.13. Schematic description of the stages of a crevice formation: (*a*) first stage; (*b*) second stage; (*c*) third stage.

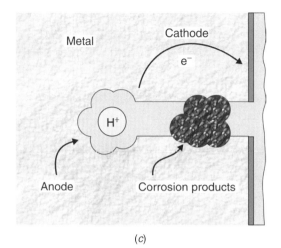

(c)   **Fig. 2.13.** (*Continued*).

Note that several metals in Table 2.1 are grouped. The potential differences within a group are not likely to be great and the metals can be combined without substantial galvanic effects under many circumstances. For the sake of comparison, Table 2.2 presents the galvanic series that was constructed for metals exposed to neutral soils and water (18).

Also note that the magnitude of the potential difference alone will not necessarily indicate the amount of galvanic corrosion. For example, metals with a potential difference of only 50 mV have shown severe galvanic corrosion problems, while metals with a potential difference of 800 mV have been successfully coupled together. This is because the potential difference gives no information about the kinetics of galvanic corrosion, which depends on the current flowing between the two metals that are coupled.

An important factor in galvanic corrosion is the area effect or the ratio of cathodic-to-anodic area. From the standpoint of practical corrosion resistance, the most unfavorable ratio is a very large cathode connected to a very small anode. This effect is illustrated in Fig. 2.14 and 2.15. Table 2.1 indicates that iron is anodic with respect to copper, and therefore is more rapidly corroded when placed in contact with it. This effect is greatly accelerated if the area of the iron is small in comparison to the area of the copper, as shown in Fig. 2.14. However, under the reverse conditions when the area of the iron is very large compared to copper, the corrosion of iron is only slightly accelerated.

If the electrical conductance of the electrolyte bridging the galvanic contact is low (e.g., in soils or during atmospheric corrosion), the effective areas taking part in galvanic cell reactions are small and the total amount of corrosion is relatively small, although it may be severe immediately adjacent to the metal junction, as shown in Fig. 2.16.

Under immersed conditions in many supply waters, which generally have a relatively low electrical conductivity, adverse effects are uncommon if the

Table 2.1. Galvanic Series of Selected Metals and Alloys in Stagnant Seawater

Active (Anodic)
Magnesium and its alloys
Zinc (hot-dip, die cast, or plated)
Beryllium (hot pressed)
Aluminum alloys: Al 7072 clad on 7075, Al 2014-T3, Al 1160-H14, Al 7079-T6
Cadmium (plated)
Uranium
Aluminum alloys: Al 218 (die cast), Al 5052-0, Al 5052-H12, Al 5456-0, H353, Al 5052-H32, Al 1100-0, Al 3003-H25, Al 6061-T6, Al A360 (die cast), Al 7075-T6, Al 6061-0
Indium
Aluminum alloys: Al 2014-0, Al 2024-T4, Al 5052-H16
Tin (plated)
Stainless steel 430 (active)
Lead
Steel 1010
Iron (cast)
Stainless steel 410 (active)
Copper (plated, cast, or wrought)
Nickel (plated)
Chromium (Plated)
Tantalum
AM350 (active)
Stainless steels (active): 310, 301, 304, 430, 17-7PH
Tungsten
Niobium (columbium) 1% Zr
Yellow brass (268)
Uranium 8% Mo
Brass, Naval brass (464), Muntz Metal (280), Brass (plated)
Nickel-silver (18% Ni)
Bronze (220), Copper (110), Red brass
Stainless steel 347 (active)
Molybdenum, Commercial pure
Copper-nickel (715), Admiralty brass
Stainless steel 202 (active)
Bronze, Phosphor 534 (B-1), Monel 400
Stainless steel 201 (active)
Alloy 20 (active)
Stainless steels (active): 321, 316, 309
Stainless steel 17-7PH (passive)
Silicone Bronze 655
Stainless steels (passive): 304, 301, 321, 201, 286, 316
AM355 (active)
Stainless steels (passive): Alloy 20, AM355, A286
Titanium alloys
AM350 (passive)
Silver
Gold
Graphite
End-Noble (Less Active, Cathodic)

Table 2.2. Practical Galvanic Series for Metals in Neutral Soils and Water[a]

Metal	Potential V (CSE)
Commercially pure magnesium	−1.75
Magnesium alloy (6% Al, 3% Zn, 0.15% Mn)	−1.6
Zinc	−1.1
Aluminum alloy (5% Zinc)	−1.05
Commercially pure aluminum	−0.8
Mild steel (clean and shiny)	−0.5 to −0.8
Lead	−0.5
Cast iron (not graphitized)	−0.5
Mild steel (rusted)	−0.2 to −0.5
Mill scale on steel	−0.2
High silicon cast iron	−0.2
Copper, brass, bronze	−0.2
Mild steel in concrete	−0.2
Platinum	0 to −0.1
Carbon, graphite, coke C	+0.3

[a] Typical potentials normally observed in neutral soils and water, measured in relation to copper sulfate reference electrode.

contacting metals are of similar area. Thus steel pipes can be used with brass connectors, but serious corrosion to the pipe end is likely to result if the contact is directly to a large area of copper, such as a tank or cylinder. Similarly, stainless steel and copper tubes can usually be joined without causing any serious problems, but accelerated corrosion of the copper tube is likely to occur if it is attached to a stainless steel tank.

Under immersed conditions in a highly conducting electrolyte, such as seawater, effective areas will be greater and severe corrosion may be encountered on small anodic areas of many metals. Extremely small anodic areas exist at discontinuities, such as cracks or pinholes, in cathodic coatings, such as magnetite (mill scale on iron) and copper plating on steel.

2.2.5. Flow Influenced Corrosion

A metal surface over which a smooth, linear flow of liquid is maintained will corrode at essentially the same rate as the metal would if the solution was stagnant. This is true because at the surface–liquid interface, there is a static film of the liquid molecules. Unless the liquid is aggressive as a corrosive, the corrosion products will form some semiprotective film over the metallic surface and remain stable.

A change in the motion of a corroding metal or alloy relative to its environment by fluid flow can increase corrosion rates by removing protective films or by increasing the diffusion or migration of corrosive agents. An increase in

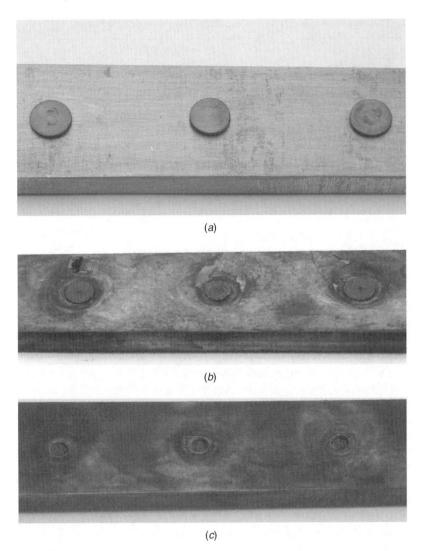

Fig. 2.14. Steel rivets on a copper bar: (*a*) at the start of the experiment; (*b*) 6 months after being submerged in 3% sodium chloride solution; and (*c*) after 10 months in the same solution.

fluid flow can also decrease corrosion rates by eliminating aggressive ion concentration or enhancing passivation or inhibition by transporting the protective species to the fluid–metal interface. In its most generic sense, erosion–corrosion is a form of corrosion that is caused or accelerated by the relative motion of the environment and the metal surface. It is characterized by surface features with a directional pattern that are a direct result of the flowing media.

The influence of flow on corrosion is most prevalent in soft alloys (i.e., copper, aluminum, and lead alloys). Alloys that form a surface film in a corrosive

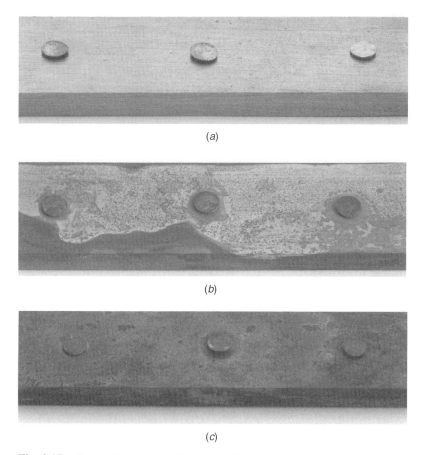

Fig. 2.15. Copper rivets on a steel bar: (*a*) at the start of the experiment; (*b*) 6 months after being submerged in 3% sodium chloride solution; and (*c*) after 10 months in the same solution.

environment commonly show a limiting velocity above which corrosion rapidly accelerates. With the exception of cavitation, flow induced corrosion problems are generally termed erosion–corrosion, encompassing flow enhanced dissolution and impingement attack. The fluid can be aqueous or gaseous, single or multiphase (19). There are several mechanisms described by the conjoint action of flow and corrosion that result in flow-influenced corrosion (20; 21):

- **Mass transport-control:** Mass transport-controlled corrosion implies that the rate of corrosion is dependent on the convective mass transfer processes at the metal–fluid interface. When steel is exposed to oxygenated water, for example, the initial corrosion rate will be closely related to the convective flux of dissolved oxygen toward the surface, and later by the oxygen diffusion through the iron oxide layer. Corrosion by mass transport will often be streamlined and smooth.

60   Chapter 2   Corrosion Detectability

Fig. 2.16 Corrosion of a water main ductile iron adjacent to a copper fitting. (Courtesy Public Works and Services, City of Ottawa.)

- **Phase transport-control:** Phase transport-controlled corrosion implies that the wetting of the metal surface by a corrosive phase is flow dependent. This may occur because one liquid phase separates from another or because a second phase forms from a liquid. An example of the second mechanism is the formation of discrete bubbles or a vapor phase from boiler water in horizontal or inclined tubes in high heat-flux areas under low flow conditions. The corroded sites will frequently display rough, irregular surfaces and be coated with or contain thick, porous corrosion deposits.
- **Erosion–corrosion:** Erosion–corrosion is associated with a flow-induced mechanical removal of the protective surface film that results in a subsequent corrosion rate increase via either electrochemical or chemical processes. It is often accepted that a critical fluid velocity must be exceeded for a given material. The mechanical damage by the impacting fluid imposes disruptive shear stresses or pressure variations on the material surface and/or the protective surface film. Erosion–corrosion may be enhanced by particles (solids or gas bubbles) and impacted by multiphase flows. The morphology of surfaces affected by erosion–corrosion may be in the form of shallow pits, horseshoe shapes, or other local patterns related to the flow direction.
- **Cavitation:** Cavitation sometimes is considered a special case of erosion–corrosion and is caused by the formation and collapse of vapor bubbles in a liquid near a metal surface. Cavitation removes protective surface scales by the implosion of gas bubbles in a fluid. Calculations have shown that the implosions produce shock waves with pressures >400 MPa.

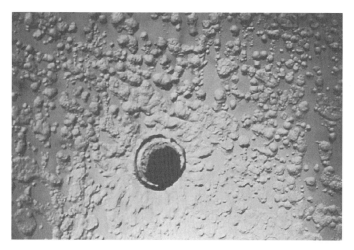

Fig. 2.17 Typical cavitation on the surface of a boiling seawater deaerator. (Courtesy Defence R&D Canada-Atlantic.)

The subsequent corrosion attack is the result of hydromechanical effects from liquids in regions of low pressure where flow velocity changes, disruptions, or alterations in flow direction have occurred. Cavitation damage often appears as a collection of closely spaced, sharp-edged pits or craters on the surface (Fig. 2.17). Cavitation also occurs in areas of high vibration, such as on engine pistons and piston liners (Fig. 2.18). Impingement attack is related to cavitation damage, and has been defined as "localized erosion–corrosion caused by turbulence or impinging flow." Entrained air bubbles tend to accelerate this action, as do suspended solids. This type of corrosion occurs in pumps, valves, orifices, on heat-exchanger tubes, and at elbows and tees in pipelines.

In general terms, as the flow of a liquid phase becomes turbulent the random liquid motion impinges on the surface to remove the naturally formed thin protective film. Additional oxidation then occurs by reaction with the liquid. This alternate oxidation and removal of the film will accelerate the rate of corrosion. The resulting erosive attack may be uniform, but quite often produces pitted areas over the surface that can result in full perforation (Fig. 2.19). Obviously, the presence of solid particles or gaseous bubbles in the liquid can accentuate this corrosion attack. Also, if the fluid dynamics is such that impingement or cavitation attack is developed, even more severe corrosion can occur.

2.2.6. Fretting Corrosion

Fretting corrosion refers to corrosion damage at the asperities of contact surfaces (Fig. 2.20). Such damage is induced under load and in the presence of repeated relative surface motion, as induced, for example, by vibration. Pits or grooves

Fig. 2.18 Cavitation damage of a diesel engine piston liner on the return stroke. (Courtesy Defence R&D Canada-Atlantic.)

Fig. 2.19 Erosion–corrosion of a brass tube carrying out seawater. (Courtesy Defence R&D Canada-Atlantic.)

Fig. 2.20 Fretting corrosion at the interface of two highly loaded surfaces that are not designed to move against each other. (Courtesy Mike Dahlager Pacific Corrosion Control Corp.)

and oxide debris characterize this damage, typically found in machinery, bolted assemblies, and ball or roller bearings. Contact surfaces exposed to vibration during transportation are exposed to the risk of fretting corrosion.

Damage can occur at the interface of two highly loaded surfaces that are not designed to move against each other. The most common type of fretting is caused by vibration. The protective film on the metal surfaces is removed by the rubbing action and exposes fresh, active metal to the corrosive action of the atmosphere. Conditions necessary for the occurrence of fretting corrosion are (*1*) the interface must be under load, and (2) vibratory or oscillatory motion of small amplitude must result in the surfaces striking or rubbing together.

The mechanism of fretting corrosion involves both chemical or electrochemical and mechanical factors. The relative motion scrubs off tiny high points or surface irregularities on the metal surface, and these, in turn, are chemically converted to oxide particles that, once dried, can be quite abrasive. They also may be sites for galling, in which case the mechanical action results in tearing out small bits of metal from one or the other of the interfaces and these oxidize to form debris.

2.2.7. Intergranular Corrosion

Intergranular corrosion is a form of localized attack in which a narrow path is corroded out preferentially along the grain boundaries of a metal. It initiates on the surface and proceeds by local cell action in the immediate vicinity of a grain boundary. Although the detailed mechanism of intergranular corrosion varies with the metal system, its physical appearance at the microscopic level is quite similar for most systems. The effects of this form of attack on mechanical properties may be extremely harmful.

The driving force is a difference in corrosion potential that develops between a thin grain boundary zone and the bulk of the immediately adjacent grains. The difference in potential may be caused by a difference in chemical composition between these two zones. Such a difference may develop as a result of migration of impurities or alloying elements in an alloy to the grain boundaries. If the concentration of alloying elements in the grain boundary region becomes sufficient, a second phase or constituent may separate or precipitate. This may have a corrosion potential different from that of the grains (or matrix) and cause a local cell to form.

In mild cases, the penetration rate may be extremely slow and even after a considerable time may penetrate to a depth of only a dozen grains or so. In more severe cases, penetration may go quickly through much thicker sections. Shallow intergranular corrosion may escape visual inspection, but is seen easily under a microscope. As attack proceeds, it can be recognized more readily by visual inspection.

Intergranular corrosion causes a drop in the ductile property of a metal, which can become appreciable before any significant loss of tensile stress or yield stress can be detected. In severe cases, there can be a marked loss of tensile properties, even though only a small volume of metal has corroded. A good example of intergranular corrosion was described in the Material Factor section of Chapter 1 when the sensitization of austenitic stainless steels was discussed.

2.2.8. Dealloying

Another type of localized corrosion involves the selective removal by corrosion of one of the elements of an alloy by either preferential attack or by complete dissolution of the matrix. Various kinds of selective dissolution have been named after the alloy system that has been affected, usually on the basis of the element dissolved (except in the case of graphitic corrosion).

Dezincification, for example, is usually associated with the corrosion of brasses containing >15% Zn, during which a porous copper surface zone develops as a result of the zinc depletion. Often the gross appearance and size of the part that has suffered dezincification is unchanged except for the development of a copper hue. The part, however, will have become weak and embrittled, and therefore subject to failure without warning. To the trained observer, dezincification is readily recognized under the microscope, and even with the unaided eye, because the red copper color is easily distinguished from the brass yellow color.

Conditions generally conducive to dezincification include the presence of an electrolyte (e.g., seawater), slightly acid conditions, the presence of carbon dioxide, and appreciable oxygen. There are two general types of dezincification. The more common is the layer type in which dezincification proceeds uniformly, as shown in Fig. 2.21. The second type is referred to as the plug type and occurs

Fig. 2.21 Layer dezincification of a brass fitting. (Courtesy of Defence R&D Canada-Atlantic.)

at localized areas. The site of plugs may be recognized by the presence above them of a deposit of brownish-white zinc-rich corrosion product.

Graphitic corrosion is a form of selective leaching that is specific to the deterioration of metallic constituents in gray cast iron, and that leaves the graphitic particles intact. Cast iron pipe may suffer graphitic corrosion as a result of the selective dissolution of the ferrite phase. Graphitic corrosion usually progresses uniformly inward from the surface, leaving a porous matrix of the remaining alloying element, that is, carbon in the form of graphite. Graphitic corrosion occurs in salt waters, acidic mine waters, dilute acids, and soils, especially those containing chlorides from deicing salts or seawater and sulfate reducing bacteria (SRB).

During graphitic corrosion, there is no outward appearance of damage, but the affected metal loses weight, and becomes porous and brittle. Depending on alloy composition, the porous residue may retain appreciable tensile strength and have moderate resistance to erosion. For example, a completely graphitized buried cast iron pipe may continue to hold water under pressure until jarred by a workman's shovel.

2.2.9. Environmental Cracking

Mechanical forces (e.g., tensile or compressive forces) will usually have little if any effect on the overall corrosion of metals. Compressive stresses do not cause cracking. In fact, shot peening is often used to reduce the susceptibility of metallic materials to fatigue, SCC, and other forms of cracking. However, a combination of tensile stresses and a specific corrosive environment is one of the most important causes of sudden cracking-type failures of metal structures.

Stress corrosion cracking and other types of environmental cracking are also the most insidious forms of corrosion because environmental cracks are microscopic in their early stages of development. In many cases, they are not evident on the exposed surface by normal visual examination, and can be detected only by special techniques. Optical or scanning electron microscopy (SEM) of failed sections may be required to fully identify them. As the cracking penetrates farther into the material, it eventually reduces the effective supporting cross-section to the point where the structure fails by overload or, in the case of vessels and piping, escape (seepage) of the contained liquid or gas occurs.

Environmental cracking is defined as the brittle fracture of a normally ductile material in which the corrosive effect of the environment is a causative factor. Environmental cracking can occur with a wide variety of metals and alloys and includes all of the types of corrosion failures listed below:

- Corrosion fatigue.
- Hydrogen embrittlement.
- Hydrogen-induced cracking (HIC), a cathodic process.
- Hydrogen stress cracking.
- Liquid metal cracking (LMC), usually a physicochemical process.
- Stress corrosion cracking (SCC), an anodic process.
- Sulfide stress cracking (SSC).

Cracking usually is designated as either intergranular (intercrystalline) as shown in Fig. 2.22 and 2.23 or transgranular (transcrystalline) as illustrated in Fig. 2.24. Occasionally, both types of cracking are observed in a failure. Intergranular cracks follow the grain boundaries in the metal. Transgranular cracks cross the grains without regard for the grain boundaries. The morphology of the cracks may change with the same material in different environments.

Failures generally do not result from the ordinarily applied engineering loads or stresses. Engineering loads, however, are often additive to the residual stresses already present in the structure. These residual stresses result from fabricating processes (e.g., deep drawing, punching, rolling of tubes into tubesheets, mismatch in riveting, spinning, welding).

Residual stresses will remain in a structure unless it is annealed or otherwise thermally stress relieved following fabrication, a practice that becomes increasingly impractical as a system gets larger or more complex. Cooling from the high temperatures required may also induce internal stresses because of nonuniform cooling. In fact, very slow, controlled cooling is a prerequisite for effective stress relief.

Corrosion products from general corrosion or other forms of corrosion may build up between mating surfaces and, because they occupy so much more volume than the metal from which they are produced, generate sufficient stresses to cause SCC. In the example shown in Fig. 2.25, moisture working down the steel rod in combination with the galvanic corrosion due to the contact with the bronze support caused enough rust build-up to generate high stresses and induce SCC.

Fig. 2.22 Intergranular SCC in a nut machined from a 2024 aluminum alloy extrusion. Stresses were in the transverse direction. Note how cracks follow a series of boundaries. Etched 2.5% HNO_3, 1.5% HCl, 1% HF. X100. (Corrosion Basics: An Introduction, 2nd edition, NACE International, by permission.)

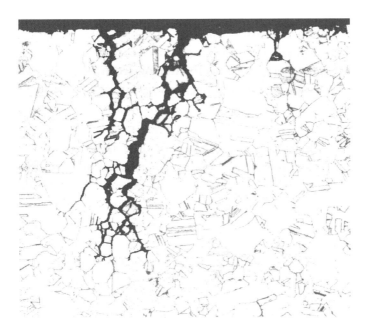

Fig. 2.23 Typical intergranular stress corrosion cracks in cartridge brass (70 Cu, 30 Zn). Etched 30% H_2O_2, 30% NH_4OH, 40% H_2O. X75. (Corrosion Basics: An Introduction, 2nd edition, NACE International, by permission.)

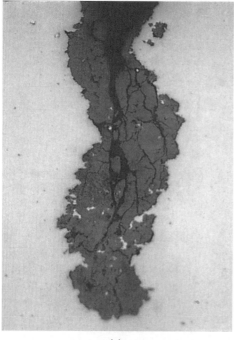

(a)

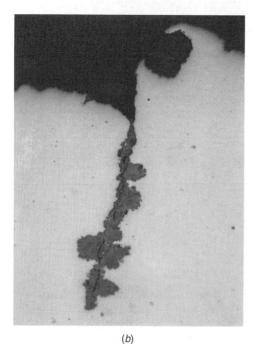

(b)

Fig. 2.24 Axial or transgranular cracks filled with oxide or scale (*a*) observed on the external surface of a pipeline. These cracks may occur in association with pits and general corrosion (*b*). (Courtesy of MACAW's Pipeline Defects, published by Yellow Pencil Marketing Co.)

Fig. 2.25 Stress corrosion crack in a bronze monument caused by build-up of rust around a decorative steel rod. (Courtesy Kingston Technical Software.)

2.2.9.1. Stress Corrosion Cracking

It is not uncommon for growing cracks to be considered acceptable (Fig 2.26) so long as they are substantially less than a critical crack size [Fig. 2.26a and Eq. (2.2)] and can be repaired at the next outage or shutdown. Unless it is found during an inspection that the crack growth rate might end up in a failure before the next outage (with some safety factor), remedial actions are usually not taken (3).

$$K_{I_c} = \sigma_m \sqrt{C \pi a} \qquad (2.2)$$

where σ_m = mean stress; a = depth of defect; C = geometric constant; K_{I_c} = critical stress intensity at which catastrophic fracture occurs.

Stress corrosion cracking is a mechanical-chemical process leading to the cracking of certain alloys at stresses below their tensile strengths. A susceptible alloy, the proper chemical environment, plus an enduring tensile stress are required. It is likely that there are no alloy systems which are completely immune to SCC in all environments. Usually, there is an induction period, during which cracking nucleates at a microscopic level. This period of latency that may be quite long (e.g., many months or years) is followed by actual propagation.

Table 2.3 lists some environments in which SCC has been observed for at least some alloys of the systems listed. This listing does not imply that all alloys of a given material will be equally susceptible, or that there are none in the class that may be immune to the environments listed.

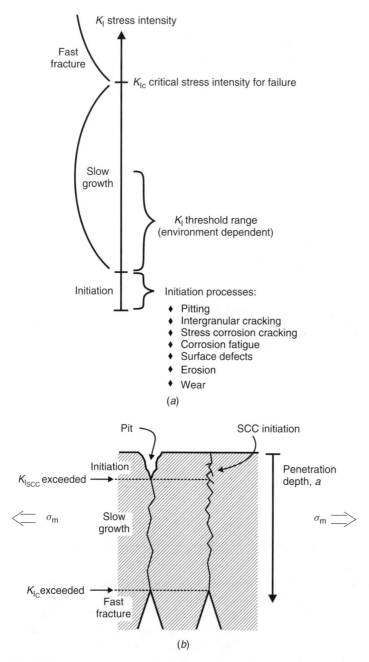

Fig. 2.26. (a) Schematic view of relative magnitudes of K_I shown with transitions and principal dependencies. The K_I parameter for threshold range is shown to be variable, (b) Morphological regimes, (c) The $\Delta a/\Delta t$ versus K for SCC and similar environmentally dependent slow growth (e.g., hydrogen), (d) $\Delta a/\Delta n$ versus ΔK for fatigue crack growth showing possible effect of SCC in increasing the slow growth (3).

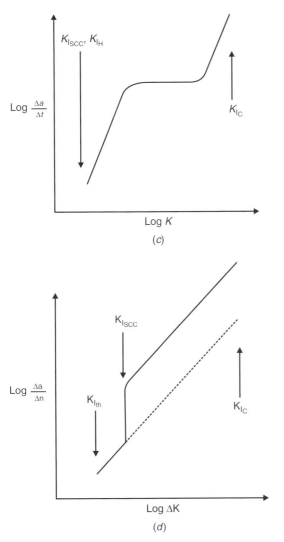

Fig. 2.26. (*Continued*).

Stress corrosion cracking is an anodic process, a fact that can be verified by the applicability of cathodic protection as an effective remedial measure, and may occasionally lead to fatigue, or vice versa. Usually, the true nature of the cracking can be identified by the morphology of the observed cracks.

In a failure by SCC, there is usually little metal loss or general corrosion associated with it. If there is severe general corrosion, SCC usually will not occur. Thus, the failure of a stress bolt rusted away until it eventually cannot sustain the applied load is not classified as SCC. However, if products from general corrosion are trapped so as to exert stress in a structure, they can cause SCC, as in the pillowing examples shown earlier.

Table 2.3. Some SCC Environments for Metals

Aluminum alloys	NaCl–H$_2$O, NaCl solutions, seawater, mercury
Copper alloys	Ammonia vapors and solutions, mercury
Gold alloys	FeCl$_3$ solutions, acetic acid salt solutions
Inconel	Caustic soda solutions
Lead	Lead acetate solutions
Magnesium alloys	NaCl-K$_2$Cr$_2$O$_7$ solutions, rural and coastal atmospheres, distilled water
Nickel	Fused caustic soda
Carbon steels	NaOH solutions; NaOH–NaSiO$_3$ solutions; CaNO$_3$, NH$_4$NO$_3$, and NaNO$_3$ solutions; mixed acids, (H$_2$SO$_4$–HNO$_3$); HCN; H$_2$S; seawater; NaPb alloy
Stainless steels	BaCl$_2$, MgCl$_2$ solutions; NaCl–H$_2$O$_2$ solutions; seawater, H$_2$S, NaOH–H$_2$S solutions
Titanium	Red fuming nitric acid

The idea, once prevalent, that only alloys and not pure metals are susceptible to SCC is quite possibly correct. The question is, 'How pure is pure?' Copper containing 0.004% phosphorous or 0.01% antimony is reported to be susceptible to SCC in environments containing ammonia or ammonium ions. Cracking has been produced in a decarburized steel containing <0.01% carbon, but still containing small amounts of manganese, sulfur, and silicon in a boiling ammonium nitrate solution. SCC has been produced in commercial titanium containing, among other constituents, 600 ppm of oxygen and 100 ppm of hydrogen. Hence, the idea that a given material cannot fail by SCC because it is commercially pure is not correct (22).

Transgranular SCC is a form of localized subsurface attack in which a narrow path is corroded randomly across grains without any apparent effect on the crack direction by the presence of the grain boundary. Transgranular SCC initiates on the surfaces and proceeds inward, presumably by local cell action. It can occur during the SCC of austenitic stainless steels and less commonly during the SCC of low-alloy steels. It can also occur in the SCC of copper alloys in certain media (e.g. ammonia). It seldom occurs in aluminum alloys.

2.2.9.2. Corrosion Fatigue

Fatigue is the failure of a metal by cracking when it is subjected to cyclic stress. The usual case involves rapidly fluctuating stresses that may be well below the tensile strength. As stress is increased, the number of cycles required to cause fracture decreases. There is usually a stress level below which no failure will occur, even with an infinite number of cycles, and this is called the endurance limit. In practice, the endurance limit is defined as the stress level below which no failure occurs in 1 million cycles. A fatigue curve, commonly known as an

S–N curve, is obtained by plotting the number of cycles required to cause failure against the maximum applied cyclic stress. A typical S–N curve is shown by the dotted line in Fig. 2.27.

When a metal is subjected to cyclic stress in a corrosive environment, the number of cycles required to cause failure at a given stress will be reduced below that shown in Fig. 2.27. This acceleration of fatigue by corrosion is called corrosion fatigue. A corrosion fatigue S–N curve can also be drawn, as shown by the solid line in Fig. 2.27. It is lower than the normal fatigue curve obtained for the same metal in air. In both cases, the frequency of stressing should be reported because this factor influences the endurance limit.

The solid S–N curve obtained in this case is the result of both corrosion and fatigue. The solid curve indicates that metal life under such conditions can be much lower than the reference curve established in air. The S–N curve with corrosion tends to keep dropping, even at low stresses, and thus does not level off, as will the ordinary fatigue curve.

A marked drop in or elimination of the endurance limit may occur even in a mildly corrosive environment, especially in the case of a film-protected alloy. For example, deionized water, which ordinarily will produce only film growth on an aluminum alloy in which it is immersed, will appreciably reduce the endurance limit of the same alloy subjected to cyclic stressing. This is because stress reversals cause repeated cracking of the otherwise protective surface film, and this allows access of the water to the unprotected metal below with resultant corrosion.

Failures that occur on vibrating structures (e.g., taut wires or stranded cables) exposed to the weather under stresses below the endurance limit are usually

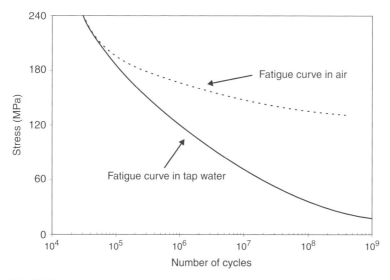

Fig. 2.27. Fatigue and corrosion fatigue curves for an aluminum alloy (22).

caused by corrosion fatigue. Corrosion fatigue also has been observed in steam boilers, due to alternating stresses caused by thermal cycling (Fig. 2.28).

The petroleum industry regularly encounters major trouble with corrosion fatigue in the production of oil. The exposure of drill pipe and of sucker rods to brines and sour crudes encountered in many producing areas results in failures that are expensive both from the standpoint of replacing equipment and from loss of production during the time required for "fishing" and rerigging.

For uniaxial stress systems, there will be an array of parallel cracks that are perpendicular to the direction of principal stress. Torsion loadings tend to produce a system of crisscross cracks at roughly 45° from the torsion axis. Corrosion fatigue cracks found in pipes subjected to thermal cycling usually will show a pattern made up of both circumferential and longitudinal cracks.

(a)

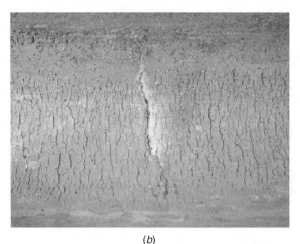

(b)

Fig. 2.28 (*a*) As-received appearance of boiler tube section with crack failure. Note remnants of tack weld to strap on the back of the tube and that the top of tube is marked establishing orientation. (*b*) Close-up ID surface of failed tube section showing band of numerous parallel partial through-wall cracks in line with through-wall crack (arrow). (2X Original Magnification.) (*c*) Photomicrograph showing as-polished longitudinal metallographic specimen through cracked area on failed tube section. Cracks originate on ID surface. Note wedge or needle-shaped, nonbranching cracks typical of corrosion fatigue. Faint black line near bottom of photo is an inclusion not related to the failure. (125X Original Magnification.) (Courtesy of Corrosion Testing Laboratories, Newark, Delaware.)

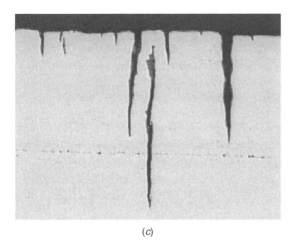

(c)   **Fig. 2.28** (*Continued*)

2.2.9.3. Hydrogen-Induced Cracking (HIC) and Hydrogen Embrittlement (HE)

There are several categories of hydrogen phenomena that are localized in nature. Atomic hydrogen, and not the molecule, is the smallest atom and it is small enough to diffuse readily through a metallic structure. When the crystal lattice is in contact or is saturated with atomic hydrogen, the mechanical properties of many metals and alloys are diminished. Nascent atomic hydrogen can be produced as a cathodic reaction when certain chemical species are present that act as negative catalysts (i.e., poisons) for the recombination of atomic to molecular hydrogen as shown in Eq. (2.3).

$$2H^{\circ} \rightarrow H_{2(g)} \tag{2.3}$$

If the formation of molecular hydrogen is suppressed, the nascent atomic hydrogen can diffuse into the interstices of the metal instead of being harmlessly evolved as a gaseous reaction product. There are many chemical species that poison this recombination (e.g., cyanides, arsenic, antimony, or selenium compounds). However, the most commonly encountered species is hydrogen sulfide (H_2S), which is formed in many natural decompositions, and in many petrochemical processes (22).

Processes or conditions involving wet hydrogen sulfide (i.e., sour services) and the high incidence of sulfide-induced HIC, has resulted in the term sulfide stress cracking (SSC). The SSC of medium strength steels has been a continuing source of trouble in the oil fields, and from these troubles has evolved in international standards (23). However, similar problems are encountered wherever wet hydrogen sulfide is encountered (e.g., acid gas scrubbing systems, heavy water plants, and waste water treatment).

Failures have occurred in the field when storage tank roofs have become saturated with hydrogen by corrosion and then been subjected to a surge in pressure, resulting in the brittle failure of circumferential welds. In rare instances, even copper and Monel 400 (N04400) have been subjected to HIC. More resistant materials, such as Inconels and Hastelloys often employed to combat HIC, can become susceptible under the combined influence of severe cold work, the presence of hydrogen recombination poisons, and a direct current from the galvanic couple due to electrical contact with a more anodic member.

The mechanism of HIC has not been definitely established. Various factors are believed to contribute to unlocking the lattice of the metal, such as hydrogen pressure at the crack tip, the competition of hydrogen atoms for the lattice bonding electrons, the easier plastic flow and dislocation formation in the metal at the crack tip in the presence of hydrogen, and the formation of certain metal hydrides in the alloy. The following phenomena have also been commonly reported in relation to hydrogen weakening of metallic components.

2.2.9.3.1. Loss of Ductility and Strength All metals will lose some portion of their ductility and to a lesser extent their strength when exposed to hydrogen. Copper, aluminum, and austenitic stainless steels are less affected than iron-or nickel-base alloys. The effect can be noted by conducting a slow strain rate tensile test of the materials when exposed to a hydrogen environment. Removing the alloy from the source of the hydrogen and baking the metal ~200°C to remove any absorbed hydrogen should restore its original mechanical properties.

2.2.9.3.2. Hydrogen Blistering When atomic hydrogen enters the metal structures, nonmetallic inclusions can catalyze the formation of molecular hydrogen

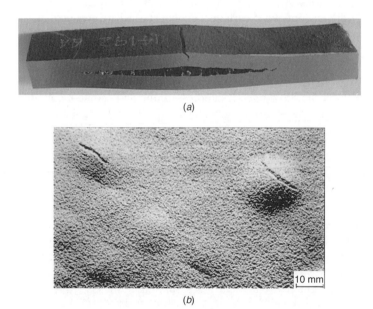

Fig. 2.29 (*a*) Hydrogen induced cracking with midwall cracks running parallel to the pipeline wall. (*b*) Surface blisters may also contain cracks. (Courtesy of MACAW's Pipeline Defects, published by Yellow Pencil Marketing Co.)

within the metal lattice, generating tremendous internal pressures and causing splits, fissures, and even blisters on the metal surface (Fig. 2.29). Normally three centimeters or so in diameter, some blisters have been observed larger than one meter in diameter in special cases. The tendency to blister can be combated to some extent by using steels of the same grain size and cleanliness as are specified for low-temperature service.

2.2.9.3.3. High-Temperature Effects At elevated temperatures and high partial hydrogen pressures, the tendency for molecular hydrogen to split into hydrogen atoms is greater and the atoms themselves are more active and mobile. Above $\sim 400°C$, cuprous oxide inclusions in metallic copper can be reduced, generating steam within the metal lattice and causing internal fissuring.

Of more importance from the engineering standpoint is the methanation of steel. At temperatures above $\sim 220°C$, atomic hydrogen will react with iron carbides in steel, forming methane within the structure and causing localized decarburization and fissuring.

REFERENCES

1. HORNE DH. Catastrophic Uncommanded Closures of Engine Feedline Fuel Valve from Corroded electrical connectors. CORROSION 2000, Paper No. 719. 2000. Houston, TX, NACE International.
2. NOWLAN FS. and HEAP HF. Reliability Centered Maintenance. AD-A066-579. 1978. Washington, DC, National Technical Information Service.
3. STAEHLE RW. Lifetime Prediction of Materials in Environments. In: Revie RW, ed. *Uhlig's Corrosion Handbook*. New York: Wiley-Interscience, 2000; pp. 27–84.
4. WEIR D. *The Bhopal Syndrome: Pesticides, Environment and Health*. San Francisco, CA: Sierra Club Books, 1987.
5. Natural Gas Pipeline Rupture and Fire Near Carlsbad, New Mexico, August 19, 2000. NTSB/PAR-03/01. 2003. Washington, DC, National Transportation Safety Board.
6. ECKERT R. *Field Guide for Investigating Internal Corrosion of Pipelines*. Houston, TX: NACE International, 2003.
7. MALO JM, SALINAS V, and URUCHURTU J. Stray current corrosion causes gasoline pipeline failure. *Materials Performance* 1994; **33**: 63.
8. 4X-AXG Boeing 747-258F El Al 04.10.92 Bijlmermeer. Accident report 92-11. 1992. Amsterdam, the Netherlands, Netherland Aviation Safety Board.
9. HAMER M. Clampdown on the Rust Buckets. *New Scientist* 1991; **146**: 5.
10. RUMMEL WD and MATZKANIN GA. *Nondestructive Evaluation (NDE) Capabilities Data Book*. 3rd edn. Austin, TX: Nondestructive Testing Information Analysis Center (NTIAC), 1997.
11. DILLON CP. *Forms of Corrosion: Recognition and Prevention*. Houston, TX: NACE International, 1982.
12. SPEIDEL MO and FOURT PM. Stress Corrosion Cracking and Hydrogen Embrittlement in Industrial Circumstances. In: Staehle RW, Hochmann J, McCright RD, and Slater JE, eds. *Stress Corrosion Cracking and Hydrogen Embrittlement of Iron Base Alloys*. Houston, TX: National Association of Corrosion Engineers, 1977; pp. 57–60.
13. FONTANA MG and GREENE ND. *Corrosion Engineering*. New York: McGraw-Hill, 1967.
14. Standard Guide for Examination and Evaluation of Pitting Corrosion. ASTM G46-94. 1999. West Conshohocken, PA, American Society for Testing of Materials.

15. HAYNIE FH. Statistical Treatment of data, Data Interpretation, and Reliability. In: Baboian R, ed. *Corrosion Tests and Standards, MNL 20*. Philadelphia, PA: American Society for Testing of Materials, 1995; pp. 62–7.
16. GUMBEL EJ. Statistical Theory of Extreme Values and Some Practical Applications. 33. 1954. Washington, DC, National Bureau of Standards. Mathematics Series.
17. Dissimilar Metals. MIL-STD-889B(3). 1993. Aberdeen, MD, Army Research Laboratory.
18. PEABODY AW and BIANCHETTI RL. *Peabody's Control of Pipeline Corrosion*. 2nd Houston, TX: NACE International, 2001.
19. POULSON BS. Erosion Corrosion. In: SHREIR LL, JARMAN RA, and BURSTEIN GT, eds. *Corrosion Control*. Oxford, UK: Butterworths Heinemann, 1994; pp. 1:293-1:303.
20. SHIFLER DA Environmental Effects in Flow Assisted Corrosion of Naval Systems. CORROSION 99, Paper No. 619. 1999. Houston, TX, NACE International.
21. ROBERGE PR. *Corrosion Testing Made Easy: Erosion–Corrosion Testing*. Houston, TX: NACE International, 2004.
22. ROBERGE PR. *Corrosion Basics—An Introduction*. 2nd edn. Houston, TX: NACE International, 2005.
23. NACE MR0175/ISO 15156, Petroleum and natural gas industries—Materials for use in H_2S-containing environments in oil and gas production. 2001. Houston, TX, NACE International.

Chapter 3

Maintenance, Management, and Inspection Strategies

3.1. The Cost Of Poor Maintenance
3.2. Corrosion Management Strategies
3.3. Maintenance Strategies
    3.3.1. Corrective Maintenance
    3.3.2. Preventive Maintenance
    3.3.3. Predictive or Condition-Based Maintenance
    3.3.4. Reliability-Centered Maintenance
3.4. Life Cycle Asset Management
    3.4.1. Life Cycle Costing
    3.4.2. Condition Assessment
    3.4.3. Prioritization
3.5. Improvement In Computer And Communication Technologies
3.6. Inspection Strategies
    3.6.1. What to Inspect?
        3.6.1.1. Anticipated Failures 'Hot Spots'
            3.6.1.1.1. Cracking Locations in Reactors
            3.6.1.1.2. Cracking Locations in Columns
            3.6.1.1.3. Cracking Locations in Heat Exchangers
            3.6.1.1.4. Cracking Locations in Piping
        3.6.1.2. Corrosion-Based Design Analysis
    3.6.2. When to Inspect?: Key Performance Indicators
        3.6.2.1. Cost of Corrosion KPI
        3.6.2.2. Corrosion Inhibition Level KPI
        3.6.2.3. Completed Maintenance KPI
        3.6.2.4. Selecting KPIs
            3.6.2.4.1. Asset Performance Metrics
            3.6.2.4.2. Tactical Perspectives
    3.6.3. Corrosion Monitoring or Corrosion Inspection?
    3.6.4. Risk-Based Inspection
        3.6.4.1. Probability of Failure Assessment
        3.6.4.2. Consequence of Failure Assessment
        3.6.4.3. Application of RBI

Corrosion Inspection and Monitoring, by Pierre R. Roberge
Copyright © 2007 John Wiley & Sons, Inc.

            3.6.4.3.1. Optimizing Inspection–Monitoring
            3.6.4.3.2. Materials of Construction Changes
            3.6.4.3.3. Key Process Parameters
   3.6.5. Risk Assessment Methodologies
        3.6.5.1. HAZOP
        3.6.5.2. FMEA and FMECA
        3.6.5.3. Risk Matrix Methods
        3.6.5.4. Fault Tree Analysis
        3.6.5.5. Event Tree Analysis
3.7. Failure Analysis Information
   3.7.1. Conducting a Corrosion Failure Analysis
        3.7.1.1. Planning the Analysis
        3.7.1.2. Conditions at the Failure Site
        3.7.1.3. Operating Conditions at Time of Failure
        3.7.1.4. Historical Information
        3.7.1.5. Sampling
        3.7.1.6. Evaluation of Samples
        3.7.1.7. Assessment of Corrosion-Related Failure
   3.7.2. Information to Look For
        3.7.2.1. Temperature Effects
        3.7.2.2. Fluid Velocity Effects
        3.7.2.3. Impurities
        3.7.2.4. Presence of Microbes
        3.7.2.5. Presence of Stray Currents
3.8. Industrial Examples
   3.8.1. Transmission Pipelines
        3.8.1.1. External Corrosion Damage Assessment
            3.8.1.1.1. Pre-assessment
            3.8.1.1.2. Indirect Inspection
            3.8.1.1.3. Direct Examinations
            3.8.1.1.4. Post-assessment
        3.8.1.2. Internal Corrosion Damage Assessment
            3.8.1.2.1. Pre-assessment
            3.8.1.2.2. Indirect Inspection
            3.8.1.2.3. Detailed Examination
            3.8.1.2.4. Post-assessment
        3.8.1.3. Hydrostatic Testing
        3.8.1.4. In-Line Inspection
            3.8.1.4.1. Metal Loss (Corrosion) Tools
            3.8.1.4.2. Crack Detection Tools
            3.8.1.4.3. Geometry Tools
   3.8.2. Offshore Pipeline: Risers
   3.8.3. Process Industry
   3.8.4. Power Industry
        3.8.4.1. Corrosion Product Activation and Deposition
        3.8.4.2. Pressurized Water Reactor Steam Generator Tube Corrosion
        3.8.4.3. Boiler Tube Waterside–Streamside Corrosion

- 3.8.4.4. Heat Exchanger Corrosion
- 3.8.4.5. SCC and CF in Turbines
- 3.8.4.6. Fuel Cladding Corrosion
- 3.8.4.7. Corrosion in Electric Generators
- 3.8.4.8. Flow Accelerated Corrosion
- 3.8.4.9. Corrosion of Raw Water Piping
- 3.8.4.10. IGSCC of BWR Piping and Internals
- 3.8.5. The RIMAP Project
  - 3.8.5.1. Process Industry
  - 3.8.5.2. Offshore Industry
  - 3.8.5.3. Power Industry
- 3.8.6. Aircraft Maintenance
  - 3.8.6.1. Corrosion Definition
    - 3.8.6.1.1. Level One Corrosion
    - 3.8.6.1.2. Level Two Corrosion
    - 3.8.6.1.3. Level Three Corrosion
  - 3.8.6.2. Maintenance Schedule
  - 3.8.6.3. Corrosion Management Assessment
  - 3.8.6.4. Maintenance Steering Group System
- 3.8.7. Water Utilities
  - 3.8.7.1. Corrosion Impact
    - 3.8.7.1.1. Health and Regulations
    - 3.8.7.1.2. Aesthetics and Customer Perception
    - 3.8.7.1.3. Premature Piping Deterioration and Economic Impacts
    - 3.8.7.1.4. Environmental Concerns
  - 3.8.7.2. Corrosion Management
    - 3.8.7.2.1. Short-Term Corrosion Management
    - 3.8.7.2.2. Long-Term Corrosion Management
    - 3.8.7.2.3. Necessity of Long-Term Corrosion Planning
    - 3.8.7.2.4. Framework for Water Pipeline Management
  - 3.8.7.3. Condition Assessment Techniques
    - 3.8.7.3.1. Water Audits
    - 3.8.7.3.2. Sonic and Acoustic Leak Detection
    - 3.8.7.3.3. Soil Corrosivity
- 3.8.8. Reinforced Concrete Infrastructure
  - 3.8.8.1. Electrochemical Corrosion Measurements
  - 3.8.8.2. Chloride Content
  - 3.8.8.3. Petrographic Examination
  - 3.8.8.4. Permeability Tests

References

3.1. THE COST OF POOR MAINTENANCE

Maintenance costs represent a significant portion of operating budgets in most industrial sectors, particularly where aging structures and equipment are involved.

Within any industrial sector, there is a range of practices when dealing with corrosion, from using old "proven" technologies to the most modern state-of-the-art methods. While some of these practices attempt to be cost effective, others could be improved in that respect.

Modern approaches to maintenance management are designed to minimize costs while improving the reliability and availability of plant and equipment. In many industries, maintenance activities are treated as an investment and, during overall rationalization, the maintenance function often has to be accomplished with shrinking technical and financial resources. In many cases, traditional corrective maintenance and time-based preventive maintenance practices are inadequate to meet modern demands. The consequences of poor maintenance practices and/or inadequate investment in the maintenance function could lead to serious consequences, such as the following:

- Direct reduced production capacity due to downtime or lack of optimal performance during uptime.
- Increased production costs and cost penalties for lack of optimal performance.
- Lower quality products and services with client dissatisfaction and possible lost sales.
- Safety hazards and failures leading to loss of life, injuries, and possibly major liabilities.

3.2. CORROSION MANAGEMENT STRATEGIES

Some of the issues involved in deciding on a cost-effective solution for combating corrosion are generic to sound system management. Others are specifically related to the impact of corrosion damage on system integrity and operating costs. Corrosion management includes all activities throughout the lifetime of a structure or system that are performed to mitigate corrosion, to repair corrosion induced damage, and to replace the structure or system that has become unusable as a result of corrosion.

Repair and rehabilitation activities are established to restore damaged structures to their original or required service level and correct the deficiencies that might have resulted in corrosion deterioration. These activities are performed at different times throughout the lifetime of a system. Maintenance is considered to be a regular and necessary activity that is characterized by an annual cost. Inspections are scheduled periodic activities, and repair is performed on an as-needed basis. Repair can involve the replacement of parts, but not the replacement of the basic structure. Rehabilitation of structures, such as bridges, is usually done only once or twice during the lifetime of the structure, generally at a high cost. The goal of corrosion management is to achieve the desired level of service at the least cost.

3.3. MAINTENANCE STRATEGIES

There are basically four general types of maintenance philosophies or strategies, namely, corrective, preventive, predictive, or condition-based, and reliability centered maintenance. Figure 3.1 illustrates the various maintenance relationships and activities associated with these strategies (1). Predictive maintenance is the most recent development. In practice, all these strategies are used in maintaining engineering systems. The challenge is to optimize the balance between the chosen strategies for maximum profitability. In general, corrective maintenance is the least cost effective option when maintenance requirements are high.

3.3.1. Corrective Maintenance

Corrective maintenance refers to actions taken only when a system or component failure has occurred. It is thus a retroactive strategy. The task of the maintenance team in this scenario is usually to effect repairs as soon as possible. Costs associated with corrective maintenance include repair costs (e.g., replace components, labor, and consumables), lost production and lost sales. To minimize the effects of lost production and speed up repairs, actions, such as increasing the size of maintenance teams, the use of back-up systems, and implementation of emergency procedures, might be considered. Unfortunately, such measures are relatively costly and/or only effective in the short term. For example, if heat exchanger tubes have leaked due to pitting corrosion and production must proceed as a matter of urgency, it may be possible to plug the leaking tubes on a short-term basis. Obviously, the longer term performance of a heat exchanger is less than assured with such measures.

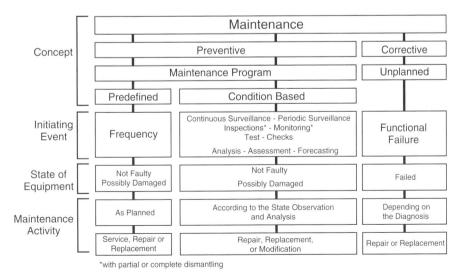

Fig. 3.1. Strategic maintenance relationships (1).

3.3.2. Preventive Maintenance

In preventive maintenance, equipment is repaired and serviced before failures occur. The frequency of maintenance activities is predetermined by schedules. The more important the failure consequences are, the greater should be the level of preventive maintenance. This ultimately implies a trade-off between the cost of performing preventive maintenance and the cost to run the equipment to failure. Preventive maintenance tasking can of course also be dictated by safety, environmental, insurance, or other regulatory compliance.

Inspection assumes a crucial role in preventive maintenance strategies. Components are essentially inspected for corrosion and other damage at planned intervals, in order to identify corrective actions before any failures actually occur. Preventive maintenance performed at regular intervals will usually result in reduced failure rates. As significant costs are involved in performing preventive maintenance, especially in terms of scheduled downtime, good planning is vital. To maximize asset value and performance, the basic aim is to perform preventive maintenance just before serious damage would set in otherwise.

The level of preventive maintenance activity needs to be driven by the importance of the equipment to the process and the desired level of reliability. In complex systems, computerized preventive maintenance is used to accomplish these objectives for most plant sizes. A preventive maintenance system also needs to be dynamic and incorporate some built-in feedback mechanism to ensure that the tasking is still valid or need to be replaced with a predictive task.

3.3.3. Predictive or Condition-Based Maintenance

Predictive maintenance refers to maintenance based on the actual condition of a component. Maintenance is not performed according to fixed preventive schedules, but rather when a certain change in characteristics is noted. Corrosion sensors supplying diagnostic information on the condition of a system or component play an important role in this maintenance strategy. Preventive maintenance aims to eliminate unnecessary inspection and maintenance tasks, to implement additional maintenance tasks when and where needed, and to focus efforts on the most critical items.

A useful analogy can be made with automobile oil changes. Changing the oil every 5000 km to prolong engine life, irrespective of whether the oil change is really needed or not, is a preventive maintenance strategy. Predictive maintenance would entail changing the oil based on changes in its properties, such as the build-up of wear debris. When a car is used exclusively for long-distance highway travel and driven in a very responsible manner, oil analysis may indicate a longer critical service interval (2).

Some of the resources required to perform predictive maintenance will be available from the reduction in breakdown maintenance and the increased utilization that results from proactive planning and scheduling. Good record keeping is very important to identify repetitive problems, and the problem areas with the highest potential impact.

3.3.4. Reliability-Centered Maintenance

Reliability-centered maintenance (RCM) involves the establishment or improvement of a maintenance program in the most cost-effective and technically feasible manner. It utilizes a systematic, structured approach based on the consequences of a failure. As such, it represents a shift away from time-based maintenance tasks and emphasizes the functional importance of system components and their failure–maintenance history. Reliability-centered maintenance does not refer to a particular maintenance strategy, such as preventive maintenance; rather it can be employed to determine whether preventive maintenance is the most effective approach for a particular system component.

The concept of RCM finds its roots in the early 1960s mentations for commercial aircraft in the late 1960s, when wide-body jets were introduced in commercial service (3). A major concern of airline companies was that existing time-based preventive maintenance programs would threaten the economic viability of larger, more complex aircraft. In the time based maintenance approach, components are routinely overhauled after a certain amount of flying hours. In contrast, as pointed out above, RCM determines maintenance intervals based on the criticality of a component and its performance history. The experience of airlines with the RCM approach was that maintenance costs remained roughly constant but that the availability and reliability of their planes improved (3). Reliability-centered maintenance is now standard practice for most of the world's airlines.

The initial development work was done by the North American civil aviation industry through "Maintenance Steering Groups" or MSG. The MSGs were established to reexamine everything currently practiced to keep aircraft airborne. These groups consisted of representatives of the aircraft manufacturers, the airlines and the Federal Aviation Administration (FAA). The first attempt at a rational, zero-based process for formulating maintenance strategies was promulgated by the Air Transport Association in Washington, DC, in 1968. The first attempt is now known as MSG 1. A refinement, now known as MSG 2, was promulgated in 1970.

In the mid-1970s the U.S. Department of Defense commissioned a report on the subject of RCM in the aviation industry. This report was written by Stanley Nowlan and Howard Heap of United Airlines. The report was published in 1978, and it is still one of the most important documents in the history of physical asset management (4). Nowlan and Heap's report represented a considerable advance on MSG 2 thinking. It was used as a basis for MSG 3, which was promulgated in 1980. MSG 3 has since been revised twice. Revision 1 was issued in 1988 and revision 2 in 1993. It is used to this day to develop prior-to-service maintenance programs for new aircraft types, for example, Boeing 777 and Airbus 330/340.

Following the application of RCM in commercial aviation and defense, these methodologies have also been applied to maintenance programs in the nuclear power industry, chemical processing, fossil fuel power generation and other industries. Potential benefits of RCM include maintaining high levels of system reliability and availability; minimizing unnecessary maintenance tasks; providing a documented basis for maintenance decision making; identifying the most cost-effective inspection, testing and maintenance methods.

3.4. LIFE CYCLE ASSET MANAGEMENT

There are significant improvements that can be made in terms of both costs and efficiency by implementing asset management and maintenance practices. From a materials performance point of view, it is critical that assets are fit for purpose, perform safely, and with respect for environmental integrity, and most of all, deliver what the users want, when and where they want it. Asset management refers to the effective management of assets, from the time of planning and acquisition to eventual disposal.

In life cycle asset management, the aim is to maximize a return on the investment in assets by providing comprehensive information about their condition and value throughout their life. The emphasis is not on short-term costs of an asset but rather on the total value (performance) through its entire life. The optimum value of an asset is dependent on an optimum level of investment. Both the asset value and the available investment levels are a function of time, a variable that assumes major importance in life cycle asset management.

The pressure to make sound decisions with respect to construction, inspection and maintenance methods, and priorities is never greater than during difficult economic times. The main challenge for maintenance managers is to ensure that available and potentially scarce resources are applied optimally to maintenance requirements. The most critical requirements should be addressed first, followed by prioritization of remaining maintenance needs. Life cycle asset management can go a long way toward providing solutions to this challenge. It can be used to justify maintenance budgets, prioritize maintenance expenditures, and predict the need to acquire new assets. Life cycle asset management focuses on the application of three basic facility management tools: life cycle costing, condition assessment, and prioritization.

3.4.1. Life Cycle Costing

Life cycle costing utilizes universally accepted accounting practices for determining the total cost of asset ownership or projects over the service life. The economic analysis is usually performed for comparing competing alternatives. Since the initial capital outlay, support, and maintenance over the service life and disposal costs are considered, the time value of money assumes major importance in life cycle costing. Discounting future cash flows to present values essentially reduces all associated costs to a common point in time for objective comparison.

In practice, defining and controlling life cycle costs may be difficult. The future behavior of materials is often uncertain, as are the future use of most systems, the environmental conditions to which they may be exposed, and the financial and economic conditions that influence relationships between present and future costs. An effective life cycle cost analysis depends on having a reasonable range of possible alternatives that are likely to deliver equally satisfactory service over a given service life.

A generalized economic analysis Eq. (3.1) was developed that is particularly adapted to corrosion engineering problems. Given the numerous uncertainties associated with most corrosion problems this equation can provide fairly good estimates for different corrosion prevention and control alternatives. This equation takes into account the influence of taxes, straight-line depreciation, operating expenses, and salvage value in the calculation of present worth and annual cost. By using this equation, a problem can be solved merely by entering data into the equation with the assistance of compound interest data (5).

$$PW = -P + \left(\frac{t(P-S)}{n}\right)\left(\frac{P}{A}, i\%, n\right) \\ - (1-t)(X)\left(\frac{P}{A}, i\%, n\right) + S\left(\frac{P}{F}, i\%, n\right) \quad (3.1)$$

where A represents the annual end-of-period cash flow; F represents a future sum of money; $i\%$ represents the interest rate; n is the number of years; PW is the present worth referred also as Net Present Value (NPV); P is the cost of the system at time 0; S represents salvage value; t is the tax rate expressed as a decimal; X represents the operating expenses.

1. **First term [$-P$]:** This term represents the initial project expense, at time zero. As an expense, it is assigned a negative value. There is no need to translate this value to a future value in time, as the PW approach discounts all money values to the present (time zero).

2. **Second term [$(t(P-S)/n)(P/A, i\%, n)$]:** The second term in this equation describes the depreciation of a system. The portion enclosed in braces expresses the annual amount of tax credit permitted by this method of straight-line depreciation. The portion in parentheses translates these as equal amounts back to time zero by converting them to present worth.

3. **Third term [$-(1-t)(X)(P/A, i\%, n)$]:** The third term in the generalized equation consists of two terms. One is $(X)(P/A, i\%, n)$ that represents the cost of items properly chargeable as expenses, such as the cost of maintenance, insurance, and the cost of inhibitors. Because this term involves expenditure of money, it also comes with a negative sign. The second part, $t(X)(P/A, i\%, n)$, accounts for the tax credit associated with this business expense and because it represents a saving it is associated with a positive sign.

4. **Fourth term [$S(P/F, i\%, n)$]:** The fourth term translates the future value of salvage to present value. This is a one time event rather than a uniform series and therefore it involves the single payment present worth factors. Many corrosion measures, such as coatings and other repetitive maintenance measures, have no salvage value, in which cases this term is zero.

Present worth (PW) can be converted to equivalent annual cost (A) by using the following formula: $A = (PW)(A/P, i\%, n)$

One can calculate different options by referring to interest tables or by simply using the formula describing the various functions. The capital recovery function (P/A) or how to find P once given A:

$$\left(\frac{P}{A}, i\%, n\right) \quad \text{where} \quad P_n = A\frac{(1+i)^n - 1}{i(1+i)^n}$$

The compound amount factor (P/F) or how to find P once given F:

$$\left(\frac{P}{F}, i\%, n\right) \quad \text{where} \quad F_n = P(1+i)^{-n}$$

The capital recovery factor (A/P) or how to find A once given P:

$$\left(\frac{A}{P}, i\%, n\right) \quad \text{where} \quad A_n = P\frac{i(1+i)^n}{(1+i)^n - 1}$$

Several examples of the application of the generalized equation to corrosion engineering problems can be found in the literature (2, 5).

3.4.2. Condition Assessment

A second major component of life cycle asset management is systematic condition assessment surveys (CAS). The objective of CAS is to provide comprehensive information about the condition of an asset. This information is imperative for predicting medium- and long-term maintenance requirements, projecting remaining service life, developing long-term maintenance and replacement strategies, planning future usage, and determining the available reaction time to damage. Therefore, CAS is in direct contrast to a short-term strategy of "fixing" serious defects as they are found. As mentioned previously, such short-sighted strategies are often ultimately not cost-effective and will not provide optimum asset value and usage in the longer term. Condition assessment surveys include three basic steps (2, 6):

1. The facility is divided into its systems, components and subcomponents, forming a work breakdown structure (WBS).
2. Standards are developed to identify deficiencies that affect each component in the WBS and the extent of the deficiencies.
3. Each component in a WBS is evaluated against the standard or reference component.

Condition assessment surveys provides the maintenance managers the analytical information needed to optimize financial resource allocation for repair, maintenance and replacement of assets. Through a well-executed CAS program, information will be available on the specific deficiencies of a system or component, the extent and coverage of that deficiency, and the urgency of repair. The following scenarios, many of which will be all too familiar to readers, indicate a need for CAS in corrosion control strategies:

- Assets are aging, with increasing corrosion risks.
- Assets are complex engineering systems, although they may not always appear to be (e.g., "ordinary" concrete is actually a highly complex material).
- Assets fulfilling a similar purpose have variations in design and operational histories.
- Existing asset information is incomplete and/or unreliable.
- Previous corrosion maintenance/repair work was performed but poorly documented.
- Information on the condition of assets is not transferred effectively from the field to management, leaving the decision makers ill-informed.
- Maintenance costs are increasing, yet asset utilization is decreasing.
- There is great variability in the condition of similar assets from poor to excellent. The condition appears to depend on local operating microenvironments, but there is no knowledge on where the next major problem will appear.
- The information for long-term planning is very limited or nonexistent.
- Commitment within an organization to conduct long-term strategies and plans for corrosion control is limited or lacking.

A requirement of modern condition assessment surveys is for the data and information to be ultimately stored and processed using computer database systems. As descriptive terms are unsuitable for these purposes, some form of numerical coding is required to describe the condition of engineering components. Such condition codes will tend to decrease with the age of the system, while maintenance work will have the effect of upgrading these codes. The overall trend in condition code behavior will thus indicate whether maintenance is keeping up with environmental deterioration.

3.4.3. Prioritization

Prioritizing maintenance activities is central to a methodical, structured maintenance approach, in contrast to merely addressing maintenance issues in a reactive, short-term mode. From the preceding sections, it should be apparent that life cycle asset management can be used to develop a prioritization scheme that can be employed in a wider set of funding decisions, not only maintenance "go–no/

go" decisions. This entails the methodical evaluation of each maintenance action against preestablished values and attributes. Prioritization methodologies usually involve a numerical rating system, to ensure that the most important work receives the most urgent attention. The criticality of equipment is an important element of some rating systems. Such an unbiased, "unemotional" rating will ensure that decisions are made for the best overall performance of an engineering system, rather than overemphasizing one of its parts.

3.5. IMPROVEMENT IN COMPUTER AND COMMUNICATION TECHNOLOGIES

The investment in computer and communication tools can obviously be considerable in terms of software, hardware, and associated maintenance (e.g., upgrades, firewall, and technical glitches). While the computing technology should ideally be up to date and leading edge for optimum efficiency, an important consideration is how adaptable the computerized tools are and how easily they can be upgraded to avoid major unnecessary reinvestments in the future. The progress in portable computing power and data storage in recent years has been paralleled by similar spectacular advances in communication, global positioning, and other information systems that can greatly support field personnel in doing tasks from the most tedious or repetitive to the most complex.

Monitoring techniques have advanced significantly as well. Measurements can be taken more rapidly using small sensors and data can be stored in handheld electronic devices or portable computers, or be sent to the control room or central office by wire or using wireless technology. These computerized sensors allow an operator to monitor local corrosion conditions in real-time and almost anywhere.

In the following example, the main functions and features of a commercial software system designed for cathodic protection (CP) data management are briefly described to illustrate how the work of a field engineer or operator can be greatly facilitated. Similar systems are available for most corrosion inspection and protection applications for which data gathering and parameters monitoring are important (e.g., water chemistry management, ultrasonic piping inspection, chemical process operations).

In this example, the software system is used to store and maintain the data from bimonthly rectifier readings and annual test-point surveys. The software operates over the corporate network as well as on stand-alone laptop computers. Using the two-way replication feature, a copy of the database can be loaded onto a laptop computer and taken into the field, giving the field worker access to all historical cathodic protection data. New readings can be added in the field. On returning to the office, new readings can be added to the master database making the new data accessible to all users over the corporate network (7).

In addition, routes[1] can be defined and downloaded onto a rugged waterproofed field computer with an integrated global positioning system (GPS) unit

---

[1] A route consists of an order list of test locations where data is to be collected including the previous or last reading collected at the site.

for capturing the location of new test points. The software system also contains a variety of standard reports, including reading history and compliance reports, as well as flexible ad hoc reporting capabilities, which are used to generate custom reports as needed. Such a data management system provides a number of advantages that can facilitate the CP system commissioning:

1. **Field access to CP history data:** The entire CP database is available on each field crew's laptop computer. As questions are encountered in the field, the historical operating data can be reviewed directly to analyze the cause of any low readings and determine if additional testing is required.
2. **Two-way replication:** By using the two-way replication capabilities of the system, field crews can synchronize their remote databases with the master copy of the databases. This synchronization process provides the pipeline operations staff with access to the latest field data.
3. **Route files:** Route files containing the list of test station locations can be downloaded into the field computer for each compressor station and for pipeline segments between compressor stations.
4. **Ad hoc reporting:** The ability to generate custom reports, either in the field or after work is one of the more useful features of a laptop-based data management system.

3.6. INSPECTION STRATEGIES

Inspection normally refers to the evaluation of the quality of some characteristic in relation to a standard or a specification. Over the last decades, the inspection process has grown in complexity in parallel with the growth in complexity of systems and their production processes. A flow diagram is useful for showing the various materials, components, and processes that collectively or sequentially make up the system. Inspection consists of the following series of actions: interpretation of the specifications; measurement and comparison with a specification; judging conformance; classification of conforming cases; classification of non-conforming cases; and recording and reporting the data obtained.

When several inspection techniques can be used, the choice of a specific schedule will depend on the accuracy and cost of the inspection, balancing the money spent on safety measures with the business return of the system being maintained (Fig. 3.2) (8).

3.6.1. What to Inspect?

In overall tight budgets, the selection of the components, parts, or systems that should be inspected is of paramount importance. A good knowledge of intricate designs is an important asset since corrosion factors are often related to the geometry of systems and components. This selection should also be based on a

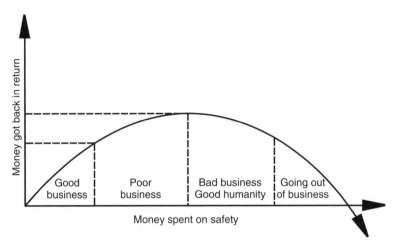

Fig. 3.2. The effects of increasing expenditure on safety and inspection.

thorough knowledge of process conditions, materials of construction, geometry of the system, external factors, and historical records.

3.6.1.1. Anticipated Failures 'Hot Spots'

Historical data gathered during previous inspection and repairs cycles can be quite useful to determine the particular locations where future maintenance actions should be focused. Figure 3.3 illustrates some of the 'hot spots' found for the P-3, a maritime patrol and antisubmarine warfare aircraft having fifty years of service.

Similarly, there is a lot of knowledge in process operations. In the process industry, units that start up and shut down frequently are more susceptible to stress corrosion cracking (SCC) than identical units in the same service that operate for long periods at a steady state. This is because start-up and shut-down produce thermal stresses and internal pressure surges that add a low-rate cyclic component to the static stress of normal operation (9).

In a single-train unit, obviously the greatest concern will be for capital intensive items, such as major stills and heat exchangers, and for items whose failure might affect production. Any items within a given train that run hotter, or could give rise to more concentrated solutions than associated equipment, are worthy of extra attention. Reboilers, for example, frequently crack first when a corrosion problem surfaces in a unit, because reboilers normally run hotter than their associated stills, and their tubes are thinner and more highly stressed than the vessel walls.

3.6.1.1.1. Cracking Locations in Reactors All weldments contain high residual stress and, unless they are suitably stress relieved, they are focal points for environmental cracking. However, the circumferential welds by which nozzles

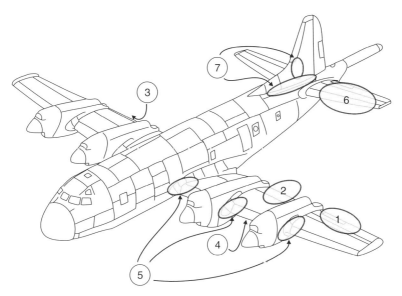

Fig. 3.3. Corrosion areas found on P-3 maritime patrol aircraft over decades of service: (1) aileron bay bonded panels, (2) flap bay bonded panels, (3) main landing gear jack pad, (4) inboard nacelle structure, (5) fillet fairing, (6) horizontal stabilizers, and (7) vertical stabilizers.

are attached to the vessels are particularly vulnerable. The radius of a dished head also contains residual stress from the forming operation, and the top head is often the first point of failure when volatile species (e.g., traces of hydrogen or ferric chloride) are involved (Fig. 3.4) (9). External jackets can cause problems, unrelated to the process, that arise from the use of cooling water or steam in the jacket that causes external SCC of the vessel proper.

3.6.1.1.2. Cracking Locations in Columns Weldments, especially circumferential nozzle welds, are a primary area for corrosion, as are the radii of dished heads. Nozzles themselves, if in the horizontal plane, may accumulate and hold chlorides or other corrosive agents, making them particularly susceptible to SCC (Fig. 3.5) (9).

Packing is especially susceptible to SCC. Expanded metal packing or mesh contains a great deal of cold-work and very high stress levels, rendering it particularly susceptible to SCC. Although both types of packing can be stress relieved, alone or with the column itself, a preferred method is to manufacture them from crack resistant alloys.

Trays, especially valve trays and bubble caps that may have high residual stress are also a primary area for environmental cracking. Tray posts, clips, fasteners, and other hardware may be highly stressed and susceptible to SCC. Sieve trays are usually manufactured by cold-punching, and bubble caps and valves are cold formed. The latter particularly, unless stress relieved or of resistant-alloy

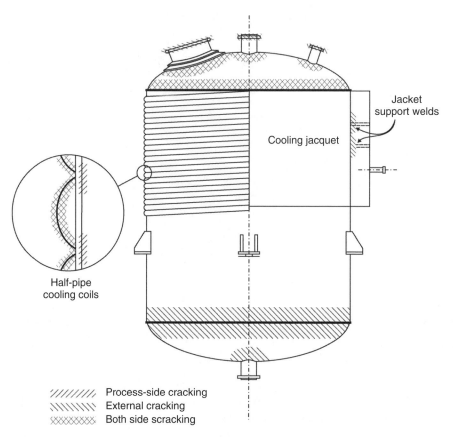

Fig. 3.4. Typical cracking locations in chemical process reactors (9).

construction, may lose their restraining legs and fall off the trays, leaving only an extremely large "sieve" tray of inefficient characteristics.

3.6.1.1.3. Cracking Locations in Heat Exchangers Heat exchangers share with other equipment the inevitable problems of stress associated with weldments, nozzles, and cold forming especially in heads or water boxes. But peculiar to tube and shell exchangers are the stresses associated with the bending of tubes for U-bundles, and the stresses imposed by rolling the tubes into the tubesheet. The areas immediately adjacent to the tubesheets, and the U-bends themselves, are primary areas for environmental cracking.

Plate–and frame exchangers have residual stresses associated with the method of manufacture of the component parts. Spiral exchangers have stresses associated with the forming operation. Panel-coil exchangers have residual stresses inherent in the forming and welding operations. All such high-stress areas are immediately suspected.

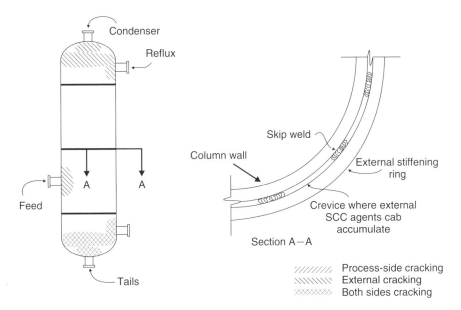

Fig. 3.5. Typical cracking locations in distillation columns (9).

3.6.1.1.4. Cracking Locations in Piping There may be a multiplicity of welds in a piping system that may be susceptible to aggravated attack unless full penetration welds are achieved. In addition, cold work due to bending operations, and especially that associated with flared pipe, provides areas highly susceptible to environmental cracking (Fig. 3.6).

3.6.1.2. Corrosion-Based Design Analysis

An analogy to the detailed design of a complete plant has been proposed to decide on maintenance and inspection activities on the basis on first principles. The Corrosion-based design analysis (CBDA) approach to predicting performance is a series of knowledge elicitation steps that require detailed consideration on a multitude of topics (10).

The two most important of these steps are described in Figs. 3.7 and 1.15 for, respectively, the environment and the material definitions. Each of the numbers in square brackets in Fig. 3.7 identifies an explicit action that needs to be considered for the environmental definition. The end point of the process is an input to a location for analysis (LA) matrix that is illustrated in Fig. 3.8 for the locations in a steam generator. A brief explanation of the individual elements in Fig. 3.7 follows:

  1. "Nominal Chemistry" refers to the bulk chemistry. For components exposed to ordinary air atmospheres, the "Major" elements mean humid air. The "Minor" elements refer to industrial contaminants, such as SO_2 and NO_2.

96 Chapter 3 Maintenance, Management, and Inspection Strategies

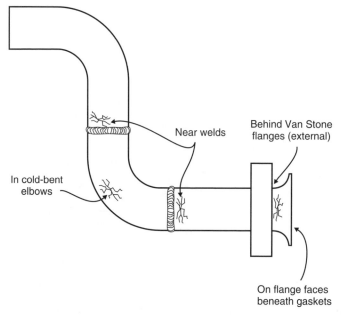

Fig. 3.6. Typical cracking locations in piping (9).

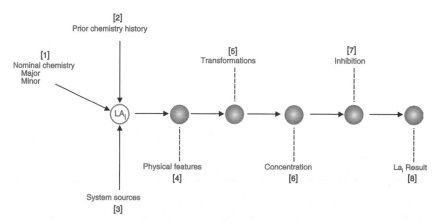

Fig. 3.7. Analysis sequence for determining environment at a location for analysis.

2. "Prior Chemistry History" refers to exposures to environmental species that might still reside on the surfaces or inside crevices.
3. "System Sources" refers to those environments that do not come directly from a component, but from an outside source.
4. "Physical Features" includes occluded geometries, flow, and long range electrochemical cells.

3.6. Inspection Strategies

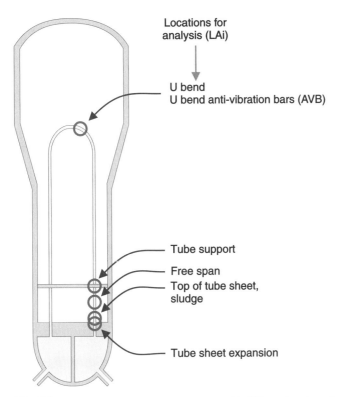

Fig. 3.8. Schematic view of steam generator with different locations for analysis.

5. "Transformations" refers, for example, to microbial actions that can change relatively innocuous chemicals, such as sulfates into very corrosive sulfide species that may accelerate hydrogen entry and increase corrosion rates.

6. "Concentration" refers to accumulations much greater than that in the bulk environment due to various actions of wetting and drying, evaporation, potential gradients, and crevices actions that prevent dilution.

7. "Inhibition" refers to actions taken to minimize corrosive actions. This usually involves additions of oxygen scavengers or other chemicals that interfere directly with the anodic or cathodic corrosion reactions.

The LA template of the locations that correspond to most likely failure sites along tubes in a steam generator of a pressurized water nuclear power plant (Fig. 3.8) is detailed in Table 3.1 for the main failure modes and submodes considered in such analysis. Maintenance and inspection actions can be decided upon by following developing trends monitored in each LA matrix thus produced.

Table 3.1. Matrix for Organizing Mode-Location cells[a]

Locations for analysis LA_i	ID	OD	LPSCC ($j=1$)	HPSCC ($j=2$)	Submodes of SCC			AkSCC ($j=5$)	PbSCC ($j=6$)	HPIGC ($j=7$)	AcIGC ($j=8$)	AkIGC ($j=9$)	Submodes of IGC		Pitting ($j=11$)	Fatigue ($j=12$)	Wear ($j=13$)
					AcSCC ($j=3$)	MRSCC ($j=4$)							Wastage ($j=10$)				
Tubular expansion ($i=1$)	x																
Tubular expansion ($i=2$)		x															
Top of tube sheet ($i=3$)	x																
Top of tube sheet ($i=4$)		x															
Sludge ($i=5$)		x															
Free span ($i=6$)	x																
Free span ($i=7$)		x															
Tube support (hot leg) ($i=8$)		x															
Tube support (cold leg) ($i=9$)		x															
U-Bend ($i=10$)	x																
U-Bend AVB ($i=11$)		x															

[a] The abbreviations, LP, HP, Ac, MR, Ak, and Pb for SCC and IGC refer to low potential, high potential, acidic, mid-range pH, alkaline, and lead for stress corrosion cracking, and 'intergranular corrosion'

3.6.2. When to Inspect?: Key Performance Indicators

The importance of meaningful management information is often highlighted when operational problems arise. However, it is operationally more useful to know beforehand what detrimental processes are relevant in a given production operation, and hence be in control of that process. Key performance indicators (KPIs) may be used in this context to optimize the need and sequence of inspection and other maintenance activities.

Management information is typically required on predicted costs of problems, the risks involved, the remaining life of the affected equipment, and what can be done to improve or eradicate these problems. The KPIs described in the following sections were developed specifically to measure the effect of corrosion on the technical and financial performance of assets involved in oil and gas production facilities and to address the performance of particular critical corrosion related systems (11). The same approach can be generalized and adapted to many high risk systems.

3.6.2.1. Cost of Corrosion KPI

The cost of corrosion KPI allows converting the amount of corrosion damage sustained during a given period into a monetary figure to provide a clearer focus on corrosion management performance. Factors considered within this KPI are existing damage sustained prior to the period in question, the cost of repair or replacement, and the remaining service life of the plant. Performance can then be computed in terms of the cost of damage in the last period examined, the annual damage cost and/or lifecycle costs. The cost of corrosion damage (C_{corr}) sustained in a given period can be derived with Eq. (3.2):

$$C_{corr} = \left(\frac{N_C \cdot R_{cost}}{FL}\right)\left(\frac{D_p}{365}\right) \quad (3.2)$$

where C_{corr} = cost of corrosion damage in a specific time period; N_C = estimated number of replacement cycles to end of service life; R_{cost} = replacement cost (including lost product cost); FL = required remaining field life (years); and D_p = days in monitoring period (days).

If, however, the calculated remaining life of a component (RL_C) as defined in Eq. (3.4) is greater than the FL, C_{corr} can be assumed to be equal to 0. This is based on the assumption that the KPI is a performance indicator reflecting the effect on operating costs (O_{pex}) and does not consider depreciation against the initial capital cost (C_{apex}). The number of replacement cycles (N_C) can be estimated from Eq. (3.3):

$$N_C = \left(1 + \left(\frac{FL - RL_C}{RL_R}\right)\right) \quad (3.3)$$

where RL_C = remaining life of current component (years) and RL_R = remaining life of replacement components (years).

Both RL_C and RL_R are, respectively, derived from Eqs. (3.4) and (3.5):

$$RL_C = \left(\frac{CA - DT}{CR}\right) \qquad (3.4)$$

$$RL_R = \left(\frac{CA}{CR}\right) \qquad (3.5)$$

where CA = corrosion allowance[2] (design or fitness for purpose) (mm); DT = damage to date (mm); and CR = measured corrosion rate (mm/year).

The above formulas have been developed from an actual pipework monitoring and replacement program as a result of internal corrosion effects. Both replacement cost (R_{cost}), component remaining life (RL_C), and field life (FL) are key factors in this method and reflect the need to understand accurate costing of installation and replacement activities. Required FL is not necessarily the difference between installation and design lives, but more likely the time remaining until end of field or production life.

3.6.2.2. Corrosion Inhibition Level KPI

This KPI is a measure of the availability of corrosion inhibitors to provide protection against corrosive processes. The inhibitor efficiency[3] itself should have been determined from a combination of previous laboratory and field testing to determine the optimum concentration the chemical inhibitor should be in the produced fluids. Of course, there may be applications where it is justified to apply a degree of over injection to provide protection to downstream facilities where it is not practical to inject.

The KPI itself is derived from a measure of the produced fluids including water and hydrocarbon phases and the inhibitor injected in the produced fluid stream to provide a correlation between how much inhibitor should be in the produced fluid stream versus actual injected inhibitor concentrations. The KPI % inhibitor availability (Inhibitor$_{AV}$) function is described by Eq. (3.6):

$$\text{Inhibitor}_{AV} = \left(\frac{C_{actual}}{C_{required}}\right) \times 100 \qquad (3.6)$$

where C_{actual} = actual concentration of corrosion inhibitor (ppm) and $C_{required}$ = required concentration of corrosion inhibitor (ppm).

---

[2]Corrosion allowance depends on the type of defect anticipated that needs to be identified by inspection. Once the defect geometry is known and the process parameters identified, the maximum allowable defect size may be calculated using fitness for purpose criteria to ensure that failure does not occur.

[3]The efficiency of an inhibitor is expressed as a measure of the improvement in lowering the corrosion rate of a system:

$$\text{Inhibitor efficiency}(\%) = \left(\frac{CR_{uninhibited} - CR_{inhibited}}{CR_{uninhibited}}\right) \times 100$$

where $CR_{uninhibited}$ = corrosion rate of the uninhibited system and $CR_{inhibited}$ = corrosion rate of the inhibited system.

This KPI when directly correlated with the corrosion cost KPI (C_{corr}) provides a clear indication of the corrosion performance of the asset and identifies clearly where effective action can be taken to improve performance if, for example, damage costs are seen to increase. A correlation between the cost of damage and inhibition level KPI may indicate how the cost of damage and inhibitor dosage level trend in actual performance. The example shown in Fig. 3.9 illustrates how the availability of corrosion inhibitors relates to the cost of operation, that is the cost is down when the availability is high while it may become quite high when the availability decreases >90%.

3.6.2.3. Completed Maintenance KPI

This KPI, which provides a measure of the reliability of the corrosion monitoring equipment, is determined from the assets maintenance performance in repairing equipment faults reported during routine corrosion inspection visits. This measure reflects the recognized importance that equipment reliability is critical to the performance of corrosion inhibition systems, and hence is also a key cost factor. The indicator is derived from a ratio of the number of maintenance actions raised versus the number of actions completed in a given monitoring period [eq. (3.7)].

$$\%\text{Completed maintenance} = \frac{\text{Maintenance actions completed}}{\text{Maintenance actions raised}} \times 100 \quad (3.7)$$

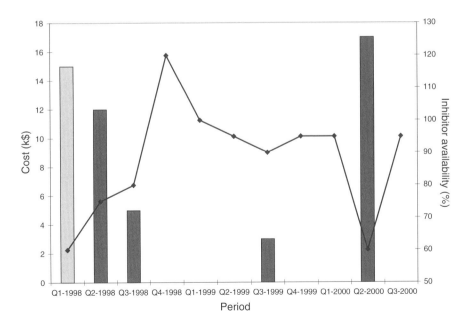

Fig. 3.9. Example of corrosion cost and inhibitor availability KPIs for estimating corrosion management performance (11).

3.6.2.4. Selecting KPIs

Measures of performance have been used by management for centuries to review current operational capabilities. Such measures have been used to assess both departmental and corporate performance, as well as trend performance achieved against plan. In many industrial facilities, these measurements are related to safety (e.g., number of incidents–accidents), environmental (e.g., number of releases), costs, and productivity. These measures are needed in order to determine not only if resources and costs have been managed for the production achieved, but also whether the assets or plant remain in good health (12).

In order to define a complete set of performance measures, companies must ensure that simple, workable measures are in place. The real challenge is not only to select those indicators that satisfy budgetary goals, but also to build the activities needed to meet the levels of asset performance required. Selecting the right measures is vital for effectiveness. Even more importantly, the metrics must be built into a performance measurement system that allows individuals and groups to understand how their behaviors and activities are fulfilling the overall corporate goals.

3.6.2.4.1. Asset Performance Metrics An asset performance management initiative is comprised of business processes, workflows, and data capture that enable rigorous analysis to help define strategies based on best practices, plant history, and fact-based decision support. Similarly to many other management issues, the key to building a set of performance metrics is to do it in stages. Clear corporate goals are important at this point, otherwise vague objectives will create impractical perspectives and metrics. By contrast, well-organized metrics and scorecards provide operational measures that have clear cause-and-effect relationships with the desired outcomes (12).

Each of these outcomes will build toward the goals of the perspective. Also, these metrics, if well chosen, will be the catalysts for change, providing warning signals to identify ineffective or failed asset performance strategies.

3.6.2.4.2. Tactical Perspectives Table 3.2 illustrates a high-level map developed for a chemical company using operational excellence goals of managing risks and improving profitability. From this strategic goal, perspectives have been defined that are specific to four functions and their associated goals:

- **Operations:** reduce operating costs–risks and maximize output.
- **Reliability:** maximize uptime, preserve plant, and asset integrity.
- **Work management:** minimize corrective work and restore asset condition.
- **Safety and environmental:** controlled–audited environmental, safe–audited operational capabilities.

Within these perspectives, each discipline has been able to take charge of factors under its control by choosing the right metrics to measure its progress

Table 3.2. Performance Indicators for Managing Risk and Improving Profitability of a Chemical Plant

Operations perspective	Reliability perspective	Work management perspective	Safety and environmental perspective
Strategic KPI			
Plant availability	Plant availability	Planning compliance	Incident rate
Number LPO events	Proactive work orders (%)	Work order complete (within 20% of planned costs)	Safety performance index
Time operating outside deterioration limits (%)	Emergency work orders on high critical systems (%)	Proactive work orders (%)	PHA/reviews completed (total)
Plant uptime (%)	Significant deterioration mechanisms improvements	Scheduling compliance	PSM compliance audits (total)
Production target compliance	Inspection compliance	Assessments of work order complete (total)	Significant environmental aspects defined/quantified (total)
	Protective device schedule compliance	Quantified availability targets (total)	
	Quantified reliability target (total)		
	Predictive maintenance compliance		

Table 3.2. (continued)

Operations perspective	Reliability perspective	Work management perspective	Safety and environmental perspective
Operational KPI			
Process availability variance	MTBF by equipment type	Emergency work orders (%)	Outstanding items from monthly safety inspection report (total)
Utility variance	MTBR by equipment type	Reactive work orders (%)	A incidents (total)
Product transfer indicator	MTBM by equipment type	Backlog work orders (total)	B incidents (total)
Quality limit excursions (total)	MTBF growth	Overtime hours (%)	C incidents (total)
Actual counter measures (total)	Cumulative nonavailability of critical assets	Work orders planned (total)	Total days lost days due to injury
Startup indicators	Unscheduled maintenance events (total)	Cumulative maintenance costs for standing order	Process hazard analysis action items
Shutdown indicators	Completed work order records on significant failures (total)	Average direct cost per maintenance event	
Offspec product	Bad actor count	Work orders scheduled (total)	
Scrap value	Current mechanical availability	Rework (%)	
Inventory	Mechanical availability trend	Closed work orders within 2 days of schedule (%)	

toward achieving the collective goal. Taking a methanol producing chemical plant as example, the Operations objectives would be to focus on delivering reduced operating costs, and managing the risks inherent in the process and in operational activities, while maximizing methanol output.

In the Safety and Environmental perspective, the focus would be on providing the systems, procedures, and training that build operational awareness, skills, functional systems and capabilities to prevent, manage, and eliminate safety and environmental incidents.

In addition, in the Work Management perspective, the focus would be on efficiently completing maintenance work while minimizing the potential for future breakdowns and restoring assets to their operating condition. Finally, for the Reliability perspective, the focus would be to build the analytics and skills required to increase and improve plant uptime while preserving the integrity and life of plant assets.

From each of these perspectives, tactical metrics can be set to stimulate new outcomes, build new processes, and build skill development and learning with clear links to the goals of each individual perspective.

3.6.3. Corrosion Monitoring or Corrosion Inspection?

The dividing line between corrosion inspection and corrosion monitoring is not always clear. Usually, inspection refers to short-term "once-off" measurements taken according to maintenance and inspection schedules while corrosion monitoring describes the measurement of corrosion damage over a longer time period and often involves an attempt to gain a deeper understanding of how and why the corrosion rate fluctuates over time. Corrosion inspection and monitoring are most beneficial and cost-effective when they are utilized in an integrated manner since the associated techniques and methods are in reality complementary rather than substitutes for each other.

Inspection techniques for the detection and measurement of corrosion range from simple visual examination to nondestructive evaluation (see Chapter 5 for more details). Significant technological advances have been made in the last decade. For example, the combined use of acoustic emission (AE) and ultrasonics (UT) can, in principle, allow for an entire structure to be inspected and for growing defects to be quantified in terms of length and depth. In parallel, modern corrosion monitoring methods have been developed for both their on-line capability and their ability to detect problems at an early stage (see Chapter 4 for more details).

The selection of a specific technique or inspection tool should be made by carrying out a cost benefit analysis in order to produce a figure of merit, such as shown in Fig. 3.10 (13). In the case of an inspection technique, the tool should have sufficient precision to detect defects considerably smaller than those that may result in a failure because these defects will probably grow in size between inspections. A cheaper and less accurate technique used frequently could be

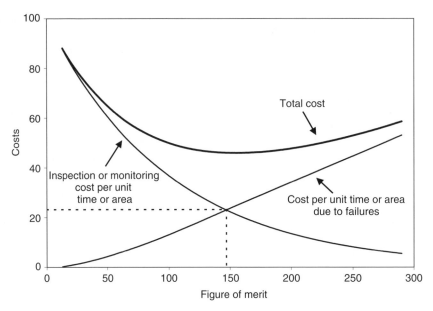

Fig. 3.10. Cost of operating a system as a function of expected cost per unit time associated with failures and inspection.

equivalent costwise to a more expensive and more accurate technique used less frequently leading, however, to greater reliability.

A considerable catalyst to the advancement of corrosion inspection and monitoring technology has been the exploitation of oil and gas resources in extreme environmental conditions. Work in these conditions has necessitated enhanced instrument reliability and the automation of many tasks, including inspection. In addition to the usual uncertainty of the onset or progression of corrosion of equipment, the oil industry has to face ever changing corrosivity of processing streams. The corrosivity at a well head can oscillate many times, during the life of an exploitation system, between being benign to becoming extremely corrosive (14). Such changes require more corrosion vigilance in terms of inspection and monitoring. Many techniques that have been accepted in the oil and gas industries for years are only now beginning to be applied in other industries, such as transportation, mining, and construction.

3.6.4. Risk-Based Inspection

Risk analysis refers to techniques for identifying, characterizing, and evaluating hazards. Risk-based inspection (RBI) is the application of risk analysis principles to manage inspection programs for plant equipment. The identification of risk, defined in Eq. (3.8), and risk analysis found their way into many applications, where they can add value in prioritization and management processes. The

Table 3.3. Examples of Risk Criteria and Their Units

Risk criteria	Expressed in
Financial risk (business impact)	Outage cost/day
Investment risk (asset damage)	Equipment cost/m^2
Safety	Injury cost/year
Environment	Cost/year
Potential of loss of life (PLL)	Events/year
Probability/likelihood of failure	Events/year

application of general risk analysis principles to help prioritize and manage the inspection program for plant equipment, now commonly referred to as RBI, is one of the newest applications of risk principles (15). Some examples of risk criteria and associated units are shown in Table 3.3.

$$\text{Risk} = \text{probability of failure (POF)} \times \text{consequence of failure (COF)} \qquad (3.8)$$

In this equation the POF is either based on failure frequency or on remaining lifetime, while COF is usually related to safety, health, environment, and economics issues.

The application of RBI to refinery and petrochemical plant equipment inspection was started by several companies in the late 1980s and was first reported in the open literature in 1993. Since that time, RBI has become a fast-developing methodology. A 1995 survey of member companies from the Materials Technology Institute (MTI) found that most corrosion engineers were routinely involved with establishing the frequency of inspection, and about one-half established the inspection frequency based on the results of risk identification.

A joint industry project sponsored by 21 refining and petrochemical companies was set in 1993 working in conjunction with the American Petroleum Institute (API) to develop RBI methodologies for application within their industry. At the same time, the API was developing an industry consensus standard for RBI (16).

Understanding the potential deterioration mechanisms that can lead to equipment failures, their likelihood of occurrence, and the potential consequences of the failures are key elements in an RBI program. The challenge faced by corrosion engineers is to provide plant inspection personnel with a sound technical understanding of potential deterioration mechanisms for use in developing a practical and effective strategy to limit the risk of potential equipment failures. The RBI program output may then be used to demonstrate the value of proactive corrosion control and thereby serve as a communication tool to influence the decision makers and stakeholders in plant equipment integrity and reliability (15).

The main objective of RBI is to use risk as a basis for prioritizing and managing the inspection program for plant equipment. In an RBI program, the risk associated with the continued operation of each piece of plant equipment is ranked by assessing the likelihood of its failure versus the severity of failure consequences.

There are several levels of risk-based assessments, usually described as qualitative, semiquantitative, and quantitative, as described later in this chapter. These vary considerably in the amount of effort and input needed and the accuracy of the resultant assessment. The qualitative assessment is usually performed to determine risk associated with whole or large portions of process units. The semiquantitative and quantitative assessments are usually performed to determine risk associated with individual equipment items.

The end result of risk assessment is a risk rating for each equipment item, which may vary from low to high. More often than not, a large percentage of the risk (>80%) is found to be associated with a small percentage of the equipment items (<20%). Once identified, the higher risk equipment become the focus of inspection and maintenance to reduce the risk, while opportunities may be found to reduce inspection and maintenance of the lower risk equipment without significantly increasing risk. The main drivers for RBI in specific industries are (17) **nuclear power industry:** increased safety and availability; **fossil fired power industry:** life extension and cost reduction; **offshore petroleum industry:** risk management (e.g., safety, environmental, and economics), plus cost reduction of the inspection costs; **petrochemical industry:** cost reduction, extension of inspection periods, and improvement of availability.

3.6.4.1. Probability of Failure Assessment

Two fundamental issues must be considered to determine the probability of a failure (POF). First, the different forms of corrosion and their rate. Second, the effectiveness of inspection. The input of corrosion experts is obviously required to identify the relevant forms of corrosion in a given situation and to determine the key variables affecting their propagation rate. It is also important to realize that full consensus and supporting data on the variables involved is highly unlikely in real-life complex systems and that simplification will invariably be necessary.

One semiquantitative approach for ranking process equipment is based on internal POF. The procedure is based on an analysis of equipment process and inspection parameters, following which the equipment is categorized on a scale of one to three, with "one" being the highest priority. The procedure requires a fair degree of engineering judgment and experience and, as such, is dependent on the background and expertise of the analyst. The procedure is designed to be both practical and efficient. The POF is intended to be a convenient and reproducible means for establishing equipment inspection priorities. As such, it facilitates the most efficient use of finite inspection resources when and where 100% inspection is not practical.

The POF approach is thus based on a set of rules heavily dependent on detailed inspection histories, knowledge of corrosion processes, and knowledge of normal and upset conditions. The equipment rankings may have to be modified and updated as additional knowledge is gained, process conditions change and equipment ages. Maximum benefits of the procedure depend on fixed equipment inspection programs that permit the capture, documentation, and retrieval of inspection, maintenance, and corrosion–failure mechanism information.

3.6.4.2. Consequence of Failure Assessment

It is important to obtain the input from experts in process engineering, safety, health, and environmental engineering to assess the consequences of a failure (COF). By considering the example of an uncontrollable fluid release, the following three factors would play a dominant role in the consequence assessment:

1. The type of fluid and associated hazards that may be released.
2. Its inventory available for release.
3. The rate of release.

A determinant factor in the rate of release is the size of the breech of containment. Fortunately, there are many breeches of containment that are small leaks which, once detected, can be readily contained and mitigated without significant incident. These are often the result of pinholes caused by localized corrosion or small, tight cracks that only allow minor fluid seepage. On the other hand, some breeches of containment are the result of major ruptures that allow large quantities of hazardous fluid to escape in a short period of time. It is difficult and sometimes impossible to react quickly enough to contain and mitigate such releases without enduring significant incidents.

Corrosion and materials engineering expertise is required to estimate the size and nature of damage that could result in a plant item. As described in Chapter 2, different corrosion mechanisms can produce different morphologies of damage. The difference in impact on the release rate created at a pinhole leak versus that of a large rupture is a good example of this aspect of consequence sensitivity.

Another important field covered by corrosion engineering is that of materials properties. For example, the risk of a catastrophic explosion from cracks in a brittle material associated with high release rates is obviously greater than in a material with higher fracture toughness. The toughness of a material is a key parameter in determining so-called "leak-before-break" safety criteria and the general tolerance toward defects. An understanding of how the toughness of a material can be reduced in service over time is thus obviously important (15).

3.6.4.3. Application of RBI

The application of RBI becomes most effective when involving a multidiscipline team including operations, process engineering, equipment specialists, and maintenance in addition to inspection and corrosion engineers. As a team, these experts can exchange valuable information and use the team synergy to arrive at a mutually agreeable approach to risk reduction that may involve approaches other than increased inspection. The following sections described three approaches to reducing the risk of operating plant equipment that are incorporated into the API RBI program (16).

3.6.4.3.1. Optimizing Inspection–Monitoring Once the risk assessment has been completed, the results are used to evaluate the effectiveness of the present inspection–monitoring strategy while looking for ways to optimize the strategy to reduce the overall risk of continued operation. For equipment identified as high risk, changes in the detailed inspection plan that would reduce the risk should be considered. For equipment identified as low risk, changes in the detailed inspection plan that would reduce costs while not increasing risk appreciably should be considered (15).

The RBI assessment identifies the potential deterioration mechanisms that can lead to failure in each piece of equipment. It is very important to determine whether the existing inspection plan is addressing all of the potential deterioration mechanisms that were identified in that service. Are appropriate inspection methods being used? For example, it is unlikely that spot ultrasonic thickness measurements will find highly localized corrosion. Or perhaps there has been no inspection performed to detect a cracking mechanism that is highly likely to occur in that service. The fundamental understanding of a deterioration mechanism may be valuable input when evaluating the likelihood that an inspection method will actually detect the deterioration.

3.6.4.3.2. Materials of Construction Changes The RBI approach can be used to evaluate the risk reduction associated with alternative alloys. Coupled with appropriate alloy cost information, the replacement alloy selection can be optimized on a risk reduction versus cost basis. Used in this manner, the RBI program becomes a valuable tool for the corrosion engineer. This type of output from an RBI program can be used when communicating with management about the justification of the added cost of a material upgrade (15).

Although it may not be readily evident, there is a great opportunity to use an RBI program when selecting materials for new construction projects. The corporate pressure for improving profitability has placed intense pressure to achieve the lowest initial cost for installation of new plants. Unfortunately, as a result, it is not uncommon to see low-cost materials with rather high expected corrosion rates used for original construction, just to save on initial plant cost, assuming that the plant's inspection program will maintain mechanical integrity. This heavy reliance on inspection more often than not comes with considerable risk that may not have been adequately quantified or addressed by the project leaders. When the new plant becomes operational, high risks require mitigation to ensure mechanical integrity, and this usually involves additional expenditures of inspection resources. As pointed out earlier, suitable risk reduction may not be achievable by increased inspection alone, and materials upgrades may be required sooner than expected. The risks associated with alternative materials selection for new projects should be quantified and considered, and an RBI program can provide an effective means of demonstrating the tradeoff between initial cost and risk.

3.6.4.3.3. Key Process Parameters A real benefit derived from the development of an RBI program is the identification of the key process parameters that most influence the equipment deterioration rate. These key process parameters often include fluid composition, temperature, pH, and fluid velocity. Once identified as such, a strong case can be made for routine monitoring of these highest impact process parameters and maintaining them within prescribed limits. It is disturbing to find that process unit operating personnel sometimes do not fully understand the relevancy of certain process parameters to mechanical integrity, and therefore are not routinely monitoring them. This is often realized when there is a loss of containment incident and root cause investigation reveals that the failed equipment has been operating beyond one or more of the key process parameter limits (15).

An RBI program can serve as a helpful tool for quantifying the risk associated with a change in the value of a key process parameter. This can be an invaluable resource to the corrosion engineer when addressing management of change issues in an operating process unit. It also can serve as an aid to reaching agreement on the desirable process monitoring and limits on operating conditions, and in communicating with process unit operators and management about the impact of changes on risk.

3.6.5. Risk Assessment Methodologies

Risk-based inspection procedures can be based on either qualitative or quantitative methodologies. Qualitative procedures provide a ranking of equipment based largely on experience and engineering judgment. Quantitative risk-based methods use several engineering disciplines to set priorities and develop programs for equipment inspection. Some of the engineering disciplines include nondestructive examination, system and component design, and analysis, fracture mechanics, probabilistic analysis, failure analysis, and operation of facilities.

The level of detail in a risk assessment should be proportionate to the level of the intrinsic hazards. In general, the greater the magnitude of the hazards under consideration and the greater the complexity of the systems being considered, the greater the degree of rigor, robustness, and enough level of detail to show that risks have been reduced "as low as is reasonably practicable" (ALARP). The level of risk arising from the undertaking should therefore determine the degree of sophistication needed in the risk assessment.

In general, qualitative approaches are easiest to apply, but provide the least degree of insight. Conversely, quantitative risk analysis (QRA) approaches are most demanding on resources and skill sets, but potentially deliver the most detailed understanding and provide the best basis if significant expenditure is involved. Semiquantitative approaches lie in between these extremes. The following sections will describe briefly some of the techniques that have been used in the context of corrosion risk assessment.

3.6.5.1. HAZOP

A hazard and operability (HAZOP) study is a method of identifying hazards that might affect safety and operability based on the use of guidewords. A team of experts in different aspects of the installation, under the guidance of an independent HAZOP leader, systematically considers each subsystem of the process typically referring to process and instrumentation diagrams (P&IDs). These experts use a standard list of guidewords to prompt them to identify deviations from design intent. Guidewords are simple words or phases used to qualify or quantify the intention and associated parameters in order to discover deviations (Table 3.4). For each credible deviation, they consider possible causes and consequences, and whether additional safeguards should be recommended. The conclusions are usually recorded in a standard format during the sessions.

Figure 3.11 illustrates how the HAZOP process is conducted on specific nodes of interest. In the HAZOP context, a node is a location on a process diagram (usually P&IDs) at which process parameters are investigated for deviations. Nodes are also points where the process parameters have identified design intent. Nodes are usually pipe sections or vessels. Plant components (e.g., pumps, compressors, exchangers) are found within nodes. In the same fashion, a parameter is an aspect of the process that describes it physically, chemically, or in terms of what is happening (e.g., flow, level, temperature). Parameters are usually classified as specific or general to, respectively, describe aspects of the process or aspects of design intent remaining after the specific parameters have been removed

Table 3.4. Standard and Auxiliary Guidewords Used to Conduct HAZOP Studies

Guide word	Meaning
Standard guidewords	
No	Negation of the design intent
More	Quantitative increase
Less	Quantitative decrease
As well as	Qualitative increase
Part of	Qualitative decrease
Reverse	Logical opposite of the intent
Other than	Complete Substitution
Auxiliary guidewords	
How	How is the step to be accomplished
Why	Is the step or operation really needed
When	Is timing of the step or operation important
Where	Is it important where the step is performed

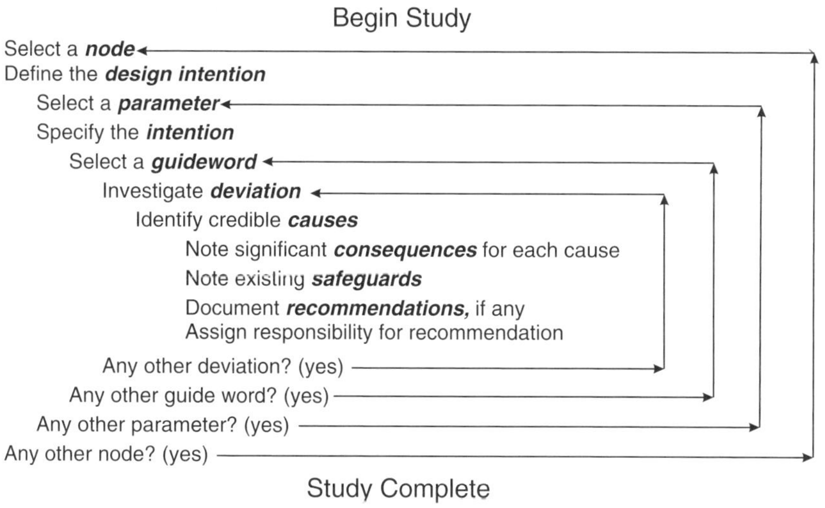

3.6. Inspection Strategies

```
                          Begin Study
Select a node ◄─────────────────────────────────────────────┐
Define the design intention                                  │
   Select a parameter ◄────────────────────────────────┐    │
   Specify the intention                                │    │
      Select a guideword ◄───────────────────────┐     │    │
         Investigate deviation ◄──────────┐      │     │    │
            Identify credible causes      │      │     │    │
               Note significant consequences for each cause
               Note existing safeguards
               Document recommendations, if any
               Assign responsibility for recommendation
         Any other deviation? (yes) ──────┘      │     │    │
      Any other guide word? (yes) ───────────────┘     │    │
   Any other parameter? (yes) ──────────────────────────┘    │
Any other node? (yes) ───────────────────────────────────────┘
                          Study Complete
```

Fig. 3.11. A HAZOP process iterative loops.

The HAZOP procedure is a powerful tool for hazard analysis and its methodical approach ensures that weaknesses in the design intent are detected and acted upon. HAZOP is widely used in RBI processes. It is also applied at an early design stage through corrosion risk assessments as a documented process for materials selection and corrosion prevention. The strengths of HAZOP are (18):

- It is widely used and its advantages and disadvantages are well understood.
- It uses the experience of operating personnel as part of the team.
- It is systematic and comprehensive, and should identify all hazardous process deviations.
- It is effective for both technical faults and human errors.
- It recognizes existing safeguards and develops recommendations for additional ones.
- The team approach is particularly appropriate to operations requiring the interaction of several disciplines or organizations.

The weaknesses of the HAZOP method are

- Its success depends on the facilitation of the leader and the knowledge of the team.
- It is optimized for process hazards, and needs modification to cover other types of hazards.
- It requires development of procedural descriptions that are often not available in appropriate detail. However, the existence of these documents may benefit the operation.
- Documentation is lengthy (for complete recording).

3.6.5.2. FMEA and FMECA

Failure modes, effects, and criticality analysis (FMECA) (or its simpler form, FMEA) are systematic methods of identifying a system failure mode. FMEA is implemented by considering each item of equipment and associated systems in the plant, detailing the possible failure modes (e.g., leak or break in the case of pressure equipment), and determining their resulting effect on the rest of the system. The analysis is more concerned with specifying the likely effects and criticality of different modes of failure rather than the mechanisms or events leading to the failure (19).

It is a simple method that is easy to apply, yet it is a powerful tool that may be used to improve the quality of products and processes. It can lead to focusing on consequences and additional safeguards to mitigate the effects of the failure. It is common for individuals familiar with system functionality to perform FMEA, but teams of experts can produce greater insight into the mechanisms and wider range consequences. The analysis uses a form that begins with a systematic list of all components in the system, and typically includes the following: component name; function of component; possible failure modes; causes of failure; how failures are detected; effects of failure on primary system function; effects of failure on other components; necessary preventative/repair action; rating of failure frequency; and rating of severity (i.e., consequence) of failure.

Failures are rated as critical if they have high frequency or severity ratings. In these cases, special protection measures may be considered. The strengths of FMECA are (18): it is widely used and well understood; it can be performed by a single analyst; it is systematic and comprehensive, and should identify hazards; and it identifies safety-critical equipment where a single failure would be critical for the system.

The FMECA weaknesses are

- Its benefit depends on the experience of the analyst.
- It requires a hierarchical system drawing as the basis for the analysis, which the analyst usually has to develop before proceeding with the analysis.
- It is optimized for mechanical and electrical equipment, and does not apply to procedures or process equipment.
- It copes difficultly with multiple failures and human errors.
- It does not produce a simple list of failure cases.

Most accidents have a significant human contribution, and FMECA is not well suited to identifying these. As FMECA can be conducted at various levels, it is important to decide before commencing what level will be adopted as otherwise some areas may be examined in greater details than others. If conducted at too deep a level, FMECA can be time consuming and tedious, but it can also lead to great understanding of the system.

3.6.5.3. Risk Matrix Methods

Risk matrices provide a framework for an explicit examination of the frequency and consequences of hazards. This may be used to rank them in order of significance, screen out insignificant ones, or evaluate the need for risk reduction of each hazard. A risk matrix uses a matrix dividing the dimensions of frequency (POF) and consequence (COF) into typically three to six categories (A–E in Fig. 3.12). There is little standardization in matters, such as the size of the matrix or the labeling of the axes.

Sometimes risk matrices use quantitative definitions of the frequency and consequence categories. They may also use numerical indices of frequency and consequence (e.g., 1–5) and then add the frequency and consequence pairs to rank the risks of each hazard or each box on the risk matrix. The strengths of the risk matrix approach are (18):

- It is easy to apply and requires few specialist skills, and for this reason it is attractive to many project teams.
- It allows risks to people, property, environment, and business to be treated consistently.
- It allows hazards to be ranked in priority order for risk reduction effort.

However, there are several problems with this approach, which are less apparent:

- Many judgments are required on likelihood and consequence and unless properly recorded, the basis for risk decisions will be lost.

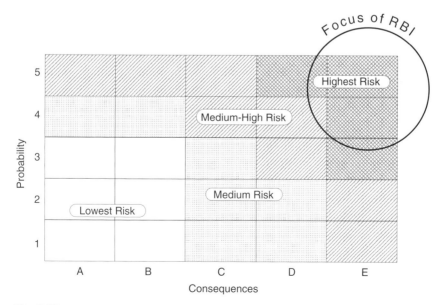

Fig. 3.12. A 5 × 5 risk matrix.

- Judgments must be consistent among different team members, a condition difficult to achieve whether qualitative or quantitative definitions are used.
- Where multiple outcomes are possible (e.g., the consequence of a fall on a slippery deck can range from nothing to a broken neck), it can be difficult to select the "correct" consequence for the risk categorization.
- A risk matrix looks at hazards "one at a time" rather than in accumulation, whereas risk decisions should really be based on the total risk of an activity. Potentially, many smaller risks can accumulate into an undesirably high total risk, but each smaller one on its own might not warrant risk reduction. Consequently, risk matrix has the potential to underestimate total risk by ignoring accumulation.

3.6.5.4. Fault Tree Analysis

Fault tree analysis (FTA) is a logical representation of the many events and component failures that may combine to cause one critical event (e.g., pipeline explosion). It uses "logic gates" to show how "basic events" may combine to cause the critical "top event". The top event would normally be a major hazard, such as "Pipeline SCC" as in the example shown in Fig. 3.13. Figure 3.14 presents, in their graphical form, the most commonly used tree symbols and gates used in the construction of fault trees (20). A brief description of these symbols is given in the following list:

- **Fault event (rectangle):** System-level fault or undesired event.
- **Conditional event (ellipse):** Specific condition or restriction applied to a logic gate (mostly used with inhibit gate).
- **Basic event (circle):** Lowest event of examination that has the capability of causing a fault to occur.
- **Undeveloped event (diamond):** Contains a failure that is at the lowest event of examination by the fault tree, but can be further expanded.
- **Transfer (triangle):** The transfer function is used to signify a connection between two or more sections of the fault tree.
- **AND gate:** The output occurs only if all inputs exist (multiply probabilities on the input, therefore decreasing resulting probability).
- **OR gate:** The output is true only if one or more of the input events occur (add probabilities on the input, therefore increasing resulting probability).
- **Inhibit gate (hexagon):** One input is a lower fault event and the other input is a conditional qualifier or accelerator [direct effect as a decreasing (<1) or increasing factor (>1)].

The FTA emphasizes the lower level fault occurrences that directly or indirectly contribute to a major fault or undesired event. The technique is one of "reverse thinking" where the analyst begins with the final undesirable event that is to be avoided and identifies the immediate causes of that event (19).

3.6. Inspection Strategies 117

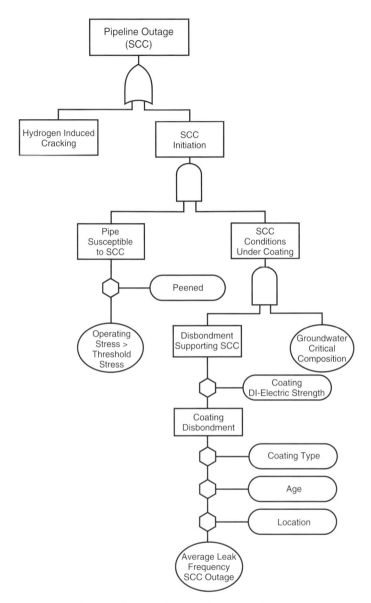

Fig. 3.13. Fault tree for natural gas pipeline outage due to SCC.

By developing the lower level failure mechanisms necessary to produce higher level occurrences, a total overview of the system is achieved. Once completed, the fault tree allows an engineer to fully evaluate a system safety or reliability by altering the various lower level attributes of the tree. Through this type of analysis, a number of variables may be visualized in a cost-effective manner.

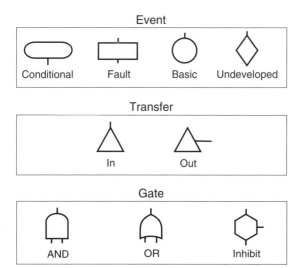

Fig. 3.14. Fault tree symbols for gates, transfers and events.

Tracing the chain of events leading to the final outcome can indicate where extra monitoring, regular inspection and protective schemes (e.g., temperature and pressure sensors, and alarms) could protect and forewarn impending failure. Fault tree analysis is a very useful tool for studying the routes by which an accident can occur, and is particularly effective at identifying accident scenarios due to secondary and tertiary causes. However, it requires a great deal of skill and effort to implement. For this reason, it is expected to be used only by industries where the consequences of failure might be very severe. The FTA has several potential uses in QRA:

- In frequency analysis, it is commonly used to quantify the likelihood of the top event occurring, based on estimates of the failure rates of each component. The top event may be an individual failure case, or a branch probability in an event tree.
- In risk presentation, it may also be used to show how the various risk contributors combine to produce the overall risk.
- In hazard identification, it may be used qualitatively to identify combinations of basic events that would be sufficient to cause or trigger the top event, known as "cut sets".

The strengths of fault tree analysis are (18):

- It is a widely used and well-accepted technique.
- It is suitable for many hazards in QRA that arise from a combination of adverse circumstances.
- It is often the only technique that can generate credible likelihoods for novel, complex systems.

- It is suitable for technical faults and human errors.
- It provides a clear and logical form of presentation.

Its weaknesses are

- The diagrammatic format discourages analysts from stating explicitly the assumptions and conditional probabilities for each gate. This can be overcome by careful back-up text documentation;
- FTA can soon become complicated, time consuming, and difficult to follow for large systems.
- Analysts may overlook failure modes and fail to recognize common cause failures (i.e., a single fault affecting two or more safeguards) unless they have a high level of expertise and work jointly with the operator.
- All events are assumed to be independent.
- FTA may easily lose its clarity when applied to systems that do not fall into simple failed or working states (e.g., human error, adverse weather).

Figure 3.13 illustrates how a major gas transmission pipeline company adopted FTA for the risk assessment of SCC corrosion on its 18,000 -km gas pipeline network (2, 21). The rupture risk FTA was normally performed for the review and analytical examination of systems or equipment to emphasize the lower level fault occurrences. These results also served to schedule maintenance operations, conduct surveys, and plan research and development efforts.

Each element of the branch in Fig. 3.13 contains numerical probability information related to technical and historical data for each segment of the complete pipeline network. In some cases, it was simpler to assume some probability values for an entire system. The probabilities of operating at maximum permitted pressure and the presence of electrolyte were both set at value unity in Fig. 3.13, therefore forcing the focus on worst case scenarios. Other more verifiable variables can be fully developed as is shown in Fig. 3.15 for two basic events describing the probable impact of a cathodic protection deficiency on the pipeline network.

3.6.5.5. Event Tree Analysis

Event tree analysis (ETA) is a logical representation of the various events that may be triggered by an initiating event (e.g., a component failure). It uses branches to show the various possibilities that may arise at each step. It is often used to relate a failure event to various consequence models. It may also be used to quantify system failure probabilities, where several contributory causes can only arise sequentially in time.

Like FTA, event tree analysis is also a logic-based methodology for identifying accident scenarios, but unlike FTA it is a "forward thinking" method. The analysis begins with a given initiating failure event and develops the resulting sequence of events, normally over a short time interval, making assumptions about the availability or otherwise of safeguards and back-up systems, such as

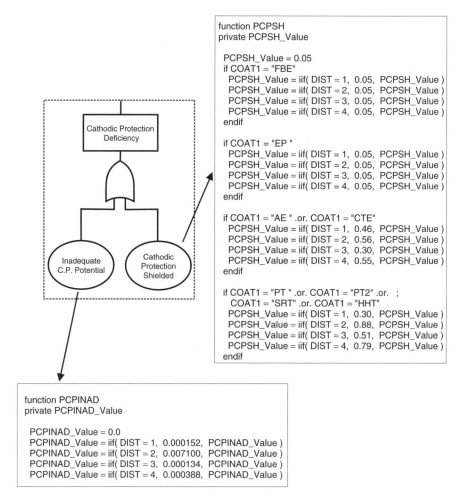

Fig. 3.15. Detailed code for the basic events leading to a natural gas pipeline cathodic protection deficiency.

protective devices. Event tree analysis is an extension of FMEA to encompass a complete system (19).

Event trees are valuable for examining the consequences of failure. They are less effective for the analysis of the causes of system failure. The short time scale over which events are considered may mask longer term consequences, such as the gradual deterioration of equipment due to faults elsewhere.

Construction starts with the initiating event and works through each branch in turn. A branch is defined in terms of a question (e.g., Protective device fails?). The answers are usually binary (e.g., yes or no), but there can also be multiple outcomes (e.g., 100, 20, or 0% in the operation of a control valve). Each branch is conditional on the appropriate answers to the previous ones in the tree.

3.6. Inspection Strategies

Usually, an event tree is presented with the initiating events on the left and the outcomes on the right. The questions defining the branches are placed across the top of the tree, with upward branches signifying "yes" and downward ones for "no". A probability is associated with each branch, being the conditional probability of the branch (i.e., the answer yes or no to the branch question) given the answers of all branches leading up to it. In each case, the sum of the probabilities of each branch must be unity. The probabilities of each outcome are the products of the probabilities at each branch leading to them. The sum of the probabilities for all outcomes must be unity as well. This provides a useful check on the analysis. The strengths of event tree analysis are (18)

- It is widely used and well accepted.
- It is suitable for many hazards in QRA that arise from sequences of successive failures.
- It a clear and logical form of presentation.
- It is simple and readily understood.

Its weaknesses are

- It is not efficient where many events must occur in combination, as it results in many redundant branches.
- All events are assumed to be independent.
- It loses its clarity when applied to systems that do not fall into simple failed or working states (e.g., human error, adverse weather).

Figure 3.16 shows an event tree analysis that was performed on each process system of a fluid catalytic cracking unit (FCCU) gas plant using actual probabilities and consequences that are particular to that process system. In a span of 7 months, the refinery had experienced 23 leaks in piping located in the fractionator overhead and the wet gas compression sections of the FCCU. Follow-up with intensive ultrasonic (UT) shear wave inspection located an additional 73 carbonate cracking like indications in the gas recovery section of the FCCU. As with many forms of SCC, radiography was not considered a suitable inspection technique for carbonate cracking (22).

In petroleum refining, carbonate SCC can occur in FCCU gas plants where the process environment contains a significant amount of carbonate–bicarbonate ions, H_2S, free water, and ammonia. The equipment most likely to be affected in the FCCU gas plant are the main fractionator overhead condensers, fractionator overhead accumulator, wet gas compressor knockout drums and condenser, deethanizer (or other similar light end fractionators), and all associated piping including the sour water streams originating from these areas. In this example, probabilities were based on the occurrence of the event in any given year. Values assigned for the probabilities were determined using the number of welds in each system that cracked divided by the total number of welds in the system. Consequences included leaks/breaks; using clamps to contain the leak; shut down of the unit (SID); fire; and vapor cloud explosion (VCE).

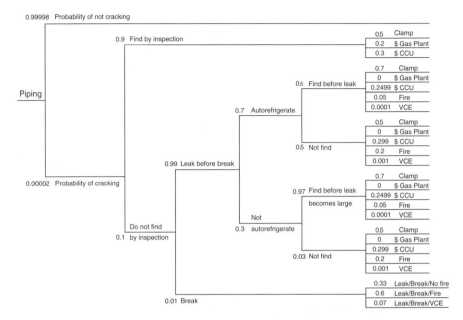

Fig. 3.16. Event tree analysis that was performed on each process system of a FCCU gas plant. ($ = cost of damage, CCU = catalytic cracking unit, VCE = vapor cloud explosion).

Dollar values for each consequence were determined in agreement with operations, engineering, and inspection personnel based on costs of past failures and loss of production. The results of the risk assessment were then used to identify which equipment and piping were to be replaced during the next refinery FCCU turnaround.

3.7. FAILURE ANALYSIS INFORMATION

Data required for developing a risk assessment program are often acquired during the analysis of failed components and systems. However, conducting a failure analysis is not an easy or straightforward task. Early recognition of corrosion as a factor in a failure is critical, since much important corrosion information can be lost if a failure scene is altered or changed before appropriate observations and tests can be made.

To avoid these pitfalls, certain systematic procedures have been proposed to guide an investigator through the failure analysis process. For example, MTI has produced an "Atlas of Corrosion and Related Failures" that maps out the process of a failure investigation from the request for the analysis to the submission of a report (23). The section that relates the origin(s) of failure to plant or component geometry in this atlas is illustrated in Figs. 3.17 and 3.18. The depth of the failure analysis into the roots of a failure is key to accurately unearthing all the failure sources. Looking at machinery failures one finds that there are (24)

3.7. Failure Analysis Information

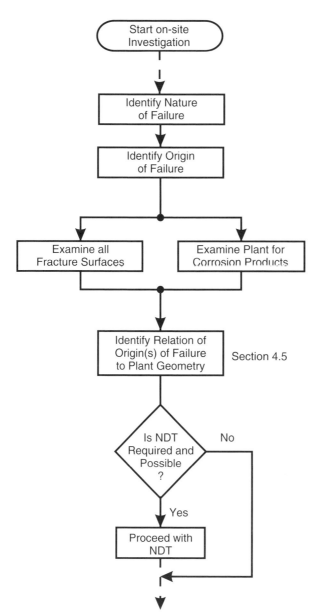

Fig. 3.17. Decision tree to guide on-site investigations dealing with corrosion damage.

Physical Roots: The physical reasons why the parts failed.

Human Roots: The human errors of omission or commission that resulted in the physical roots.

Latent Roots: The deficiencies in the management systems or the management approaches that allow human errors to continue unchecked.

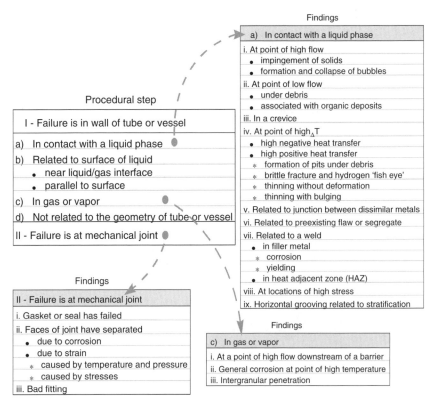

Fig. 3.18. Recommendations for relating the origin(s) of failure to plant geometry.

The more detailed the analysis, the better all the events and mechanisms that contribute as the roots of the problem can be elucidated. The analyses can be further divided into three categories in order of complexity and depth of investigation:

Component failure analysis (CFA), which looks at the piece of the machine that failed, for example, a bearing or a gear, and determines that it resulted from a specific cause such as fatigue or overload or corrosion and that there were these x, y, and z influences.

Root cause investigation (RCI) is conducted in much greater depth than the CFA and goes substantially beyond the physical root of a problem to find the human errors involved. It stops at the major human causes and does not involve management system deficiencies. The RCIs are generally confined to a single operating unit.

Root cause analyses (RCA), which includes everything the RCI covers plus the minor human error causes and, more importantly, the management system problems that allow the human errors and other system weaknesses to exist. An RCA can sometimes extend to sites other than the one involved in the original problem.

Although the cost of an analysis increases as it becomes more complex, its benefits generally increase as well. Using a CFA to find the causes of a component failure answers why that specific part or machine failed and the results of the analysis can be used to prevent the occurrence of similar failures in the future. The cost of progressing to a RCI is 5–10 times that of a CFA, but the RCI adds a detailed understanding of the human errors contributing to the breakdown and can be used to eliminate groups of similar problems in the future. However, conducting a RCA and correcting the major roots has the potential to eliminate huge classes of problems.

3.7.1. Conducting a Corrosion Failure Analysis

The first step in a failure analysis is to learn as much as possible from the existing failed sample. Care must be taken to avoid destroying valuable evidence. Much can be deduced from the physical appearance of the corrosion as viewed with the unaided eye, or with a low power magnifying glass or microscope. The geometry of attack, color, and form of corrosion products all provide valuable clues.

Other important information can be obtained only by altering the shape of or partially destroying the sample. For example, metallographic observation of type and depth of attack usually will require cutting and polishing the sample. In planning the sequence of investigation, all visual examinations, collection of corrosion products for chemical and X-ray analysis should be completed before cutting the sample for metallographic examination.

The following sections discuss the various options to consider during a corrosion failure investigation (25).

3.7.1.1. Planning the Analysis

Early recognition of corrosion as a factor in a failure analysis is critical to any such investigation. Therefore, it is desirable to conduct the analysis as soon as possible after the apparent failure. It is also desirable to protect the physical evidence until the analysis can begin. Much important corrosion information can be lost if a failure scene is altered or changed before appropriate observations can be made. A written plan for the detailed analysis should be prepared. The plan may include methods of documentation (e.g., photographs before and during analysis, sketches, and statements), responsibilities of parties, reporting needs, and scheduling.

3.7.1.2. Conditions at the Failure Site

An overall examination of the conditions at a failure site prior to cleaning, moving, or, sampling debris should be conducted. Impressions (e.g., physical arrangements, odors, colors, textures, and conditions of adjacent structures) can provide important clues as to active corrosion processes and should be reported appropriately with the use of annotated photographs, videotapes, as well as sketches and drawings. Interviews with those who were present or nearby when the failure occurred would be appropriate. Information on time, materials and supplies may be significant and should be evaluated.

3.7.1.3. Operating Conditions at Time of Failure

Special attention should be given to deviations observed in the operating conditions preceding the failure and to out-of-specifications or any other unusual upset conditions. It may be necessary to plot or track operating conditions for a long period of time prior to the detection of failure to reveal possible unusual contributory operating conditions. The operating conditions of any similar equipment should also be noted and used as reference. Any present corrosion monitoring instruments and coupons should, of course, be examined to help document operating conditions at the time of failure.

3.7.1.4. Historical Information

Historical information, when available, can be extremely useful in understanding some situations. Useful details regarding original constructions may include, for example, design drawings and specifications, material specifications, joining, and surface treatments. Details regarding modifications made subsequent to the original fabrication may also be extremely important because they often reveal less than optimum field work. Modifications may have been made for one or more reasons, including problems with the original design, changes in service requirements, corrected earlier failures, and correction of safety and environmental concerns.

3.7.1.5. Sampling

Careful sampling is critical to the successful investigation of corrosion-related failures. Sampling in corrosion investigations is similar to that used in forensic investigations by criminologists. ASTM Guide E 1459 and Practice E 1492 address issues of labeling and documenting field evidence. These standards may provide useful guidance during sampling for corrosion investigations.

Contamination should be avoided during sampling by using clean tools and by wearing gloves. Sample containers should be clean and sealable to protect samples from contamination and damage. The material of sample containers should be selected carefully to avoid undesirable interaction with samples. Each container should be dated and identified according to the sampling plan. If the location of failure initiation is apparent, it should be sampled. Samples should be cut with care to prevent heating the samples and to avoid the introduction of cutting and cooling fluids that could alter the surface and metallurgical conditions. Because of the solubility in water of many corrosion products, samples must be protected from extraneous moisture.

Corrosion products and deposits should be given special sampling treatment because they are often key elements in understanding the failure. Care should be used in the selection of tools for collecting these samples. Nonmetallic tools are often preferred because they present less chance for contamination of the sample or for damaging critical corroded surfaces. It may be desirable to obtain samples of the local environment (e.g., process stream, soils, and concrete). The most useful samples are those taken from the failure location as soon as possible after the event. Special sampling procedures are required when microbiological factors are suspected of being involved (see Chapter 4, the section on Monitoring Microbiologically Influenced Corrosion).

3.7.1.6. Evaluation of Samples

Metallic samples should be evaluated for metallurgical condition and structure. In some cases, it may be necessary to remove corrosion products from a sample to permit further evaluation. The evaluation itself may involve mechanical and physical property tests, metallographic examination of cross-sections, and specific corrosion tests.

The location and orientation of each specimen must be documented by photographs, drawings, or written descriptions. Each specimen should be labeled to aid in identifying its original location within the sample. Failure locations, such as pits, fracture surfaces, crevices, and generally attacked surfaces, should be examined, and measurements should be made to document surface chemistry, pit depths, crack dimensions, and metal losses and other modes of attack. These examinations often require the use of optical microscopes, scanning electron microscopes (SEM), X-ray powder diffraction, and other instruments.

3.7.1.7. Assessment of Corrosion-Related Failure

The type and extent of corrosion should be noted. The extent of corrosion may be determined by measurements and calculations of general corrosion rate from thickness loss, pitting penetration, or crack growth rate. From these observations and findings, the investigator should be able to identify the causes and factors involved in the failure. In many cases, more than one factor will have to be suggested as having played a role in the failure. The investigator may provide explanations and rationales for suggested corrective actions.

3.7.2. Information to Look For

It is not always be possible to anticipate the actual environment in which a metallic structure will operate. Even if initial conditions were known completely, there is often no assurance that operating temperatures, pressures, or even chemical compositions will remain constant over the expected equipment lifespan. The complexity of corrosion processes and their impact on equipment often buried or out-of-sight complicates many situations, rendering simple life prediction difficult. The following sections illustrate some of these complicating factors.

3.7.2.1. Temperature Effects

In most chemical reactions, an increase in temperature is accompanied by an increase in reaction rate. A rough rule-of-thumb suggests that the reaction rate doubles for each 10 °C rise in temperature. Although this rule has many exceptions, it is important to take into consideration the influence of temperature when analyzing why materials fail.

Changing the temperature of an environment can influence its corrosivity. Many household hot water heater tanks, for example, were historically made of galvanized steel. The zinc coating offered a certain amount of cathodic protection to the underlying steel, and the service life was considered adequate. Water tanks seldom were operated >60 °C. With the development of automatic dishwashers and automatic laundry equipment, the average water temperature was increased so that temperatures of ∼80 °C have now become common in household hot water tanks.

Coinciding with the widespread use of automatic dishwashers and laundry equipment was a sudden upsurge of complaints of short-life of galvanized steel water heater tanks. Electrochemical measurements showed that in many cases, iron was anodic to zinc ∼75 °C, whereas zinc was anodic to iron at temperatures <60 °C. This explained why zinc offered no cathodic protection ∼75 °C, and why red water and premature perforation of galvanized water tanks occurred so readily at higher temperatures. This particular problem was partly solved by using magnesium sacrificial anodes or protective coatings, and by the replacement with new alloys.

3.7.2.2. Fluid Velocity Effects

Unless otherwise protected, metals generally owe their corrosion resistance to a tightly adherent, protective film that builds up on the metal surface by corrosion processes. This film may consist of reaction products, adsorbed gases, or a combination of these. Any mechanical disturbance of this protective film can stimulate attack of the underlying metals until either the protective film is reestablished, or the metal has been corroded away. The mechanical disturbance itself can be caused by abrasion, impingement, turbulence, or cavitation.

Erosion–corrosion is encountered most frequently in pumps, valves, centrifuges, elbows, impellers, inlet ends of heat-exchanger tubes, and agitated tanks. Locations in flowing systems where there are sudden changes in direction or flow cross-section, as in heat exchangers where water flows from the water boxes into the tubes, are likely places for erosion–corrosion. Under these conditions, which stimulate some corrosion of the metal surface, the effects of flow velocity may be to displace the corrosion products, thereby exposing fresh metal to the corrosion action of the solution. This action may lead to a much increased corrosion rate.

3.7.2.3. Impurities

Impurities present in minute amounts have often more influence on the corrosion behavior of materials than substances present in much greater quantities.

Sometimes impurities in trace quantities may accelerate the corrosion attack while at other times they may act as inhibitors. The introduction of small amounts of ions of metals such as copper, lead, or mercury, can cause severe corrosion of aluminum equipment, for example, corrosion of upstream copper alloy equipment can result in contamination of cooling water. Under these circumstances, copper can plate out on downstream aluminum equipment and pipe, setting up local galvanic cells that can result in severe pitting and perhaps perforation.

3.7.2.4. Presence of Microbes

As mentioned in Chapter 1, microbes are present almost everywhere in soils, freshwater, seawater, and air. However, the mere detection of microorganisms in an environment does not necessarily indicate a corrosion problem. Nonetheless, it is well established that numerous buried steel pipes have suffered severe corrosion as the result of bacterial action. A special section of Chapter 4 is devoted to this specific topic.

In unaerated or anaerobic soils, this attack is attributed to the influence of the sulfate-reducing bacteria (SRB). The mechanism is believed to involve both direct attack of the steel by hydrogen sulfide and cathodic depolarization aided by the presence of bacteria. Even in aerated or aerobic soils, there are sufficiently large variations in aeration that the action of SRB cannot be neglected. For example, within active corrosion pits, the oxygen content becomes exceedingly low.

Bacteria, fungi, and other microorganisms can play a major part in soil corrosion. Spectacularly rapid corrosion failures have been observed in soil due to microbial action and it is becoming increasingly apparent that most metallic alloys are susceptible to some form of microbiologically influenced corrosion (MIC).

3.7.2.5. Presence of Stray Currents

The corrosion resulting from stray currents coming from external sources is similar to that from galvanic cells that generate their own current. However, stray current strengths can be much more damaging than ordinary galvanic cells and, as a consequence corrosion may be much more rapid. Stray currents causing corrosion may originate from direct current distribution lines, substations, or street railway systems, and flow into an adjacent metallic structure. Alternating currents very rarely cause corrosion.

Another difference between galvanic-type currents and stray currents is that the latter are more likely to operate over long distances since the anode and cathode are more likely to be remotely separated from one another. Seeking the path of least resistance, stray currents from a foreign installation may travel along a pipeline or any other buried metallic structure causing severe corrosion where it leaves the line or current carrier.

3.8. INDUSTRIAL EXAMPLES

As mentioned earlier, there are increasing pressures in industry to minimize costs while improving the reliability and availability of plant and equipment. In many industries, maintenance activities are now treated as an investment with a requirement to accomplish the maintenance function with shrinking technical and financial resources. In this context, risk-based inspection (RBI) as a method for prioritizing the inspection of the plant has received considerable attention over the past few years and methods have been developed, on the continent, by the API and by a number of private organizations. The recent changes in the maintenance philosophy of electric utility systems, for example, have been considered significant enough to be called a maintenance revolution. These changes have been driven by the following factors (3, 26):

- More open, competitive markets, placing emphasis on cost issues.
- Operating and maintenance costs can be directly controlled by a utility.
- Rising relative importance of operating and maintenance costs.
- Aging assets with increasing maintenance requirements.

The principles of risk-based analysis were known to the power industry since the early 1970s when a technique known as probabilistic risk assessment (PRA) was introduced. However, the concept of risk as a measure of nuclear plant safety was only established in the late 1980s (17).

In the late 1980s and early 1990s the interest shifted to economy aspects and the concepts of RCM was introduced into practice. At about the same period, the concept of RBI was introduced in the American Society of Mechanical Engineers (ASME) codes. The main development of methods has been done in the nuclear field, by API for the Petrochemical plants, and by ASME/EPRI for the power industry. A coordinated approach to these developments is also underway in Europe, that is, the RIMAP project. The following sections describe some efforts that have developed in recent years in various crucial sectors of industry.

3.8.1. Transmission Pipelines

The transmission pipeline industry is an irreplaceable component of modern infrastructures. Pipelines have historically been the safest means of transporting natural gas and hazardous liquids. However, recent pipeline failures have heightened the awareness of transmission pipeline systems. In particular, in many parts of the world cities have grown and are now located close to pipeline systems that have been operating for decades (27).

Many millions of kilometers of pipeline crisscross the globe, carrying oil and natural gas. In the United States alone there are ~3.5 million km of transmission pipeline, 525 thousands for the transmission and gathering of natural gas, 260 thousands for the transmission and gathering of hazardous liquid, and the balance for the distribution of natural gas. Each system is unique with respect to potential integrity threats and associated consequences in the unlikely event of a failure. The probability of a release can be reduced through an effective management program that addresses these integrity threats.

There are many causes and contributors to pipeline failures. The U.S. Department of Transportation (DOT) Research and Special Programs Administration, Office of Pipeline Safety (RSPA/OPS) compiles data on pipeline accidents and their causes. Tables 3.5–3.7 summarize the results collected during a 2-year period (e.g., 2002 and 2003) for, respectively, natural gas transmission and gathering, hazardous liquid transmission, and natural gas distribution.

It is notable, in Tables 3.5 and 3.6, that corrosion was the most common cause (e.g., 37%) of natural gas transmission and gathering pipeline incidents during that period, and the second most common cause (e.g., 24%) of hazardous liquid pipeline incidents. During the same period over 60% of natural gas distribution pipeline incidents were caused by outside force (e.g., excavation by the operator or other parties, damage from natural forces etc.) while only a small fraction (e.g., 0.1%) of the property damage was attributed to corrosion. Figures 3.19 and 3.20 illustrate the property damage costs associated with external and internal corrosion of natural gas (Fig. 3.19) and hazardous liquid (Fig. 3.20) transmission and gathering pipelines during a period of eight years (e.g., 1997–2004). The data in these figures clearly indicate that both external and internal corrosion are serious contributors to pipelines property damage.

The pipeline industry uses considerable resources to minimize the likelihood of failures. A study recently completed by the DOT Federal Highway Administration (FHWA) estimates that the pipeline industry spends ~$7 billion year^{-1}

Table 3.5. Natural Gas Transmission and Gathering Pipeline Incident Summary by Cause for 2002 and 2003

Reported cause	Number of incidents	% of Total incidents	Property damages (k$)	% of Total damages	Fatalities	Injuries
Excavation damage	32	17.8	4,583	6.9	2	3
Natural force damage	12	6.7	8,278	13	0	0
Other outside force damage	16	8.9	4,689	7.1	0	3
Corrosion	46	25.6	24,273	37	0	0
Equipment	12	6.7	5,337	8.0	0	5
Materials	36	20.0	12,131	18	0	0
Operation	6	3.3	2,286	3.4	0	2
Other	20	11.1	4,773	7.2	0	0
Total	180		66,351		2	13

Table 3.6. Hazardous Liquid Pipeline Accident Summary by Cause for 2002 and 2003

Reported cause	Number of accidents	% of Total accidents	Barrels lost	Property damages (k$)	% of Total damages	Fatalities	Injuries
Excavation	40	14.7	35,075	8,988	12	0	0
Natural forces	13	4.8	5,045	2,646	3.5	0	0
Other outside force	12	4.4	3,068	2,063	2.8	0	0
Materials or weld failure	45	16.5	42,606	30,682	41	0	0
Equipment failure	42	15.4	5,717	2,761	3.7	0	0
Corrosion	69	25.4	55,610	17,776	24	0	0
Operations	14	5.1	8,332	817	1.1	0	4
Other	37	13.6	20,022	9,060	12	1	1
Total	272		175,475	74,792		1	5

Table 3.7. Natural Gas Distribution Pipeline Incident Summary by Cause for 2002 and 2003

Reported cause	Number of incidents	% of Total incidents	Property damages (k$)	% of Total damages	Fatalities	Injuries
Construction–operation	20	8.1	3,086	6.7	0	16
Corrosion	3	1.2	60	0.1	2	9
Outside force	153	62.2	32,334	70	6	48
Other	70	28.5	10,618	23	13	31
Total	246		46,098		21	104

on corrosion control. This figure includes operations and maintenance activities, capital expenditures, and corrosion failure repairs (27). The tools most commonly used by operators to verify pipeline integrity include, both external and internal direct assessments, hydrostatic testing, and in-line inspection (ILI). Each of these tools can be used under various circumstances for baseline assessments and future reassessments.

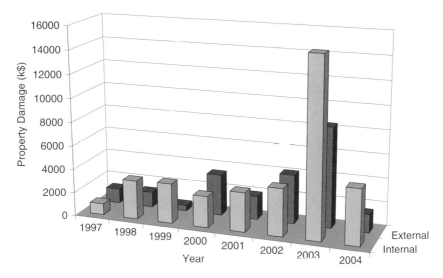

Fig. 3.19. Property damage costs attributed to external and internal corrosion of transmission and gathering pipeline during the 1997–2004 period.

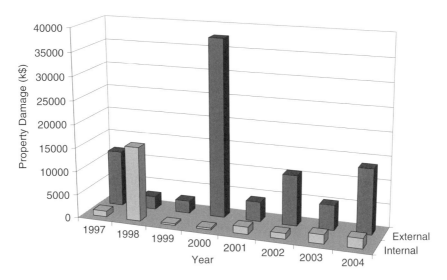

Fig. 3.20. Property damage costs attributed to external and internal corrosion of hazardous liquid transmission pipeline during the 1997–2004 period.

3.8.1.1. External Corrosion Damage Assessment

External corrosion direct assessment (ECDA) is a structured process that consists of four key steps: pre-assessment, indirect examination, direct examination, and post-assessment (Fig. 3.21). The ECDA is intended to assist pipeline operators in

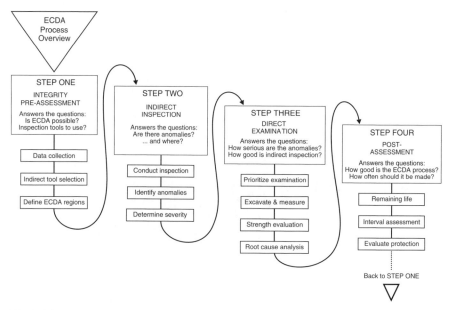

Fig. 3.21. Overview of the ECDA process.

establishing the integrity of pipelines. The process uses aboveground CP survey methods, many of which have been used in the pipeline industry for decades. The ECDA further defines the process, validation, and data integration for these survey methods (27). The ECDA process incorporates standard techniques for compiling historical information, pipeline and soil surveys, external pipeline inspections, and data analyses.

3.8.1.1.1. Pre-assessment Pre-assessment is the first step in the ECDA process. All pertinent historical information is compiled, falling generally into five categories (29): (1) pipe data; (2) construction data; (3) soil–environmental conditions; (4) corrosion protection data; (5) operating parameters–history.

The data collected for pre-assessment are similar to the information required for risk assessment, which may be performed concurrently. The feasibility of using ECDA must be evaluated based on available data. Other factors that, besides the lack of information, could preclude ECDA from being the assessment method of choice include: electrical shielding from disbonded coating; rocky terrain and backfill; paved surfaces; inaccessible locations; and paralleling buried metallic structures.

If an operator determines that ECDA is an appropriate assessment methodology, the pipeline then must be divided into ECDA regions and two indirect inspection techniques need to be selected for each region. For example, on a cathodically protected pipeline with good coating, a close-interval potential survey (CIPS) will provide information concerning the level of protection, interference, and current shielding. A direct current voltage gradient (DCVG)[4] survey performed in conjunction with the CIPS will detect specific locations of damaged coating. However, it is crucial that, independently of the two techniques selected, the ECDA be performed along the entire length of the specified section of pipeline with both techniques.

3.8.1.1.2. Indirect Inspection The purpose of indirect inspection is to identify the locations of coating faults, insufficient cathodic protection, electrical shorts, interference, geologic current shielding, and other anomalies along the pipeline. Indirect inspection also permits to identify the areas where corrosion may be occurring or has occurred. Typical pipeline survey techniques that may be used are (29):

- Close-interval on/off potential surveys (Fig. 3.22)[5].
- DCVG surveys.

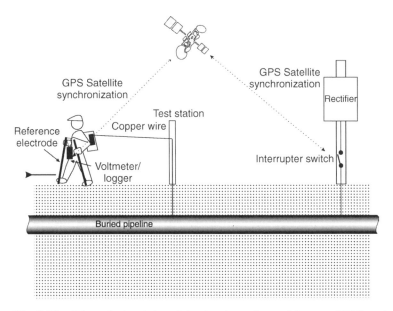

Fig. 3.22. Schematic description of the close-interval potential survey (CIPS) methodology using a global positioning system (GPS) to log the location data.

---

[4]DCVG surveys are a modern method to locate defects on coated buried pipelines and to make an assessment of their severity. The potential gradient is measured by an operator between two reference electrodes (e.g., copper sulfate electrodes) separated by a distance of approximately half a meter. A pulsed dc signal is imposed on the pipeline for DCVG measurements (29;30).

[5]The principle of a close interval potentials survey (CIPS) is to record the potential profile of a pipeline over its entire length by taking potential readings at intervals of approximately one meter. A reference electrode is connected to the pipeline at a test post and positioned in the ground over the pipeline at regular intervals (∼1 m) for the measurement of the potential difference between the reference electrode and the pipeline (30, 31).

136 Chapter 3 Maintenance, Management, and Inspection Strategies

- Electromagnetic current attenuation surveys;
- Alternating current voltage gradient surveys.

All four survey techniques are well established. Each has its advantages and disadvantages, depending on the type of anomaly that needs to be detected and the condition of the pipeline right of way. However, it is important that the inspections with the two chosen techniques be carried at approximately the same time and that the data be tied to all permanent features along the pipeline route so that the two data sets can be aligned and compared. The data should then be analyzed and indications of corrosion activity identified. The data analysis should reveal all discrepancies in the two data sets, resolve these differences, and compare the results to those obtained in the pre-assessment.

3.8.1.1.3. Direct Examinations Direct examination requires excavating the pipe and performing physical inspections and tests on the pipe surface and surrounding soil–water electrolytes. Direct examination includes

- Ranking and prioritizing indications identified during indirect inspections.
- Excavating to expose the pipe and collecting data where corrosion activity is most likely.
- Measuring coating damage and corrosion defects.
- Performing a root-cause analysis (RCA).
- Evaluating the process.

Each indication obtained with indirect inspections is then categorized for an action level as either: immediate action required; scheduled action required; or suitable for monitoring.

Subsequently, dig locations to expose the pipe are identified and given a priority. At least one excavation and direct examination is required for each ECDA region. When performed for the first time, a minimum of two direct examinations are required. Before exposing the pipe, the procedures for data collection and record keeping should be defined to ensure consistency. The procedures should include (29): photographic documentation; pipe-to-soil (P/S) potential measurements; soil and groundwater tests; coating assessment; undercoating liquid pH; mapping and measurement of corrosion defects; data for other analyses, such as MIC and SCC.

Once corrosion defects have been found, the remaining strength of the pipe should be estimated using a standard calculation routine [e.g., ASME B31G (32), RSTRENG[6], or DNV RP-F101 (33)] following which a RCA should be carried out for all significant corrosion activities before implementing an appropriate corrosion mitigation program.

---

[6]RSTRENG uses the same basic equation for predicting failure stress as B31.G, but calls for different definitions of three key variables (i.e., flow stress, Folias factor, and area of missing metal). The last has a direct bearing on the way the results of a metal loss inspection are interpreted.

3.8.1.1.4. Post-assessment The final step in the ECDA process is post-assessment, which provides an opportunity to assess the overall effectiveness of direct assessment and define reassessment intervals based on remaining-life calculations. Various methods can be used to calculate the remaining life from the corrosion growth rate, wall thickness, calculated failure pressure, yield pressure, maximum allowable operating pressure, and appropriate safety margin.

Under some conditions, corrosion rates can be measured with a corrosion monitoring technique [e.g., linear polarization resistance (LPR), corrosion coupons]. In the absence of other data, the NAGE ECDA document (34) recommends to use a pitting rate of 0.4 mm/year, which can be reduced by 24% in cases where the pipeline has consistently been cathodically protected by a minimum of 40 mV. The maximum reassessment interval is then established as one-half the calculated remaining life. For hazardous liquid pipelines, the maximum inspection interval is 10 years according to the high consequence area (HCA) regulations. The rule for gas pipelines is a maximum interval of 7 or 5 years when using direct assessment.

3.8.1.2. Internal Corrosion Damage Assessment

As previously discussed in reference to Fig. 3.19, internal corrosion is a serious problem for natural gas and transmission pipelines. Internal corrosion direct assessment (ICDA) is also a four step method (Fig. 3.23) developed to determine the presence of internal corrosion in pipelines and ensure their integrity.

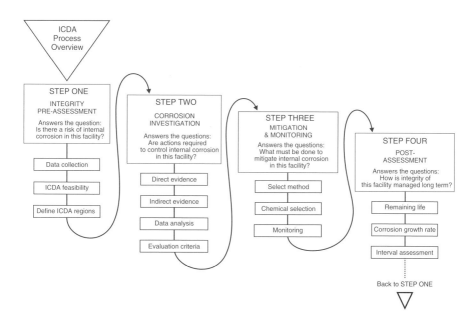

Fig. 3.23. Overview of the ICDA process.

Chapter 3 Maintenance, Management, and Inspection Strategies

The ICDA is applied to natural gas systems that normally carry dry gas, but experience intermittent dropout of liquid water. It is believed that internal corrosion occurs at regions where water accumulates along the pipeline. This observation forms the basis for the ICDA process to conduct a detailed examination at those regions (35).

3.8.1.2.1. Pre-assessment Pre-assessment involves the collection of essential data required to conduct the assessment, feasibility determination of ICDA process, and classification of ICDA regions. Data to be collected include pipe operating history, pipe design, features with inclination and elevation data, maximum flow rates, temperature and pressures, corrosion monitoring, and other internal corrosion related data. Once the required data have been collected, the requirement to carry out the ICDA is assessed based on the following conditions:

- The gas in the pipeline should contain <110 g m^{-3} of water as determined by the chemical analysis of collected fluids.
- The pipeline should not have a past history of transporting crude oil or products.
- The pipeline has a history of inefficient coating practices.
- The pipeline is not treated with corrosion inhibitors.
- Pipeline is not subjected to regular pigging.

The ICDA regions are then classified based on changes in temperature, pressure, and flow rate (e.g., in and out pressures).

3.8.1.2.2. Indirect Inspection The main objective of the indirect inspection step is to predict regions highly susceptible to internal corrosion. These regions are further selected for detailed examination. For this reason, critical degree of inclination and pipeline inclinations are calculated and compared. This step is applicable to pipelines in which stratified flow[7] is the dominant flow regime. A flow modeling calculation using Eq. (3.9) is applied to determine critical inclination angles (θ) (35):

$$\theta = \sin^{-1}\left(\frac{\rho_g}{\rho_l - \rho_g}\right) x \left(\frac{V_g^2}{g x d_{id}}\right) \times F \tag{3.9}$$

where ρ_l = liquid density; ρ_g = gas density; g = acceleration due to gravity (9.8 m s^{-2}); d_{id} = pipe internal diameter; V_g = superficial gas velocity; F = modified Froude number.

These calculations are bound by the following conditions:

- The maximum superficial gas velocity in the pipeline should be ≤ 8 m s^{-1}.
- The nominal pipe diameter should be between 0.1 m and 1.2 m.

---

[7]Stratified flow is defined as a flow pattern with liquid flowing at the bottom of the pipe and gas at the top.

- Pressures should be between 3.4 and 7.6 MPa unless flow modeling has been performed outside this range.

The actual pipeline inclination angle is calculated based on eq. (3.10):

$$\text{Inclination} = \tan^{-1}\left(\frac{\Delta \text{ elevation}}{\Delta \text{ length of pipe section}}\right) \quad (3.10)$$

Regions with water accumulation, which are most susceptible to internal corrosion, are identified based on the comparison between critical degree of inclination and pipeline inclination angles.

3.8.1.2.3. Detailed Examination This step involves comparison of field measured performance data with the pipeline historical and existing operating data to assess the effect of internal corrosion on the pipeline. Detailed examination is performed on the predicted high corrosion regions to confirm the presence of internal corrosion. If the pipeline inclination angle exceeds the critical degree of inclination, detailed examination is performed at pipe locations with inclination greater than the critical inclination within the ICDA region (35).

If there is no instance where the critical angle is exceeded, then detailed examination is performed at the angle of greatest inclination within the ICDA region. While performing a detailed examination, accurate measurements of wall thickness within wall loss areas need to be identified. The remaining strength of the corroded areas of the pipeline should also be determined.

Once the site has been exposed, corrosion monitoring devices (e.g., coupons, electrical resistance probes) can be installed to identify and monitor corrosivity in the pipeline. Inline inspection (ILI) results may also provide information to assess the downstream conditions of the pipe. Once regions most susceptible to corrosion are free of damage, it is then considered that pipeline integrity to a large extent is assured.

3.8.1.2.4. Post-assessment Post-assessment determines the effectiveness of ICDA for dry gas transmission lines based on the correlation between detected corrosion obtained during detailed examination and ICDA predicted regions. Reassessment intervals are determined depending on the remaining life of the pipe and rate of corrosion growth in the pipe.

3.8.1.3. Hydrostatic Testing

Hydrostatic testing is used to conduct strength tests on new pipes while in the manufacturing process, as well as at the completion of pipeline installation in the field prior to being placed in service. Hydrostatic testing is also used, at times, for integrity assurance after a pipeline is in operation. Hydrostatic tests are generally the preferred integrity assessment method when the pipeline is not capable of being internally inspected or if defects are suspected that may not be detectable by ILI tools.

140 Chapter 3   Maintenance, Management, and Inspection Strategies

Hydrostatic testing is an effective verification method for pipelines. It provides information about the integrity status at the time of the hydrostatic test and provides a "go/no go" answer. Although hydrostatic testing is an effective method for validating near-term integrity, it provides limited information about the status of integrity threats, creates environmental problems related to water treatment, and results in potentially significant downtime, affecting deliveries.

The hydrostatic test establishes the pressure carrying capacity of a pipeline and may identify defects that could affect integrity during operation. Testing is done to a pressure that is greater than the normal operating pressure of the pipeline. This provides a margin of safety. The test stresses the pipeline to a predetermined percentage (e.g., 93–100%) of its specified minimum yield strength (SMYS) generally for 8 h. If SCC is suspected, the test pressure may be increased to 100% of SMYS or higher for 30 min to 1 h. Axial flaws, such as SCC, longitudinal seam cracking, selective seam corrosion, long narrow axial (channel) corrosion and axial gouges, which are difficult to detect with magnetic flux leakage (MFL) pigs, are better detected with a hydrostatic test.

3.8.1.4. In-Line Inspection

In-line inspection (ILI) tools, also commonly called "smart" or "intelligent pigs", are cylinder-shaped electronic devices used by the pipeline industry to detect loss of metal and in some cases deformations in the pipeline. Inserted into the pipeline and propelled by the fluid or gas pressure, ILI tools record certain physical data about the pipeline integrity (e.g., location of reduced pipe wall thickness, dents) as it moves through the pipeline. Evaluation of the collected data allows the pipeline operator to make integrity decisions about the pipeline and to find and mitigate potential problem areas before they become a problem.

Since their development in the 1960s, these tools have undergone several generations of technological advancements. These tools now include caliper–deformation tools to characterize dents and changes in ovality, and magnetic flux leakage (MFL) and ultrasonic testing (UT) tools to characterize corrosion-caused metal loss. Other technologies, such as ultrasonic crack-detection tools, are also used by pipeline operators to address specific integrity threats. The most commonly used and established ILI tool is without any doubt MFL. The type of ILI tool to be used should be determined based upon several criteria summarized in Fig. 3.24 (36).

3.8.1.4.1. Metal Loss (Corrosion) Tools   Metal loss tools are used to detect defects that have resulted in wall thinning in a pipeline. They can discriminate to some extent between manufacturing defects, corrosion defects, and mechanical damage. There are two main types of tools:

1. Magnetic flux leakage tools use a circumferential array of MFL detectors embodying strong permanent magnets to magnetize the pipe wall to near saturation flux density. As the tool travels down the pipe it records magnetic flux leakages that are converted into information revealing anomalies

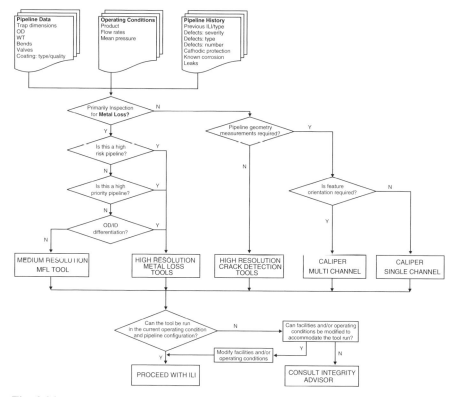

Fig. 3.24. The ILI tool selection decision tree (36).

in the pipe wall, such as corrosion pits. The leakage fluxes are detected by Hall probes or induction coils that are part of the MFL ILI tool.

Two types of MFL tools are available, that is, high resolution MFL and standard resolution MFL. The main difference between the two is in the number of sensors and the effective resolution. Most MFL tools can determine the location and o'clock position of the metal loss anomaly and detect if a corrosion anomaly is internal or external to the pipe wall. MFL tools also provide data of each corrosion anomaly including its length and maximum depth to determine the pipe remaining strength. The MFL tools are typically capable of detecting corrosion depth $>20\%$ of the pipe wall thickness. Axially oriented flaws, such as SCC, selective seam corrosion, and axial gouges are difficult to detect with MFL tools.

2. Ultrasonic (UT) wall measurement tools use the ultrasound echo time technique for measuring the remaining wall thickness. The UT tool transmits an ultrasonic pulse into the pipe wall and directly measures its thickness. Since this technology requires a clean pipe wall, it is generally not used for certain pipelines, such as crude lines with a paraffin build-up. There are also wall-thickness limitations with the UT tool. It

works well with heavy-wall pipe, but not as well with thin-wall pipe, and it is not as widely used as the MFL tool. This particular tool offers the following advantages in a pipeline environment:
- Direct measurement and high accuracy of wall thickness and defect depth.
- Precise distinction between internal and external defects with the exception of defects adjacent to welds.
- UT tools can also be used to approximate the remaining strength of affected pipe area.
- UT tools provide a more precise description of anomalies than MFL tools.

3.8.1.4.2. Crack Detection Tools Crack detection tools are typically designed for longitudinal crack detection, but they can also be adapted to circumferential crack detection.

1. AUT crack detection tool generates an ultrasonic signal into the pipe wall that is reflected off the internal and outer surfaces of the pipe. If a crack is detected, the signal reflects back along the same path of the tool. Since a liquid couplant is required between the sensors and the pipe wall, this tool works only with liquid pipelines.
2. Transverse MFL tools magnetize the pipe wall around its circumference to detect cracks, such as longitudinal seams cracks and longitudinal seam corrosion. This tool is similar to the standard MFL tool, but its induced magnetic field is in a transverse or perpendicular direction. This tool also has limitations, for example, cracks must have sufficient width, or gap, to be detected, and the severity of the crack is not determined.
3. Elastic wave tools operate by sending ultrasounds in two directions along the pipeline to locate and size longitudinally oriented crack and manufacturing defects.
4. Electromagnetic acoustic tools are particularly adapted to the detection of cracks in dry gas pipelines. This tool generates ultrasonic signals without requiring liquid couplant to transfer the ultrasound into the steel. These tools are new to the pipeline inspection industry, so their effectiveness has not yet been determined.

3.8.1.4.3. Geometry Tools Geometry tools gather information about the physical shape, or geometry, of a pipeline. Geometry tools are primarily used to find "outside force damage", or dents, in the pipeline. However, they can also detect and locate mainline valves, or fittings. As with all ILI tools, geometry tools have limitations on their use and in the usefulness of the results obtained.

1. Caliper tools use a set of mechanical fingers or arms that ride against the internal surface of the pipe or use electromagnetic methods to detect dents or deformations.

2. Pipe deformation tools operate on the same principle of caliper tools with the addition of gyroscopes to provide the o'clock position of the dent or deformation in the pipe. This tool can also provide pipe bend information.
3. Mapping tools can be used in conjunction with other ILI tools to provide GPS correlated mapping of the pipeline and other physical details (e.g. valves).

3.8.2. Offshore Pipeline: Risers

Offshore pipeline failure statistics collected for >30 years have revealed that 55% of all 4000 thousand Gulf of Mexico incidents reported to the U.S. Department of the Interior were caused by corrosion damage (Fig. 3.25) (28). External and internal corrosion contributed, respectively, to 70 and 30% of the reported cases caused by corrosion. The largest cause of failure of risers was also found to be due to corrosion, with 92% of riser corrosion failures due to external corrosion and only 8% due to internal corrosion damage. Possible causes of the high pipeline riser corrosion failure rates could be the following:

- Lack of required inspection because of ineffective regulations.
- Lack of inspection because of ineffective operator policy–procedures.
- Lack of corrosion protection because of ineffective design.
- Ineffective coatings in the splash zone.

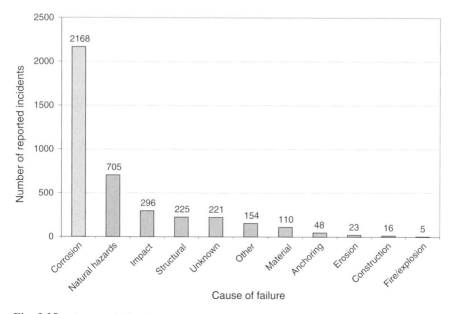

Fig. 3.25. General pipeline failure statistics for the Gulf of Mexico.

- Ineffective cathodic protection.
- Corrosive operating environment.
- Human factors.
- Internal–external inspection tool inefficiencies.

Marine riser systems form the link between the seabed pipeline and the topsides processing equipment (Fig. 3.26). Various production and export riser designs are currently used and the chosen design usually dictates or limits how a riser can be inspected. The continuing trend to produce oil and gas in deeper water is forcing the evolution towards more efficient methods for detecting and monitoring corrosion and other damages in risers. Riser inspectability or needs thereof depend on the following characteristics:

- What is the material of construction: that is, carbon steel, stainless steel, titanium, composite, etc.?
- What is the coating system used: that is, no coating, coated, but uninsulated pipe with paint or sprayed aluminum coating, or insulated pipe coated with a long-lasting protection elastomer?

External riser surfaces are very often covered with thick biomass that prevents an easy inspection. The consistent wetting and drying in the splash zone combined

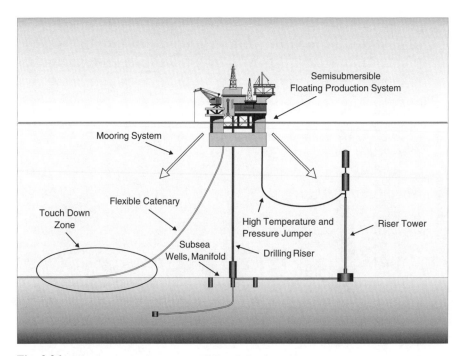

Fig. 3.26. Floating production system (FPS) and risers.

with defects in the coatings are the usual contributors to corrosion problems. The following is a list of the most common riser corrosion mechanisms: general corrosion; localized corrosion pitting; erosion corrosion; crevice corrosion; mesa corrosion[8]; SCC, sulfide, hydrogen, or chloride induced.

Below the splash zone a moderate (0.1 mm y^{-1}) corrosion rate is expected, but up to 1 -mm year^{-1} rates have been observed for local riser connections and other elements in this region. Corrosion damage can be inspected either from the outside or inside of the riser. Internal inspection using techniques other than magnetic flux leakage (MFL) technique requires bleeding down the riser and filling it with fluid resulting in significant lost production costs. A compact riser ILI tool was developed in the early 1990s to inspect risers from the inside. The system uses straight beam conventional ultrasonic multiple transducers for corrosion mapping and time-of-flight diffraction (TOFD) technique for circumferential weld inspection.

External inspection of risers with surface coatings and without casings typically involves marking the riser surface into a grid pattern, followed by point-by-point ultrasonic thickness measurements of individual grid sections using manually manipulated measuring instruments or multiple scans with single or multiple conventional ultrasonic transducers. This tedious task often results in limited measurement accuracy. The costs and difficulties of this process are compounded by the need for inspection divers to work in the hazards of the splash zone.

A summary of the technologies capable of detecting, locating, and monitoring corrosion damage in risers is shown in Table 3.8 with the respective advantages, disadvantages, and primary corrosion damage detected by each technique (28).

3.8.3. Process Industry

The public demands that process industry manufacturers operate and maintain their plants with a minimum of incidents. Bhopal and numerous other accidents involving chemical spills, explosions, fires, and other hazards have convinced many that chemicals are "highly hazardous" and that process industries cannot safely handle and manufacture their products. The public pressure to restrain process industries has lead many countries to adopt series of increasingly stringent regulations.

In process operations, where corrosion risks can be extremely high, costs are often categorized by equipment type and managed as asset loss risk (Fig. 3.27) (37). By referring to Fig 3.27, the operations department of a process plant can prioritize the maintenance functions with decreasing attention on piping, reactors,

---

[8]Mesa corrosion is one of the common types of corrosion experienced in service involving exposure of carbon or low alloy steels to flowing wet carbon dioxide conditions at slightly elevated temperatures. A protective iron carbonate surface scale may form in this type of environment. However, under the surface shear forces produced by flowing media, this scale can become damaged in a localized attack producing mesa-like features.

Table 3.8. Summary of Methods and Techniques for Detecting and Monitoring Corrosion Damage in Risers

Method and technology	Advantages	Disadvantages	Specific damage
Visual conventional	Large area of exposure and inexpensive	Limited to external damage, measurements not accurate, subjective and labor intensive	External general corrosion or pitting
Visual enhanced	Large area of exposure and very fast inspection	Requires preparation, still difficult quantification and subjective	External general corrosion or pitting through magnification or accessibility
Short-range ultrasonics (manual point-by-point measurements, single echo, or echo-to-echo)	Need access only to one side, sensitive and accurate, no coating removal	Requires couplant and clean and smooth surface for single echo and coating removal if it is thicker than 6 mm for echo-to-echo	Corrosion loss and pitting
Short-range ultrasonics (bonded array, single echo, or echo-to-echo)	Continuous local corrosion condition monitoring	Requires bonding of an array of flexible transducers strip, coating removal, clean and smooth surface	Corrosion loss and pitting
Short-range ultrasonics (semi-AUT[a] TOFD[b])	Fast inspection with good resolution	Requires couplant and clean and smooth surface and coating removal	Erosion corrosion
Short-range ultrasonics (AUT mapping with single/multiple focused probes or PA)	Fast inspection with good resolution and sensitivity	Requires couplant and clean and smooth surface and rust/coating removal	External/internal corrosion loss and pitting if internal/external surface is regular
Short-range ultrasonics (AUT pigging with single/multiple L- or SV-waves probes or PA[c])	Fast inspection with good resolution and sensitivity	Requires couplant and clean and smooth surface, riser opening	Pitting, corrosion loss, and SSC
Long-range ultrasonics (guided waves and dry-couplant transducers)	Global screening technique, fast inspection, requires no couplant	Sensitive to both internal and external damage, no absolute measurements	General corrosion loss

Table 3.8. (*continued*)

Method and technology	Advantages	Disadvantages	Specific damage
Long-range ultrasonics (guided waves and MsS)	Global screening technique, fast inspection, requires no couplant	Sensitive to both internal and external damage, no absolute measurements	General corrosion loss
Long-range ultrasonics (guided/SH-waves and electromagnetic acoustic transducers)	Global screening technique, fast inspection, requires non couplant	Sensitive to both internal and external damage, no absolute measurements	General corrosion loss, SSC
ET conventional RFEC	Good resolution, multiple layer capability portability	Low throughput, operator training	Surface and subsurface flaws
		Sensitive to both internal and external damage	Surface and subsurface flaws
Pulsed eddy current	Deep penetration	Large footprint	General corrosion loss
MFL	Through coatings penetration	Thickness limitations	General thinning, pitting
Alternating current field measurement (ACFM)	Through coatings penetration	Low throughput, operator training	Surface and subsurface flaws
Field signature method (FSM)	Continuous local corrosion monitoring	Small area, expensive	Surface flaws
Digital radiography	Good resolution and image interpretation	Radiation safety	Pitting and general corrosion
Tangentional radiography	Portable	Radiation safety	General loss
Acoustic emission (AE)	Global monitoring technique	Prone to false indication from wave motions	SSC
Infrared thermography	Large area scan	Complex equipment, layered	Surface corrosion
Magnetic particles	Easy, portable	Clean surface	Surface cracks

[a] AUT = automated ultrasonic testing.
[b] TOFD = time-of-flight diffraction.
[c] PA = phased array

148 Chapter 3   Maintenance, Management, and Inspection Strategies

tanks, and process towers. Some examples of process equipment, where SCC damage has been historically reported have been shown in the previous section on "Hot Spots".

The primary purposes of fixed equipment (e.g., pressure vessels, piping, tanks) are to contain the process under all normal service conditions and to store or process various substances. During process excursions various controls and pressure relieving devices are employed to relieve extreme conditions while at the same time controlling any loss of containment in a safe manner. One of the primary functions of other process equipment (e.g., pumps, valves, instruments) is also to maintain containment (38).

Process equipment must be strong enough to maintain containment while pressurized and subjected to other mechanical loads under service conditions. Process containment is stated public policy that is most likely to become even more demanding during the new millennium. There will thus be an increasing duty to achieve and constantly maintain even higher levels of mechanical integrity than are acceptable today. Many chemical companies are using a form of RBI to heighten the mechanical integrity of their plants. A variety of portable and fixed detection and data collection instruments are used to handle inspection requirements.

Probably the most important inspection function related to plant reliability is focused on process piping systems. Piping systems not only connect all other equipment within the unit, but also interconnect units within the operation. The fitness for service of piping can thus be considered to be a good barometer of conditions occurring within a given process. It has been demonstrated repetitively

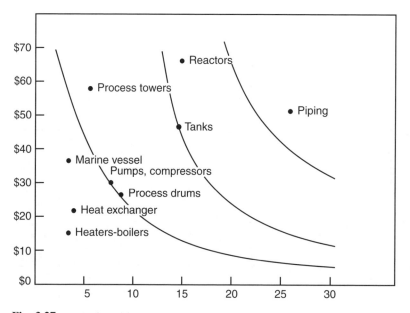

Fig. 3.27. Asset loss risk as a function of equipment type.

that, if an inspection department has control over the condition of piping within a unit, the condition of the remaining equipment will also be known with a relatively high degree of confidence (37). It is rare that corrosion or other forms of deterioration found in major components of process equipment are not found in the interconnecting piping. The latter is generally more vulnerable to corrosion and subject to initial failure because

- Corrosion allowance on piping generally is only one half that provided for other pieces of refinery equipment.
- Fluid velocities are often higher in piping, leading to accelerated corrosion rates.
- Piping design stresses normally are higher and the piping system may be subject to external loading, vibration, and thermal stresses that are more severe than those encountered in other pieces of equipment.
- A larger number of inspection points in a piping system represent a bigger control and monitoring task.

Leaks in pressurized piping systems are extremely hazardous and have led to several catastrophes. Components requiring close attention include

- Lines operating at temperatures below the dew point.
- Lines operating in an industrial marine atmosphere.
- Points of entry and exit from a building, culvert, and so on, where a break in insulation could occur.
- Pipe support condition, fireproofing.
- Piping alignment, provision for thermal expansion, position of pipe shoes on supports.
- Welded joints representing elevated stress levels from residual stress effects and stress concentrations, geometrical discontinuities, complex metallurgical structures and possible galvanic cells or preferential weld corrosion.
- Flanged or screwed joints for evidence of leakage.
- Geometrical changes that affect fluid flow characteristics (e.g., bends, elbows, section changes), with a resulting risk of erosion–corrosion.

Lines handling corrosive materials, such as salt-water ballast, acids, bases, and brine are subject to internal corrosion and require frequent inspection until a satisfactory service history is developed. Frequency and degree of inspection must be individually developed with regard to rate of deterioration and seriousness of an unpredicted leak. Testing of piping systems is typically done with a variety of techniques including pressure testing, radiography, dye-penetrant, magnetic-particle, ultrasonic, and eddy current testing.

150   Chapter 3   Maintenance, Management, and Inspection Strategies

3.8.4. Power Industry

A detailed example of cost data for corrosion problems affecting the following broad sectors of the electric power industry has recently been published for the United States (39, 40):

- Fossil steam generation.
- Combustion turbine generation.
- Nuclear power generation.
- Transmission and distribution.

The compiled and sorted costs are presented in Table 3.9 with the highest cost listed first. These dollar amounts include corrosion-related costs associated with both O&M and depreciation. They also include both direct and indirect corrosion-related costs. The total cost of the listed items is ~11.7 billion or 76% of the total $15.4 billion cost of corrosion for 1998. The balance of the corrosion cost ($3.7 billion) likely stems from many miscellaneous less costly corrosion problems.

As shown in Table 3.9, corrosion costs in the nuclear power and fossil steam power sectors dominate corrosion costs in the electric power industry. The very large cost problems in the nuclear and fossil sectors at the top of the list warrant serious attention. Some special monitoring techniques are available to supplement normal water chemistry control in these important industrial applications.

These tools are designed to help operators carry out specific monitoring tasks, for example, scale and deposits, composition of moisture droplets and liquid film in two-phase regions, *in situ* corrosion potential, at-temperature pH, and exfoliation in the superheater and reheater. Table 3.10 provides a brief description of the results that can be obtained with such devices and Fig. 3.28 illustrates where, for example, the measurement points would be for monitoring a PWR steam generator (41).

Two factors that are not monitored often enough in the power producing industry are thermal stresses in heavy sections, which can lead to low-cycle corrosion fatigue (CF), and vibration in rotating machinery, which can lead to high-cycle CF. Table 3.11 summarizes the applications of inspection and corrosion monitoring tools that can be used for field monitoring these problems.

As also revealed in Table 3.9, the corrosion costs of some distribution, transmission, and combustion turbine sector problems may be substantial and warrant attention. Detailed discussions of the items listed in Table 3.9 can be found in a special EPRI report (40). The summary given here is limited to a brief discussion of the 10 most costly corrosion problems listed in Table 3.9.

3.8.4.1. Corrosion Product Activation and Deposition

This corrosion problem is identified as the most costly to the electric power industry. It affects all nuclear power plants and occurs as a result of the accumulation and activation of corrosion products in the reactor core and the subsequent

Table 3.9. 1998 Costs of Corrosion: Problems from All Power Industry Sectors[a]

Corrosion problem	Sector	$ Million	%
Corrosion product activation and deposition	Nuclear	2,205	18.80
Steam generator tube corrosion including IGA and SCC	Nuclear	1,765	15.05
Waterside/streamside corrosion of boiler tubes	Fossil	1,144	9.76
Heat exchanger corrosion	Fossil and nuclear	855	7.30
Turbine corrosion fatigue (CF) and SCC	Fossil and nuclear	792	6.75
Fuel clad corrosion	Nuclear	567	4.83
Corrosion in electric generators	Fossil and nuclear	459	3.91
Flow-accelerated corrosion	Fossil and nuclear	422	3.60
Corrosion of service water	Fossil and nuclear	411	3.51
Intergranular SCC of piping and internals	Nuclear	363	3.10
Oxide particle erosion of turbines	Fossil	360	3.07
Fireside corrosion of waterwall tubes	Fossil	326	2.78
Primary water SCC of nonsteam-generator alloy 600 parts	Nuclear	229	1.95
Corrosion of concentric neutrals	Distribution	178	1.52
Copper deposition in turbines	Fossil	149	1.27
Fireside corrosion of superheater and reheater tubes	Fossil	149	1.27
Corrosion of underground vault equipment	Distribution	142	1.21
Corrosion of flue gas desulfurization system	Fossil	131	1.12
Corrosion in valves	Nuclear	120	1.03
Liquid slag corrosion of cyclone boilers	Fossil	120	1.02
Back-end dewpoint corrosion	Fossil	120	1.02
Atmospheric corrosion of enclosures	Distribution	107	0.91
Hot corrosion of combustion turbine (CT) blades and vanes	Combustion turbine	93	0.79
Boric acid (H_3BO_3) corrosion of CS and low-alloy steel parts	Nuclear	93	0.79
Irradiation-assisted SCC of reactor internals	Nuclear	89	0.76
Corrosion in pumps	Nuclear	72	0.61
Corrosion of tower footings	Transmission	45	0.38
Hot oxidation of CT blades and vanes	Combustion turbine	35	0.30
Corrosion of boiling water reactor control blades	Nuclear	32	0.27
Corrosion of anchor rods	Transmission	27	0.23
Corrosion of tower structures	Transmission	27	0.23
Heat recovery steam generator (HRSG) CF	Combustion turbine	20	0.17

152 Chapter 3   Maintenance, Management, and Inspection Strategies

Table 3.9. (*continued*)

Corrosion problem	Sector	$ Million	%
Conductor deterioration	Transmission	18	0.15
HRSG flow-accelerated corrosion	Combustion turbine	10	0.09
HRSG underdeposit corrosion	Combustion turbine	10	0.09
Corrosion of CT compressor section	Combustion turbine	9	0.08
Corrosion of CT exhaust section	Combustion turbine	9	0.08
Corrosion of splices	Transmission	9	0.08
Corrosion of shield wires	Transmission	9	0.08
Corrosion of substation equipment	Transmission	5	0.04

[a]Ref. 40.

distribution of these activated materials around the reactor coolant system. The costs are incurred in work activities, such as radiation protection, radioactive waste disposal, decontamination, and primary coolant chemistry control.

3.8.4.2. Pressurized Water Reactor Steam Generator Tube Corrosion

Steam generator problems, notably deterioration of the steam generator tubes have proven to be responsible for forced shutdowns and capacity losses (26). These tubes are obviously a major concern, as they represent a fundamental reactor coolant pressure boundary. In fact, corrosion of pressurized water reactor (PWR) steam generator tubing is the second most costly corrosion problem in the industry. Corrosion of steam generator tubes, especially intergranular attack (IGA) and SCC of mill-annealed N06600 alloy tubes from the secondary side, has been a major problem for many years. It has led to extensive inspections and repairs of affected plants, the need for extensive plant design and operating practice changes to significantly improve water chemistry practices, and the replacement of steam generators at many plants.

The safety issues concerning tube failures are related to overheating of the reactor core (multiple tube ruptures) and also radioactive release from a rupture in the pressurized radioactive water loop. The cost implications of repairing and replacing steam generators are enormous, at replacement costs of $100–300 million, depending on the reactor size. Forced shutdowns of a 500 MW power plant may exceed $500,000 day^{-1}. Decommissioning costs of plant due to steam generator problems run into hundreds of million dollars.

3.8.4.3. Boiler Tube Waterside–Streamside Corrosion

Waterside–streamside corrosion of boiler tubes is the most costly problem in fossil steam power plants. This specific problem it is the largest cause of boiler

Table 3.10. Monitoring Related to Water and Steam Chemistry, Scale, and Deposits[a]

Device	Applications	Monitoring results
Steam turbine deposit collector/simulator	High-pressure (HP), intermediate-pressure (IP), and low-pressure (LP) turbines	Deposit composition, morphology, and rate of deposition vs. operation
Converging–diverging nozzle for LP turbines	Fossil and nuclear LP turbines Simulates HP turbine deposition	Quantity and types of impurities depositing on LP turbine blades and corrosiveness of the environment
Converging nozzle for HP turbines	Simulates moisture drying on hot surfaces in LP turbines	Types of impurities depositing on HP turbine blades
Drying probe for wet steam stages	Boilers/turbines	Deposits of low-volatility impurities in LP turbines are collected
Boiler carryover monitors	LP turbines, boilers, condensers	Mechanical carryover
Early condensate samplers	Piping, turbines; also used to monitor effectiveness of steam blow and foreign object damage	Chemistry of water droplets formed in the final stages of the LP turbine, and 10m
Particle flow monitor for exfoliated oxides	Condensers, cooling towers Piping, heat exchangers, boilers	Number and size distribution of oxide particles in superheated and reheated steam
Biofouling monitor	Boiler tubes	Detection of biofouling, collection of organic matter
In situ pH and corrosion potential	Boiler tubes	Susceptibility to general and localized corrosion
Heat flux gauge	Deposit buildup in turbines	Value of local heat flux, type of boiling, potential for impurity concentration
Chordal thermocouples		Boiler tube temperature vs. scale
Rotor position and thrust bearing wear	Deposits and erosion in turbines	Damage to thrust bearing caused by deposit accumulation on blades
Turbine first-stage pressure		Degree of deposition or erosion of control stage

[a] Ref. 41.

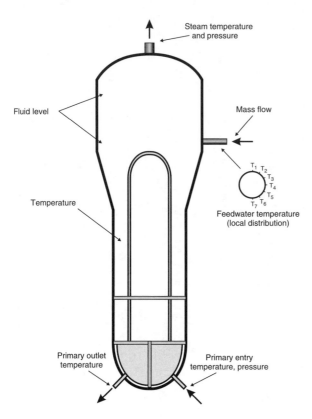

Fig. 3.28. Measurement points for monitoring of a pressurized water reactor (PWR) steam generator.

tube failures and also the largest cause of forced loss of production in these plants. The problems include the following, listed in order of decreasing importance:

- Corrosion fatigue (CF) on the waterside.
- Long-term overheat creep of superheater and reheater tubes aggravated by corrosion.
- Underdeposit corrosion of waterwalls, for example, hydrogen attack, acid phosphate corrosion, caustic gouging.
- Fireside corrosion from nitrogen oxide control.
- Pitting, attributed to improper shutdown, is another problem increasing in importance in coal-fired plants.

3.8.4.4. Heat Exchanger Corrosion

Many types of nuclear and fossil plant heat exchangers are affected by corrosion, including condensers, feedwater heaters, service water heat exchangers, lube oil coolers, component cooling water heat exchangers, and residual heat removal heat exchangers. Corrosion in heat exchangers continues to result in high costs

Table 3.11. Summary of Devices for Field Corrosion Monitoring[a]

Device	Applications	Monitoring results
Corrosion product monitor	Feedwater systems including feedwater heaters	Quantitative determination of corrosion product transport
Erosion–corrosion (flow accelerated corrosion)	Piping components	Thinning rate for materials of concern in the specific suspected areas of piping
U-bend and double U-bend specimens	LP turbines, piping, feedwater heaters, condensers, boilers	Detects general corrosion, pitting, and SCC; double U-bends simulate crevice and galvanic effects
Fracture mechanics specimens	LP turbine disks and rotors, piping, headers, deaerators	Stress corrosion and CF crack growth rate and crack incubation times
Heat exchanger or condenser test tube	Condensers, feedwater heaters, and other heat exchangers; installed within the circuit	Accumulation of scale and corrosion pertinent to plant-specific conditions
Model crevice	Studies of crevice chemistry and corrosion heat exchangers, PWR steam generators	Crevice chemistry and corrosion data for specific conditions
Corrosion hydrogen monitor	Boiler tubes, feedwater system, and PWR steam generator corrosion	Detect general corrosion versus load and chemistry; hydrogen damage, caustic gouging in boiler tubes, potential for impurity concentration
Vibration signature	Turbines and pumps-periodic monitoring	Detection of cracks and other distress
Boiler tube leak monitor	Fossil boilers	Early detection and location of a tube leak
Acoustic emission leak detector	Feedwater heaters and other heat exchangers	Early detection of tube leaks
Cavitation monitor	Feedwater piping and pumps	Early detection of cavitation noises
Stress and condition monitoring system	All types of steam cycles and major components	Actual on-line stresses, temperatures, and other conditions; used to determine damaging conditions and residual life
Turbine blade telemetry	LP turbines	Resonant frequencies and alternating stresses

[a]Ref. 41.

at both nuclear and fossil plants. There are many different corrosion mechanisms involved:

- Inlet–end erosion–corrosion of copper alloy tubes.
- Sulfide attack of copper alloy tubing in polluted waters.
- FAC or erosion–corrosion of CS tubing and supports.
- SCC of stainless steel (SS) tubing and some types of copper alloy tubing.
- Pitting, MIC, and underdeposit corrosion of copper alloy, carbon steel, and stainless steel tubing.
- Hydriding of titanium tubing.
- Galvanic corrosion of CS water boxes and tube sheets and copper alloy tube sheets.
- Dealloying of copper alloy tube sheets and graphitization of cast iron waterboxes.

3.8.4.5. SCC and CF in Turbines

Both SCC and CF continue to create large costs at both nuclear and fossil plants. The SCC of turbine disks occurs at several locations-including the bore of the disk, where it typically occurs at keyways. Other susceptible areas are the high-stress locations along the disk face and at blade attachments often called rim cracking. Both SCC and CF affect turbine blades. SCC and CF of turbine disks and blades lead to significant expenditures for periodic inspections and repairs, and for occasional rotor replacements. There has been extensive work to determine the causes of such cases of SCC and CF, including determining the sensitivity to the alloy composition and the steam system chemistry and developing inspection and remediation methods.

3.8.4.6. Fuel Cladding Corrosion

Restrictions on fuel burnup and local power levels are imposed to keep the internal and external corrosion of fuel cladding in nuclear plants at acceptable values. These limitations on burnup and local power level result in significant costs. Current boiling water reactor (BWR) fuel problems include secondary degradation, where an initial clad penetration, (e.g., due to foreign material), leads to internal hydriding, clad splitting, escape of fuel material through the splits, and consequential contamination problems. Pellet clad interactions also appear to have become more common again as a result of changes made to barrier cladding in response to the secondary degradation problem. Another costly problem is crud-induced fuel failures.

3.8.4.7. Corrosion in Electric Generators

There are two significant corrosion problems in large, water-cooled electric generators. One is clip-to-strand crevice corrosion of older generators that were fabricated using a copper 5% phosphorus bronze alloy. The second problem affecting watercooled electric generators is flow restrictions, that is, plugging the hollow strands or clogging the strainers, in the stator cooling water system.

Several decades ago, SCC of electric generator end rings was a major problem. This problem has largely been brought under control by replacement of the more susceptible end ring materials of 18Mn-5Cr steel with 18Mn-18Cr steel and by periodic inspection and repair programs. Nevertheless, the need for continuing inspections and occasional repairs result in some continuing significant costs.

3.8.4.8. Flow Accelerated Corrosion

Carbon steel piping and components in nuclear reactors can experience thinning of up to several millimeters per year because of FAC. This problem is most severe in two-phase flow areas, for example, in the steam drains and moisture separator reheaters, but also can occur in single-phase condensate and feedwater systems. It has caused serious pipe bursts, which has led to significant inspections, repairs, and replacements. The problem is relatively well understood. It is known to be a function of several variables, such as material composition, flow velocity, temperature, pH, and oxidation-reduction potential. Water chemistry changes, such as increasing pH, are known to be useful remedial approaches. Flow accelerated corrosion applies both to fossil and nuclear plants, but it is much more important from a cost standpoint at nuclear power plants because of the higher costs of downtime, inspections, and repairs at these facilities. In addition, feedwater chemistry optimization in fossil units, such as the use of oxygenated water treatment, has reduced the rate and extent of FAC in these plants.

3.8.4.9. Corrosion of Raw Water Piping

A myriad of corrosion-related problems affect raw water service piping and components. Mortar-lined carbon steel and cast iron piping carrying seawater or brackish water often have often been damaged by corrosion and leakage at field-installed joints where the mortar was applied locally after the field structural joint was made. Carbon steel piping used for service water and circulating water piping often is unlined when used in freshwater service and, as such, is susceptible to several types of corrosion problems, including MIC, pitting, erosion–corrosion, and underdeposit attack. Piping and heat exchanger tubing made from copper alloys, carbon steel, stainless steel, and other alloys suffer underdeposit attack often exacerbated by MIC. Fouling by Asiatic clams and zebra mussels is a frequent problem that can stimulate underdeposit corrosion. Localized corrosion occasionally occurs at flaws (holidays) in the external coating used on buried carbon steel piping. The reinforcing steel in some large-diameter reinforced concrete pipe has also suffered from corrosion.

3.8.4.10. IGSCC of BWR Piping and Internals

With normal water chemistry containing ~200 ppb oxygen, a boiler water reactor (BWR) coolant is oxidizing enough to cause intergranular stress corrosion cracking (IGSCC) of sensitized austenitic stainless steel and sensitized N06600 alloy. In response to the piping IGSCC problem, an array of remedies was developed in the late 1970s and 1980s. Countermeasures include replacing susceptible piping, using induction heat stress improvement, performing heat sink welding, adopting hydrogen water chemistry, and reducing impurity levels in the reactor coolant. As a result of the application of these remedies, the occurrence of IGSCC in BWR piping is now relatively rare. In addition, improved inspection practices have allowed industry to detect and correct IGSCC in BWR piping before it causes serious problems.

3.8.5. The RIMAP Project

The motivation behind the Risk-based Inspection and MAintenance Procedures (RIMAP) European project is that current inspection and maintenance planning practices in most industries in Europe are based on tradition and prescriptive rules, rather than being an optimized process where risk measures for safety and economy are integrated. However, the trend is common across industries to move from time-based inspection to RBI as new technologies for making risk-based decisions are emerging in a broad range of sectors (17). The following sections describe the status of RBI in various European industrial sectors.

3.8.5.1. Process Industry

It is widely accepted in the European process industry that the traditional time-based approach to plant inspection is too conservative. However, the limited use of modern maintenance methodologies or analysis tools in most companies is in contrast with the abundance of publications on the subject that can be found in the open literature. One explanation for this discrepancy is that 99% of the published literature is focused on the most advanced maintenance strategies that are only applicable to a small fraction of existing equipment. The following observations are telling:

- Inspection planning and maintenance planning are still mostly not integrated, albeit some companies strive explicitly for integration of inspection and maintenance.
- RBI is being adopted by certain sectors of the process industry; however, it has mostly been by larger companies.
- Present RBI is mostly focused on static equipment.
- For most companies, the RBI methodologies are "black boxes" and no quantitative criteria are available to express safety. In other words, RBI analyses and/or methods may result in different levels of safety and the questions "How safe is safe enough?" or "What is sufficient safety?" are left unresolved.

There is a variety of approaches in the current RBI methodologies to estimate a POF, which may be derived from historical probabilistic values or from an assessment of the residual lifetime for a given component. These two philosophies make use of historical inspection data, but they differ in the appreciation of the weight given to such data.

One of the most influencing factors in estimating a POF is the effectiveness of inspection, an expression of both the performance of inspection techniques and the inspection coverage. However, both parameters are poorly described in quantitative terms. This missing knowledge is a dominant weakness in RBI.

Properly defined and estimated probabilities are a precondition for successful application of RBI, this is one of the most critical parts of a RBI analysis. Therefore the following questions regarding this issue still have to be resolved:

- How to deal with the differences between trendable and nontrendable degradation mechanisms.
- How can nontrendable degradation mechanisms be predicted.
- How the "human factors" should be included in the RBI methodology.
- How to deal with the evaluation of inspection effectiveness and its influence on the capacity to predict equipment integrity.

3.8.5.2. Offshore Industry

In the European offshore industry, economic, safety, health, and environmental risks are generally estimated to prioritize the various maintenance and inspection activities. Risk and condition-based concepts are implemented to focus maintenance and inspection on the high-risk items. Different methods are used in this industry to assess remaining life and plan maintenance and inspection for various types of equipment.

However, it is not common to compute the overall risk level as part of the inspection and maintenance planning. In the offshore industry, inspection and maintenance planning are often considered as two separate activities. Furthermore, there is a need for better models and decision rules for updating the risk-based plans based on new information. In particular, methods that model the risk reduction when only a limited number of "hot spots" are inspected needs to be developed.

The results from structural reliability analysis are often used to prioritize inspection and maintenance activities on structures. RBI is used for static equipment and to some extent for pipelines. RCM has also been used with mixed success for rotating equipment and electrical or instrumental systems. One serious difficulty is the need to perform an initial screening to consider in the RCM analysis only the equipment that contributes to the key performance indicators.

The overall risk level is not commonly considered as a component of the inspection and maintenance planning. The analysis often starts by determining the overall acceptance criteria. However, these criteria may not be linked to the overall safety level and the risk determining factors may vary with the system in focus, (e.g., topside, process, and structures). For production systems safety is often the dominating factor. However, for the rest of the equipment possible economic losses are usually the dominating factor.

Offshore RBI typically covers vessels, piping, and structures. Usually, internal and external degradation is considered, taking into consideration materials and environments. Risk matrices are quite commonly used to combine POF and COF. The POF is usually determined using degradation data and expert judgment. Both qualitative methods based on expert judgment and quantitative methods based on mathematical models, (e.g., QRA), are used.

3.8.5.3. Power Industry

Risk-based life management (RBLM) is a concept that is gaining support primarily in the power plant industry. Current codes are still a limitation for broader application of RBI/RBLM because they are mainly based on prescriptive requirements, (e.g., statements on how often to open a process system for inspection). This is usually not cost efficient, and surely not the optimal solution for the safety of plants.

Modern plant management calls for integration of these departments and further integration of the efforts to minimize downtime and reduce production cost. In the RIMAP perspective, the reasons for applying an RBLM analysis are to achieve the following goals (17):

- Improved overall safety.
- Improved strategies for risk mitigation.
- Better management of inspection scheduling issues.
- Improved linking of inspection effort to the system or component condition.
- Optimized flow of maintenance and life management information.
- Simplified and transparent decision processes;
- Cost savings in both operating expenses and risk exposure by addressing the high risk items or systems.
- Increased awareness of the risk exposure, damage mechanisms, and associated uncertainties in testing–inspection and maintenance actions.

Corrective maintenance, mainly for "noncritical" components, still dominates the daily practice in this sector. However, due to reductions in personnel, the practice of inspecting components at fixed time intervals can impair plant safety by diluting inspection resources. Screening is often achieved by performing a risk-based assessment with minimal information gathering. As regards inspection planning, emphasis has been put mostly on the following issues:

- Opportunity driven planning and performing of inspections, that is when components are available.
- Use of nonintrusive inspection techniques that do not require opening vessels.

- Use of monitoring techniques, for example temperature, stress, and material damage due to aging and other factors.
- Quantification of the reliability of information provided by inspection and monitoring techniques.

3.8.6. Aircraft Maintenance

Despite the intense media coverage of air tragedies, flying remains one of the safest modes of transportation. The reliability and safety record of aircraft operators is indeed enviable by most industrial benchmarks.

The current operating fleet can be divided into three generations. The first generation airplanes include the B-707, DC-8, DC-9, B-727, L-1011, and the earlier production models of the B-737 (-100, -200), B-747 (-100, -200, -300, SP), and the DC-10. These airplanes are characterized by a design that primarily addressed strength and fail-safe criteria while little or no attention was paid to incorporating corrosion protection into the design (42). Some of the first generation jet transport airplanes that were designed in the 1950s and 1960s are still in operation.

The second generation of jet transport airplanes, which were designed in the 1970s and the 1980s, include the B-737 (-300, -400, -500), B-747 (-400), B-757, B-767, MD-81, -82 and—83, MD-88, MD-11, and F-100. In addition to the strength and fail safety requirements, these airplanes are characterized by the incorporation of durability and damage tolerance standards into the design. It was realized that corrosion in aircraft was becoming an economic burden and could possibly become detrimental to the structural integrity of the airplane. Thus, as part of the durability standards, airframe manufacturers started to use corrosion-inhibiting primers and sealants for the faying surfaces of lap joints and fastener holes.

The third generation of jet transport airplanes include the B-777 and the new generation B-737 (-600, -700, and -800) and B-747 (-400). In addition to the key characteristics of the first and second generation airplanes, these airplanes are characterized by the incorporation of significant improvements in corrosion prevention and corrosion control in design.

3.8.6.1. Corrosion Definition

The FAA has issued Airworthiness Directives (AD) related to corrosion control in design and maintenance. The corrosion control program in the AD defines three levels of corrosion. Note that the various modes of corrosion are not included in these definitions. Only the total loss of material, which affects the load-carrying capacity of a structure, is defined (42).

3.8.6.1.1. Level One Corrosion

- Corrosion damage occurring between successive inspections that is local and can be reworked or blended out within the allowable limits, as defined by the manufacturer.

- Corrosion damage that is local and exceeds allowable limits, but can be attributed to an event not typical of the operator's usage of other airplanes in the same fleet (e.g., mercury or acid spill).
- Operator experience over several years has demonstrated only light corrosion between successive inspections, and latest inspection and cumulative blend-out now exceed allowable limit.

3.8.6.1.2. Level Two Corrosion

- Corrosion occurring between successive inspections that requires rework or blend-out of structural elements as defined by the original equipment manufacturer's structural repair manual.

3.8.6.1.3. Level Three Corrosion

- Corrosion found during the first or subsequent inspections that is determined (normally by the operator) to be a potentially urgent airworthiness concern that requires expeditious action.
- In addition to the degree of corrosion, the extent of corrosion is taken into consideration. The appearance of corrosion on a single skin panel, single stringer, or single frame, where it does not affect any adjacent members, is defined as local corrosion. Widespread corrosion is defined as corrosion on two or more adjacent frames, chords, stringers, or stiffeners;
- The baseline program is designed to eliminate severe corrosion on airplanes and to control corrosion of all primary structures to Level One or better, meaning minor corrosion that never affects the airworthiness of the aircraft. Level Two and Level Three Corrosion must be reported to the airplane manufacturer, who, in principle, uses the reported data to determine any actions required to ensure continuing airworthiness and economic operation.

3.8.6.2. Maintenance Schedule

A typical maintenance program begins with nightly inspections of each airplane, which consists of a detailed visual inspection and a review of the pilot's report. There are then scheduled periodic inspections (42):

- **A. Check:** This is a more detailed visual inspection conducted every 4–5 days after 65–75 flying hours. The interior and the exterior of each airplane are visually checked for general condition and any obvious damage, with particular attention given to areas where exposure to accidental or environmental damage may have occurred.
- **B. Check:** This check occurs approximately every 30 days. Specific access panels are removed for inspection. In addition to engine servicing, other safety and airworthiness items are checked as well.

C. Check: This is performed every 12–18 months after the aircraft has flown ~5500 h. It is an in-depth, extended, heavy structural, and maintenance check.

D. Check: This is the most comprehensive inspection, conducted after 20,000–25,000 flying hours. The paint is removed from the exterior, and the interior of the airplane is completely stripped to allow for close inspection of all structural members of the fuselage.

3.8.6.3. Corrosion Management Assessment

Significant improvements have been made in the corrosion design of new airplanes. The airframe manufacturers have implemented many key design improvements over the past three decades. The improvements range from the replacement of corrosion-prone materials, such as aluminum alloy 7075-T6 (UNS A97075), to improved adhesive bonding processes, to the use of sealants in fastener holes and on faying surfaces, to the control of spillage, such as galley and lavatory fluids. Other airplane models have made similar improvements.

Maintenance practices may vary depending on the type of airline. One major U.S. airline, for example, tracks corrosion problems by aircraft tail number and trend data to determine threshold levels for maintenance actions for the fleet. Inspections are performed on letter checks (major inspections) under FAA requirements every 9 months to 1 year. Because flight profiles and utilization cycles are very close for all the aircraft and since local basing environments have little influence on corrosion and other maintenance factors, all aircraft in the fleet are considered equal. As a result, this airline company is able to predict and plan the maintenance requirements of all of their airplanes with high accuracy, while maintaining high efficiency in their maintenance operations.

3.8.6.4. Maintenance Steering Group System

The aircraft industry and its controlling agencies have additionally a top down system to represent potential failures of aircraft components, the Maintenance Steering Group (MSG). The first generation of formal air carrier maintenance programs was based on the belief that each part of an aircraft required periodic overhaul. As experience was gained, it became apparent that some components did not require as much attention as others and new methods of maintenance control were developed. Condition monitoring was thus introduced in the decision logic of the initial maintenance steering group document (MSG-1) was applied to Boeing 747 aircraft.

The experience gained with MSG-1 was used to update its decision logic and create a more universal document for application to other aircraft and powerplants (43). When applied to a particular aircraft type the MSG-2 logic would produce a list of Maintenance Significant Items (MSIs), to each of which one or more process categories would be applied, that is "hard time", "on-condition", or/and "reliability control".

164   Chapter 3   Maintenance, Management, and Inspection Strategies

The most recent major update to the MSG system was initiated in 1980. The resultant MSG-3 is based on the basic philosophies of MSG-1 and MSG-2, but prescribes a different approach in the assignment of maintenance requirements. In 1991, industry and regulatory authorities began working together to provide additional enhancements to MSG-3. As a result of these efforts, Revision 2 was submitted to the FAA in September 1993 and accepted a few weeks later (44). One particularly interesting addition in this revision is a procedure for incorporating corrosion damage in the MSG logic. The environmental deterioration analysis (EDA) involves the evaluation of the structure against probable exposure to adverse environments. The evaluation of deterioration is based on a series of steps supported by reference materials containing baseline data expressing the susceptibility of structural materials to various types of environmental damage and produces inputs for the structure maintenance program.

The MSG-3 (Revision 2) structures analysis begins by developing a complete breakdown of the aircraft systems, down to the component level. All structural items are then either classified as structure significant items (SSIs) or other structure. An item is classified as an SSI depending on consideration of COF and POF as well as material, protection and probable exposure to corrosive environments. All SSIs are then listed and categorized as damage tolerant or safe life items to which life limits are assigned (43).

Accidental damage, environmental deterioration corrosion prevention and control, and fatigue damage evaluation are then performed for these SSIs following the logic diagram illustrated in Fig. 3.29. Once the MSG-3 structure analysis is completed, each element of the structural analysis diagram (Fig. 3.29) can be

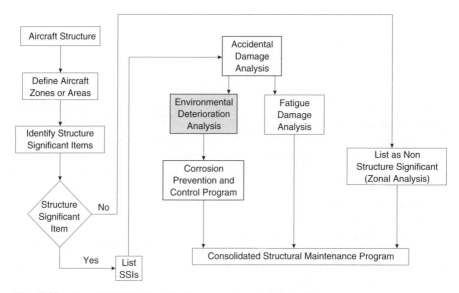

Fig. 3.29. Overall MSG-3, Revision 2, structural analysis logic diagram.

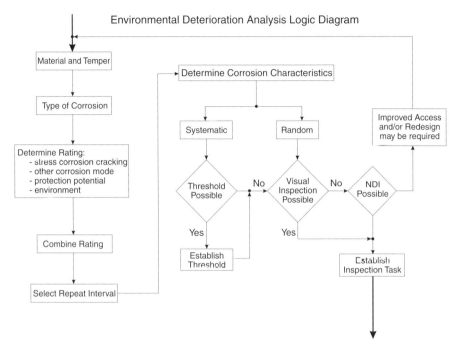

Fig. 3.30. Environmental deterioration analysis logic diagram.

expanded right to the individual components and associated inspection and maintenance tasks. The logic of the EDA, illustrated in Fig. 3.30, requires the input of a multitude of parameters such as component part number, location, material composition, and protective coating that can be actuated in practical templates, such as illustrated in Fig. 3.31.

3.8.7. Water Utilities

North Americans use a lot of water. Annually, ~63 billion m^3 of drinking water serves ~73 million customers in North America, with the average total water use rate per customer ranging between 475 and 660 L capita^{-1} day^{-1}. Recent benchmark estimates by the American Water Works Association (AWWA) on indoor water use rates, indicated an average use rate of 245 L capita^{-1} day^{-1}. The average consumer cost for clean water ranges from \$0.50–2.60 per 4 m^3 (42).

Based on AWWA data, there was ~1.4 million km of municipal water piping in the United States in 1995. Table 3.12 presents an estimated profile of the different materials that make up these water pipes. New pipes are being installed at a rate that extends the system length by 1.5 % per year, while an additional 0.5 percent is being replaced annually.

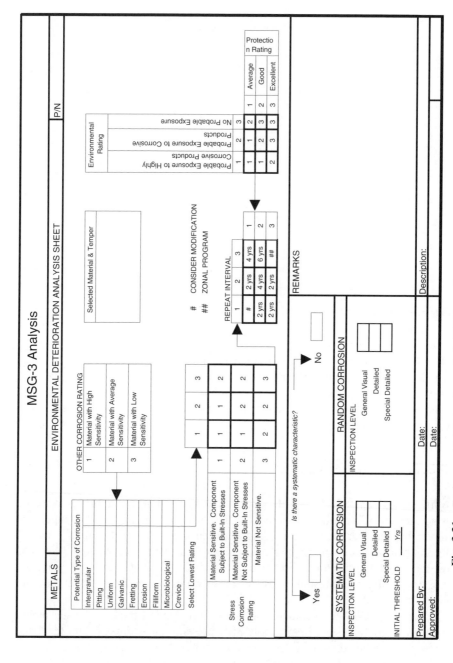

Fig. 3.31. Environmental deterioration analysis (EDA) template.

Table 3.12. Profile of Different Materials Used for U.S. Transmission Water Pipes, as Determined from the 1992 AWWA Water Industry Database[a]

Material	Percentage
Cast iron	48
Ductile iron	19
Concrete and asbestos concrete	17
PVC	9
Steel	4
Other	2

[a] Ref. 42.

3.8.7.1. Corrosion Impact

Corrosion jeopardizes system reliability by causing leaks and breaks and by affecting water quality. Corrosion can have a variety of impacts grouped into the following four categories (45).

3.8.7.1.1. Health and Regulations From a health standpoint, the most significant metal that can enter the drinking water via corrosion is lead. The contribution of drinking water to the total daily intake of lead in the United States is estimated to be ~20%. In drinking water, the primary sources of lead are lead service lines, lead plumbing, brass fixtures, and 50:50 tin-lead solder used to join copper piping. Lead enters the water when the water has been standing motionless in contact with the lead source for extended time periods.

3.8.7.1.2. Aesthetics and Customer Perception The metals of most interest from an aesthetic standpoint are copper, iron, and zinc. The corrosion products of copper piping in drinking water may cause a metallic taste, blue-green staining particles and discoloration of the water. Iron species from the corrosion of unlined cast-iron piping, unlined steel piping, and galvanized pipe may discolor drinking water, form precipitates, cause rusty water and red staining of laundry and fixtures, impart metallic taste to water, and promote the growth of iron bacteria and sulfate-reducing bacteria. Rusty or red water is one of the most common customer complaints received by water utilities.

Zinc can cause bitter or astringent tastes in drinking water at concentrations of 4–5 mg^{-1} L. High concentrations may give water a milky appearance at room temperature and a greasy appearance when boiling.

3.8.7.1.3. Premature Piping Deterioration and Economic Impacts Internal corrosion can have significant adverse economic impacts on the water transmission and distribution systems and consumer plumbing. Typically, consumer plumbing is most seriously affected by internal corrosion because piping is normally unlined and its diameter is small. Such piping will have a tendency to leak or develop tubercles that may reduce both flow and pressure (Fig. 3.32). In recent years, many owners of large-diameter transmission and distribution piping systems typically have protected their investment by installing pipe with internal linings, by cleaning and lining pipe in place, or by adding a cathodic protection system for external corrosion (Fig. 3.33). However, there are still many thousands of kilometers of unlined metallic piping in use throughout North America and Europe. Costs to maintain operability over and above the normal accrue when this piping must be replaced prematurely or cleaned and lined in place.

3.8.7.1.4. Environmental Concerns Corrosion of water distribution piping raises environmental concerns due mostly to the presence of lead, cadmium, zinc, and copper in drinking water. These metals enter the wastewater collection system and eventually accumulate in the sludge and end up in a landfill, on croplands, or other locations, depending on the disposal method. Metals corroded from water piping are the largest source of these contaminants in the wastewater of many communities.

3.8.7.2. Corrosion Management

System reliability is of the utmost importance to water suppliers and their customers. However, corrosion problems can vary greatly within a single system

Fig. 3.32. Tubercles in a small diameter water pipe. (Courtesy Public Works and Services, City of Ottawa.).

3.8. Industrial Examples 169

(a)

(b)

Fig. 3.33. (a) Holes being drilled for sacrificial anode installation; (b) thermit welding of anode connection on water main; (c) anode cluster ready to be buried and connected to fire hydrant tap. (Courtesy Public Works and Services, City of Ottawa.).

because many variables affect corrosion, for example, pipe material, pipe age, pipe wall thickness, water additives, corrosion inhibitor treatment, soil chemistry, soil moisture content and/or local groundwater level, and stray currents (42). Table 3.13 summarizes some of the physical, environmental and operational

(c)

Fig. 3.33. (*continued*).

factors that can affect the deterioration rate of water distribution systems and lead to their failure (46).

The AWWA has developed a six-step procedure shown in Fig. 3.34 for the assessment and control of internal corrosion of water distribution systems (45). Although the application of this method appears straightforward, working out the details for each system in an actual network can be quite complex. The procedure specifically focuses on older systems and does not consider corrosion prevention for new systems. It assumes that corrosion is already present and that the corrosion only occurs internally. Although the cost per system can be calculated reasonably accurately using this method, interactions with other systems are difficult to evaluate. The system size, location, population served, materials used, water quality, and soil conditions all significantly influence corrosion susceptibility.

3.8.7.2.1. Short-Term Corrosion Management Short-term corrosion problems are often indicated by customer complaints, such as the occurrence of red or yellow 'rusty' water or a sudden decrease in water pressure. A reason for rust-colored water is generally the presence of corrosion products that have flaked off of the internal pipe walls, while a water pressure drop may be caused by a leak in the transmission or distribution system. Finding a leak in an underground pipe system is often difficult because the leak may start small and go undetected for a period of time. Once the leak is so severe that water is literally coming from the ground, it may cause a local flood. In addition to the lost water, the damage can be significant and the repair work is more than what would have been needed to fix a small leak (42).

Table 3.13. Factors That Contribute to Water System Deterioration[a]

Factor	Explanation
Physical	
Pipe material	Pipes made from different materials fail in different ways.
Pipe wall thickness	Corrosion will penetrate thinner walled pipe more quickly.
Pipe age	Effects of pipe degradation become more apparent over time.
Pipe vintage	Pipes made at a particular time and place may be more vulnerable to failure.
Pipe diameter	Small diameter pipes are more susceptible to beam failure.
Type of joints	Some types of joints have experienced premature failure (e.g., leadite joints).
Thrust restraint	Inadequate restraint can increase longitudinal stresses.
Pipe lining and coating	Lined and coated pipes are less susceptible to corrosion.
Dissimilar metals	Dissimilar metals are susceptible to galvanic corrosion.
Pipe installation	Poor installation practices can damage pipes, making them vulnerable to failure.
Pipe manufacture	Defects in pipe walls produced by manufacturing errors can make pipes vulnerable to failure.
Environmental	
Pipe bedding	Improper bedding may result in premature pipe failure.
Trench backfill	Some backfill materials are corrosive or frost susceptible.
Soil type	Some soils are corrosive; some soils experience significant volume changes in response to moisture changes, resulting in changes to pipe loading.
Groundwater	Some groundwater is aggressive toward certain pipe materials.
Climate	Climate influences frost penetration and soil moisture. Permafrost must be considered in the north.
Pipe location	Migration of road salt into soil can increase the rate of corrosion.
Disturbances	Underground disturbances in the vicinity of an existing pipe can lead to actual damage or changes in the support and loading structure on the pipe.
Stray electrical currents	Stray currents cause electrolytic corrosion.
Seismic activity	Seismic activity can increase stresses on pipe and cause pressure surges.
Operational	
Internal or transient water pressure	Changes to internal water pressure will change stresses acting on the pipe.
Leakage	Leakage erodes pipe bedding and increases soil moisture in the pipe zone.
Water quality	Some water is aggressive, promoting corrosion.
Flow velocity	Rate of internal corrosion is greater in unlined dead-ended mains.
Backflow potential	Cross connections with systems that do not contain potable water can contaminate water distribution system.
O&M practices	Poor practices can compromise structural integrity and water quality.

[a]Ref. 46.

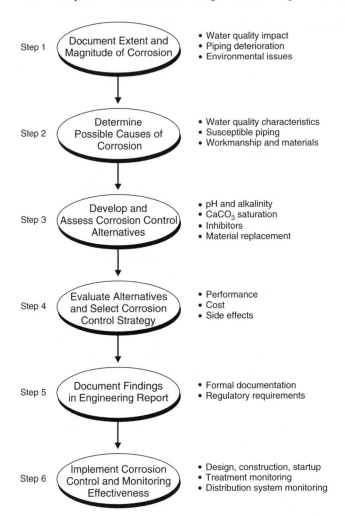

Fig. 3.34. Corrosion control program implementation flowchart.

3.8.7.2.2. Long-Term Corrosion Management Long-term corrosion impact is generally indicated by system integrity studies. Maintenance and inspection teams are tasked to find leaks and failures. Many utilities apply a specialized corrosion team to monitor the water quality, using corrosion loops in which treated water circulates over weight loss coupons. Coupons made from different materials are exposed to various water flow rates. Periodically, these coupons are measured and average corrosion rates are determined. In addition to the weight loss coupons, water samples are routinely tested to ensure that the water quality is acceptable. The test results are used to make assessments about corrosion as well. For example, the pH of the water is important both for water consumers and for system integrity. The pH is kept within a predetermined range by adding pH adjusters to the treatment process. Dedicated corrosion groups mainly focus on corrosion prevention. They generally work with a fixed annual budget prorated as a percentage of the total water utility budget (42).

3.8.7.2.3. Necessity of Long-Term Corrosion Planning Because of the long life expectancy of water systems, a long-term vision for corrosion management is required. Unfortunately, some managers give in to short-term cost savings over long-term investments. As an example, the average thickness of cast iron and ductile iron pipe has been continuously decreased over the last 100 years because thinner, higher strength pipe has become available (42). Unfortunately, corrosion rates are not significantly dependent on the strength of ductile iron or steel as was demonstrated by extensive corrosion studies (47). As a result, thinner wall pipe will have a smaller corrosion tolerance than thicker wall pipe and will show more frequent failures. The time to corrode through a pipe wall was found to be directly proportional to the square of the wall thickness, for example, for a wall thickness reduction of 50% the corrosion life will be reduced to 25% of the life of the original pipe thickness.

3.8.7.2.4. Framework for Water Pipeline Management The major tools for managing and preventing pipeline failures until recently have been simple statistical approaches based on numbers of pipe breaks per kilometer and reactive inspection techniques, such as leak detection. While, these approaches have been useful for managing pipeline failures, new technologies and knowledge about water system piping make it possible to develop more efficient and accurate approaches to maintaining pipeline integrity. The framework illustrated in Fig. 3.35 was recently proposed to help the introduction of these new techniques

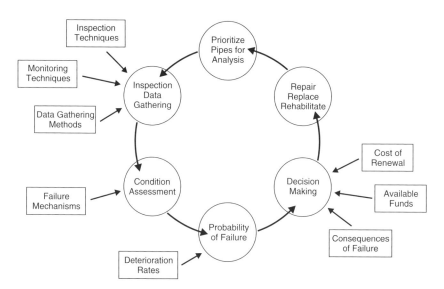

Fig. 3.35. Water pipeline management cycle (48).

in service and use them even before the required research and development has been completed (48).

A major component of this framework is the use of nondestructive evaluation techniques to provide information about the condition of the pipeline. All pipes will eventually fail, but the rate of failure will depend on both the pipe material and the actual exposure to environmental and operational conditions. The most important feature of the framework relates to the cyclical nature of pipeline management. Each pipe in the system must be examined periodically, and its condition reassessed in order to determine what action should be taken to maintain or upgrade its condition. The entry point to the cycle is the pipe selection area labeled as "Prioritize Pipes for Analysis" in Fig. 3.35. First, a pipeline must be selected for analysis before any additional work can be done on it.

A second important point is that the maintenance and inspection strategy should differ between distribution and transmission systems to reflect the much higher failure consequences for transmission pipelines compared to those of a typical distribution line. Approaches that may not be economically viable for a distribution system could be employed advantageously in a transmission system to prevent failure. Essentially, the low consequences associated with a single distribution failure also mean that the emphasis in distribution systems should be on failure management in order to minimize life-cycle costs while a more proactive failure prevention approach would be better suited for transmission systems due to the high consequences of associated failures (48).

3.8.7.3. Condition Assessment Techniques

The pipes in water distribution and transmission systems are difficult to inspect for damage due to their location below the surface of the ground. This difficulty has led water utilities to rely on techniques, such as breakage records, leak detection, and water audits to determine the health of their systems. While these techniques have been shown to be very useful in prioritizing repairs and replacements, they have the disadvantage of being reactive in nature. In each case, problems with the water system only become apparent after the pipes have failed in some manner (49).

Table 3.14 provides a comparative listing of the different techniques for inspecting metallic water pipes. Each technique is named and the relative advantages and disadvantages of the technique are given. Table 3.15 shows a similar listing for prestressed concrete pipes. It is apparent from Table 3.14 that most of the techniques available for inspecting metallic water pipes complement, rather than compete with each other. Water audits provide broad, reactive information about the condition of a wide area of a water distribution network. Acoustic leak detection methods find already broken or damaged pipes, corrosion monitoring, can, in theory, detect areas where corrosion activity on pipes is likely, while the remote field effect can inspect pipes to find damage before they fail. A complete diagnostic program is likely to use all these inspection methods.

The following sections will briefly described the main condition assessment techniques used by water utilities without including the nondestructive techniques

Table 3.14. Comparison of Diagnostic Techniques for Metallic Water Mains[a]

Technique	Advantages	Disadvantages
Zone water audits	Cheap Covers large areas of a city quickly Allows for a comparison of water losses between individual districts Useful as a screening process for other techniques Can be used to evaluate the effectiveness of repair programs	Does not give the precise location of leaks Requires isolation of zones Work must be performed at night Only gives an overview of current problems
Sonic–acoustic leak detection	Widely practiced Known to find leaks accurately Known to find leaks of different sizes Operates from outside the water line	Percentage of leaks missed by the technique is unknown Currently works best in metal water lines Only gives information on the current condition of the line (the tool has little predictive value) Background noise problems
Remote field inspection (hydroscope)	Most advanced technique currently available Detects areas of corrosion pitting, as well as through holes Can be used to give an estimate of the future life of a line	More expensive than leak detection Requires access to the inside of the water line, which may require cleaning Knowledge of the relationship between pit size and residual life of the pipe is not yet complete Limited to pits of <3 cm^3 in size.
Magnetic flux leakage	Established technology in oil and gas industry Known to be capable of detecting small defects and through holes in steel pipe	Not yet commercially available for water lines Requires access to and complete cleaning of the inside of the pipe
Ultrasound	Most versatile NDE technique Established technology in oil industry	Technique will not work through tuberculation Not yet commercially available for water lines Requires access to and complete cleaning of the inside of the pipe

Table 3.14. (*continued*)

Technique	Advantages	Disadvantages
Soil corrosivity measurements	Simple to conduct May act as a screening mechanism for more expensive methods	Cities may have similar levels of corrosivity across their region, making use difficult Do not always correlate with corrosion reality
Half-cell potential measurements	Simple to conduct May act as a screening mechanism for more expensive methods Well established as method to detect corrosion activity in buried objects	Factors, such as stray currents and soil conditions, may affect readings Accuracy of results depends on distance between readings Small areas of localized corrosion can not be detected

[a]Ref. 49.

described in Chapter 5, that is, ultrasonics, eddy current, and magnetic flux leakage testing.

3.8.7.3.1. Water Audits Water audits are the simplest way of evaluating the condition of large parts of a water system. In this type of test, a comparison over an entire city or within a district is made between the minimum nightly flow rate per person or per household and a target value that represents the background leakage, (e.g., seeping at joints and similar sources). The difference between the two is an indication of how much water is being consumed at night in the area being monitored (49, 50). The water audit therefore provides an estimate of a system's water loss.

A coarse system wide audit can be done by monitoring the flow rates through a city's transmission lines and comparing the resulting overall water consumption to regional or national averages. However, it is more effective to follow a system wide audit with district measurements. In this procedure, a municipality is divided up into individual districts or segments of pipeline that are isolated from the rest of the system except for one valve that allows water into the district and a second that allows water to leave it. Such a division lets the municipality compare individual districts and determine the areas most likely to be experiencing serious leakage problems. This type of measurement can have a number of benefits:

- Quickly quantify leakage potential, identifying good candidates for immediate acoustic surveys.
- Provide opportunities to update the city's data base on such matters as water consumption patterns, hydrant locations and valve locations.

Table 3.15. Comparison of Inspection Methods for Prestressed Concrete Pipes[a]

Technique	Advantages	Disadvantages
Half-cell potential measurements	Standard technique Results easy to interpret Simple to perform Does not require pipe entry	Experience suggests the method is ineffective for use with these lines in an urban setting Indicates the presence of corrosion activity on or nearby the pipe, but not the extent of the damage.
Visual inspection and sounding	Longest record of successfully detecting damaged pipes Examines condition of concrete	Requires man entry into pipes Does not provide direct information on wire breaks Unclear whether all wire breaks will produce noticeable concrete damage Subjective in nature and dependent on skill of inspection team
Acoustic monitoring	Works in an operating pipeline Can detect wire breaks during monitoring period and locate them Works in all types of prestressed concrete pipes	Only detects damage that occurs during monitoring period
Remote field inspection	Can detect single or multiple broken wires Inspection gives complete picture of damage to the pipeline	Currently only available for use in embedded type pipelines Requires man entry into pipes
Impact echo–spectral analysis of surface waves	Examines condition of concrete Objective measurement system	Does not give direct information on wire breaks Slower inspections than remote field effect Requires an empty pipeline

[a] Ref. 49.

- Provide an opportunity to exercise valves that are not normally used, a good way to reveal the valves that are not working properly and might cause potential flow problems during fire fighting.

While periodic city wide leak detection campaigns are typical in North America, more systematic approaches to leak detection using water audits may be taken by logging meters that are permanently installed at the entrances and exits to each district. Water audits by themselves cannot precisely locate leaks. However, they can be used to prioritize areas for leak location procedures. They are commonly used by municipalities during the beginning of a leak detection program and may be reduced in extent or dropped as the program matures.

3.8.7.3.2. Sonic and Acoustic Leak Detection Leaks produce a distinctive sound that can be used to identify and locate them. This may be done by sonic detection methods that enhance the natural hearing of a listener, or by electronic acoustic leak detection methods. These two methods are the most common in use today in North America. Typical sonic leak detection methods involve the use of stethoscopes or listening horns placed against hydrants or the ground above a pipe. Acoustic methods may use similar locations, but with geophones or hydrophones to pick up the sound, allowing it to be recorded for future reference. Typical systems use an autocorrelator to compare the signals from two different hydrants or sites. A comparison between the time of arrival of the sounds from each device is used to deduce the approximate location of the leak.

Leak detection systems have been used primarily in metallic distribution systems. However, two new developments promise to extend their use to other water lines. A system for performing leak detection on large diameter water lines has recently been developed in the United Kingdom. This system places a neutrally buoyant hydrophone into an operating pipe (Fig. 3.36) (49). A parachute like droud is then deployed so that the force of the water flow carries the hydrophone downstream for a distance of up to three km. The droud is then collapsed and the hydrophone is slowly wound back to the entry point while it listens for leaks inside the pipe. The loudest leak noises would be expected to be heard at the

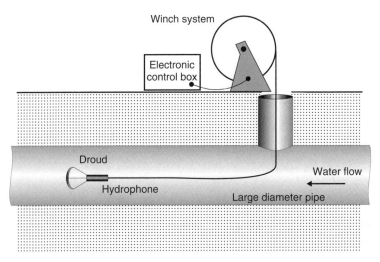

Fig. 3.36. Schematic description of the Water Research Center's large diameter pipe leak detection system (49).

point where the detector is closest to the leaks, allowing them to be located precisely. The major advantage of this system is that very few access points to the pipe are required from the surface. It can therefore be easily deployed within a deeply buried, long transmission line where surface leak detection techniques would not work.

While the effectiveness of leak detection is well recognized by municipalities, there do not appear to have been any tests carried out to determine what percentage of leaks were actually found during a survey. In addition, some municipalities may prefer to minimize the number of falsely detected leaks (false positives) in order to minimize unnecessary excavations. Consequently, operators may tend to set the threshold levels in their devices to reduce the number of false positive leaks that are detected, which would also increase the likelihood of missing real leaks with weak noise signals.

The issue of the percentage of missed leaks needs to be explored. Some cities have unaccounted for water losses of ~25% despite conducting leak detection programs. It is likely that at least some of this unaccounted for water is due to undetected leaks. Furthermore, not knowing the true number of leaks in a water line means that incorrect decisions may be made regarding choices between repairing, rehabilitating or replacing water lines. The detection of 10 leaks in 5 km of pipe may mean that the repairs are the lowest cost option for a water utility. However, if those 10 leaks are an indication that 20 more lie undetected below the ground a more extensive rehabilitation or replacement program might produce lower long-term costs.

3.8.7.3.3. Soil Corrosivity The high leak rate of an unprotected line depends on a few factors. The greatest difference is in the soils in which lines lie. An easy way to get an approximate measure of the quantity of dissolved salts in soil is by measuring the electrical resistivity of the soil. Probably the second largest factor and it can easily be argued to be the first, is the coating, if any. Pipe wall thickness, as already mentioned, will also have an effect on leak rates. Other things being equal, a thin-walled pipe will develop a leak at a given location in less time than one with a thicker wall.

Soils come in many varieties and have many different properties that can be measured. Soils may be classified by texture, organic content, dissolved chemicals, color, and origin depending on what one is interested in. Soil resistivity generally decreases with increasing water content and the concentration of ionic species. Sandy soils are high up on the resistivity scale, and therefore considered the least corrosive. Clay soils, especially those contaminated with saline water are on the opposite end of the spectrum. The generally adopted corrosion severity ratings are shown in Table 3.16 (30).

Soil resistivity is by no means the only parameter affecting the risk of corrosion damage. A high soil resistivity alone will not guarantee absence of serious corrosion. The AWWA has developed a numerical soil corrosivity scale, applicable to cast irons (Table 3.17) (51). The severity ranking is based on assigning points for different variables. When the total number of points of a soil in the

Table 3.16. Corrosivity Ratings Based on Soil Resistivity[a]

Soil resistivity (Ω cm)	Corrosivity rating
>20,000	Essentially noncorrosive
10,000–20,000	Mildly corrosive
5,000–10,000	Moderately corrosive
3,000–5,000	Corrosive
1,000–3,000	Highly corrosive
<1,000	Extremely corrosive

[a]Ref. 30.

AWWA scale equals 10 (or higher), additional corrosion protective measures (e.g., cathodic protection) should be considered or recommended.

3.8.8. Reinforced Concrete Infrastructure

Forty percent of the bridges in North America are >40-years old and concrete is the primary materials of construction (52). It was recognized by the mid-1970s that the main deterioration factor of concrete bridge structures was the corrosion of the reinforcing steel in the concrete which, in turns, was induced by the intrusion of chloride from the deicing salts into the concrete. According to a 1997 report, of the 581,862 bridges in and off the U.S. National Bridge Inventory, ~40% were either functionally or structurally deficient.

Most of these bridges were severely deteriorated with extensive loss of serviceability and reduced safety, with some of the bridges load posted so that overweight trucks would be required to take a longer alternative route (53). The estimated cost to eliminate all backlog bridge deficiencies (both structural and functional) in Canada was estimated at $10 billion (52) and in the United States to be $78 billion, increasing to as much as 112 billion, depending on the number of years it would take to meet the objective.

Reinforced concrete bridge decks presently constitute the weakest link in North America's infrastructure network. Even though the cost of maintaining bridge decks is becoming prohibitively expensive, the benefits provided by deicing salts are too great in terms of reducing vehicular accidents and minimize traffic disruption (54). Therefore its use is not likely to decrease in the future. In fact, the use of road deicing salts, which are extremely corrosive due to the disruptive effects of its chloride ions on protective films on metals, has actually increased in the first half of the 1990s after leveling off during the 1980s.

In this context, techniques that could provide an early warning of critical corrosion problems would be particularly helpful. Once rebar corrosion has proceeded to an advanced state as being visually apparent on external surfaces it

Table 3.17. Point System for Predicting Soil Corrosivity According to the AWWA C-105 standard[a]

Soil parameter	Assigned points
Resistivity (Ω cm)	
<700	10
700–1000	8
1000–1200	5
1200–1500	2
1500–2000	1
>2000	0
pH	
0–2	5
2–4	3
4–6.5	0
6.5–7.5	0
7.5–8.5	0
>8.5	3
Redox potential (mV)	
>100	0
50–100	3.5
0–50	4
<0	5
Sulfides	
Positive	3.5
Trace	2
Negative	0
Moisture	
Poor drainage continuously wet	2
Fair drainage generally moist	1
Good drainage generally dry	0

[a] Ref. 30.

is usually too late to implement effective corrosion control measures and high repair or replacement costs are inevitable. Tools for assessing the effectiveness of remedial measures in short, practical time frames are also greatly required.

A systematic comparison of condition assessment techniques has been developed for concrete bridge components under a contract of the U.S. Strategic

Highway Research Program (SHRP). The comparison scheme developed in this program compares in Table 3.18 13 conventional, well-established test methods and seven new methodologies divided in the following three survey types (55):

- The initial (baseline) evaluation surveys focus on parameters that undergo relatively little change over time. The test methods recommended for this baseline survey essentially represent tests that should form part of the acceptance testing of new concrete bridge components (Table 3.19).

Table 3.18. Conventional and New Methods for Reinforced Concrete Structures[a]

Property	Test methodology
Existing techniques in SHRP bridge assessment guide	
Depth of concrete cover	Magnetic flux devices
Concrete strength from test cylinders	ASTM C 39 for test cylinders
Concrete strength from core samples	ASTM C 42
Concrete strength from pull out tests	ASTM C 900
Concrete strength/quality from rebound hammer tests	ASTM C 805
Concrete strength/quality from penetration tests	ASTM C 803
Air void system characterization in hardened concrete	ASTM C 457
Microscopic evaluation of hardened concrete quality Alkali-silica reactivity	ASTM C 856 petrographic examination
Delamination detection	ASTM D 4580 sounding
Damage assessment by pulse velocity	ASTM C 597
Cracking damage	ACI 224.1R
Probability of active rebar corrosion	ASTM C 876 based on corrosion potential (note no corrosion rate is determined)
New techniques in SHRP bridge assessment guide	
Instantaneous corrosion rate measurement	Electrochemical measurements, applicable to uncoated steel
Condition of preformed membranes on decks using pulse velocity	
Relative effectiveness of penetrating concrete sealers with electrical resistance	
Evaluating penetrating concrete sealers by water absorption	
Chloride content in concrete by specific ion sensor	
Relative concrete permeability by surface air flow	
Other	
Chloride ion content by titration	AASHTO T-260
Rebar location	X-ray and radar
pH and depth of carbonation	Phenolphthalein solution or pH electrode in extracted pore solution
Concrete permeability with respect to chloride ions	ASTM C 1202 and ASTM C 642
Delamination, voids and other hidden defects	Impact echo, Infrared thermography, pulse echo and radar
Material properties	ASTM C 642 (density), moisture content (ASTM 642), shrinkage (ASTM C 596, C 426), dynamic modulus (ASTM C 215), modulus of elasticity (ASTM C 464)

[a] Ref. 55.

3.8. Industrial Examples   183

Table 3.19. Initial Evaluation Survey of Reinforced Bridge Components[a,b]

Assessment procedure	Comments
Air void and petrographic core samples	Applicable to structures up to 15-years old
Alkali-silica reactivity test	Applicable to structures that are between 1 and 15-years old
Concrete strength	Applicable at all ages
Relative permeability	Applicable at all ages
Rebar cover	Applicable at all ages

[a]*Note*: Test and sampling details may vary, depending on the age of the structure.
[b]Ref. 55.

- The goal of the second type of surveys in the SHRP scheme is to provide subsequent evaluation during the life of a structure. The initial step in these surveys is visual inspection. The nature of subsequent inspection techniques depends on the morphology of damage observed in the visual inspection phase. The emphasis on reinforcing steel corrosion damage is placed under concrete spalling phenomena. The recommended assessment procedures for this form of damage are presented in Fig. 3.37.

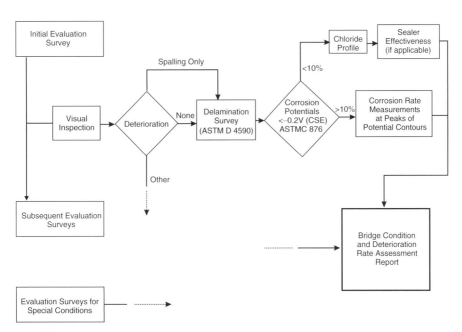

Fig. 3.37. Part of the SHRP guide to assessing concrete bridge components (applicable to uncoated steel rebar).

- The third survey category describes special surveys, applicable to asphalt covered decks, pretensioned and posttensioned concrete members and rigid deck overlays.

Traditional methods, such as visual inspection, core sampling for compressive strength tests or petrographic analysis, and chain drag sounding have formed the basis of condition assessments. However, as is apparent in Table 3.18, a host of new nondestructive evaluation (NDE) methods and corrosion monitoring techniques are available in modern engineering practice. Some of these techniques are briefly described in the following sections.

3.8.8.1. Electrochemical Corrosion Measurements

These electrochemical corrosion measurements can be performed nondestructively on the actual reinforcing steel or on sensors that are embedded in the concrete structure. However, it is necessary to expose a reinforcing bar to make an electrical connection for the portable sensors. In general, electrochemical techniques are highly sensitive and therefore can detect corrosion damage at a very early stage. There are three different types of measuring device:

- Half-cell potential measurements that can be used to map the anodic areas across a concrete surface (56).
- Corrosion rate measurements by polarization resistance measurement which may be used to evaluate the corrosion rate of the bars nearest the sensor (57, 58).
- Measurement of galvanic current flow between embedded steel sensors. Such measurements are usually embedded prior to construction. The corrosion sensors are essentially small sections of rebar steel, with shielded electrical leads attached for potential and current measurements. Preferably these sensors are embedded at different depths of cover and their use must obviously be defined in the design stage of the structure.

3.8.8.2. Chloride Content

Samples for determining the chloride level in concrete are collected in the form of powder produced by drilling or by the extraction of cores, sections of which are subsequently crushed. This latter method can provide an accurate chloride concentration depth profile. The chloride ion concentration is subsequently determined by potentiometric titration. Two distinctions are made in chloride ion concentration testing. Acid soluble chloride content (ASTM C 114) refers to the total chloride ion content, while the water soluble content represents a lower value.

3.8.8.3. Petrographic Examination

Petrographic examination is a microscopic analysis of concrete, performed on core samples removed from the structure according to the ASTM C856. This technique yields information such as the depth of carbonation, density of the cement paste, air content, freeze–thaw damage and direct attack of the concrete.

3.8.8.4. Permeability Tests

Permeability tests are either based on ponding core samples in chloride solution with subsequent chloride content analysis, or on "forced" migration of chloride ions under the influence of an external electric field. The application of the electrical field accelerates chloride ion migration and hence reduces the testing time.

REFERENCES

1. Guidance for Optimizing Nuclear Power Plant Maintenance Programmes. IAEA-TECDOC-1383. 2003. Vienna, Austria, International Atomic Energy Agency (IAEA).
2. ROBERGE PR. *Handbook of Corrosion Engineering*. New York: McGraw-Hill, 2000.
3. DOUGLAS J. The Maintenance Revolution. *EPRI Journal* 1995; **20**.
4. NOWLAN FS and HEAP, HF Reliability Centered Maintenance. AD-A066-579. 1978. Washington, DC, National Technical Information Service.
5. VERINK ED. Corrosion Economic Calculations. In: *Metals Handbook: Corrosion*. Metals Park: ASM International, 1987; pp. 369–374.
6. COULLAHAN R, and SIEGFRIED C. Life Cycle Asset Management. *Facilities Engineering Journal* 1996; **1996**.
7. BOYD L, SCHOW BL, and NICHOLAS KW. CP Data Management Software, Field Computers and Remote Monitoring Facilitate Cathodic Protection System Commissioning. CORROSION 2005, Paper No. 135. 2005. Houston, TX, NACE International.
8. KLETZ T. *HAZOP and HAZAN*. 3rd edn. Rugby, UK: Institution of Chemical Engineers, 1992.
9. McINTIRE DR and DILLON CP. *Guidelines for Preventing Stress Corrosion Cracking in the Process Industries*. Columbus, OH: Materials Technology Institute, 1985.
10. STAEHLE RW. Lifetime Prediction of Materials in Environments. In: Revie RW, ed. *Uhlig's Corrosion Handbook*. New York: Wiley-Interscience, 2000; pp. 27–84.
11. QUEEN DME, RIDD BR, and PACKMAN C. Key Performance Indicators for Demonstrating Effective Corrosion Management in the Oil and Gas Industry. CORROSION 2001, Paper No. 056. 2001. Houston, TX, NACE International.
12. McNEENEY A. Selecting the Right Key Performance Indicators. *Maintenance Technology* 2005; **18**: 27–34.
13. BRAY DE and STANLEY RK. *Nondestructive Evaluation*. New York: Mc Graw-Hill, 1989.
14. MORELAND PJ and HINES JG. *Hydrocarbon Processing* 1978; 543.
15. HORVATH RJ. The Role of the Corrosion Engineer in the Development and Application of Risk-Based Inspection for Plant Equipment. *Materials Performance,* 1998; **37**: 70–75.
16. Risk-Based Inspection. ANSI/API RP 580, 1st ed. 2002. Washington, DC, American Petroleum Institute.
17. Report on Current Practices. DNV Report No. 2004-0305. 2005. Hovik, Norway, DNV Library Services.
18. Marine Risk Assessment. Contract Research Report 2001/0631. 2002. Sudbury, UK, Health and Safety Executive (HSE).
19. WINTLE JB, KENZIE BW, AMPHLETT GJ, and SMALLEY S. Best practice for risk based inspection as a part of plant integrity management. Contract Research Report 363/2001. 2001. Sudbury, UK, Health and Safety Executive (HSE).

20. *Fault Tree Analysis Application Guide*. Rome, NY: Reliability Analysis Center (RAC), 1990.
21. ROBERGE PR. Modeling Corrosion Processes. In: Cramer DSaCBS, ed. *Volume 13A: Corrosion: Fundamentals, Testing, and Protection*. Metals Park, OH: ASM International, 2003.
22. RIVERA, M., BOLINGER, S., and WOLLENWEBER, C. Carbonate Cracking Risk Assessment for a FCCU Gas Plant. CORROSION 2004, Paper No. 639. 2004. Houston, TX, NACE International.
23. WYATT LM, BAGLEY DS, MOORE MA, and BAXTER DC. *An Atlas of Corrosion and Related Failures*. St. Louis, MO: Materials Technology Institute, 1987.
24. SACHS NW. Understanding the Multiple Roots of Machinery Failures. *Reliability Magazine* 2002; **8**: 18–21.
25. Standard Guide for Corrosion-Related Failure Analysis. ASTM G 161-00. [Annual Book of ASTM Standards, Vol 03.02]. 2000. Philadelphia, PA, American Society for Testing of Materials.
26. DOUGLAS J. Solutions for Steam Generators. *The EPRI Journal;* 1995; **20**.
27. VIETH PH. Comprehensive, long-term integrity management programs are being developed and implemented to reduce the likelihood of pipeline failures. *Materials Performance* 2002; **41**: 16–22.
28. LOZEV M, GRIMMETT B, SHELL E, and SPENCER R. Evaluation of Methods for Detecting and Monitoring of Corrosion and Fatigue Damage in Risers. Project No. 45891GTH. 11-4-2003. Washington, DC, Minerals Management Service, U.S. Department of the Interior.
29. KROON DH. External Corrosion Direct Assessment of Buried Pipelines: The Process. *Materials Performance* 2003; **42**: 28–32.
30. PEABODY AW and BIANCHETTI RL. *Peabody's Control of Pipeline Corrosion*. 2nd ed. Houston, TX: NACE International, 2001.
31. ROBERGE PR. *Corrosion Basics—An Introduction*. 2nd edn. Houston, TX: NACE International, 2005.
32. *ASME B31G*—Manual for Determining the Remaining Strength of Corroded Pipelines. 1992. New York: American Society of Mechanical Engineers (ASME).
33. DNV-RP-F101 Corroded Pipelines. 2004. Høvik, Norway, Det Norske Veritas.
34. NACE RP0502-2002—Standard Recommended Practice Pipeline External Corrosion Direct Assessment Methodology. 2002. Houston, TX, NACE International.
35. TADEPALLY VP and HENDREN ES. Internal Corrosion Direct Assessment: Refining the Method through New Decision Support Models. CORROSION 2005, Paper No. 178. 2005. Houston, TX, NACE International.
36. WAKER S, ROSCA G, and HYLTON M. In-Line Inspection Tool Selection. CORROSION 2004, Paper No. 168. 2004. Houston, TX, NACE International.
37. TIMMINS PF. *Predictive Corrosion and Failure Control in Process Operations*. Materials Park, OH: ASM International, 1996.
38. SMALLWOOD, R. Equipment Integrity in the New Millennium. CORROSION 2004, Paper No. 215. 2004. Houston, TX, NACE International.
39. SYRETT BC and GORMAN JA. Cost of Corrosion in the Electric Power Industry—An Update. *Materials Performance* 2003; **42**: 32–38.
40. Cost of Corrosion in the Electric Power Industry. EPRI 1 004 662. 2001. Palo Alto, CA, Electric Power Research Institute (EPRI).
41. JONAS O. Monitoring of Steam Plants. *Materials Performance* 2003; **42**: 38–42.
42. KOCH GH, BRONGERS MPH, THOMPSON NG, VIRMANI YP, and PAYER JH. Corrosion Costs and Preventive Strategies in the United States. FHWA-RD-01-156. 2001. Springfield, VA, National Technical Information Service.
43. Anonymous. Airworthiness Manual—Chapter 571—Maintenance of Aeronautical Products. TP6197E. 1987. Ottawa, Canada, Transport Canada Aviation.

44. NAKATA D. MSG-3 Aircraft Systems/Powerplant Analysis Method. In: *Aircraft Maintenance and reliability Seminar/Workshop Vol. 1*. Palm Harbor, FL: Transportation Systems Consulting Corp., 1997.
45. KIRMEYER GJ, WAGNER I, and LEROY P. Organizing Corrosion Control Studies and Implementing Corrosion Control Strategies. In: *Internal Corrosion of Water Distribution Systems*. Denver, CO: American Water Works Association, 1996; 487–540.
46. Deterioration and Inspection of Water Distribution Systems: A Best Practice by the National Guide to Sustainable Municipal Infrastructure. Deterioration and Inspection of Water Distribution Systems Issue No. 1.1. 2003. Ottawa, Canada, National Research Council of Canada.
47. ROMANOFF M. *Underground Corrosion*. Houston, TX: NACE International, 1989.
48. MAKAR JM and KLEINER Y. Maintaining water pipeline integrity. NRCC-43 986. 2001. Ottawa, Canada, National Research Council.
49. MAKAR JM and CHAGNON N. Inspecting systems for leaks, pits, and corrosion. *American Water Works Association* 1999; **91**: 36–46.
50. *Water Audits and Leak Detection (M36)*. 2nd edn. American Water Works Association, 1999.
51. PALMER JD. Environmental Characteristics Controlling the Soil Corrosion of Ferrous Piping. In: Chaker V and Palmer JD, eds. *Effects of Soil Characteristics on Corrosion*. Philadelphia, PA: American Society for Testing and Materials, 1989; pp. 5–17.
52. LOUNIS Z. Maintenance Management of Aging Bridges: Economical and Technological Challenges. *Canadian Civil Engineer* 2002; **19**: 20–23.
53. Corrosion Protection: Concrete Bridges. FHWA-RD-98-088. 1998. Washington, DC, U.S. Department of Transportation.
54. Highway Deicing: Comparing Salt and calcium magnesium acetate. Special Report 235. 1991. Washington, DC, Transportation Research Board, National Research Council.
55. KESSLER R. Concrete Deterioration Manual. SHRP 2032. 1991. Washington, DC, 1991, Strategic Highway Research Program, National Research Council.
56. Standard Test Method for Half-Cell Potentials of Uncoated Reinforcing Steel in Concrete. Annual Book of ASTM Standards. C 976-91. 1997. West Conshohocken, PA, American Society for Testing and Materials.
57. FLIS J PICKERING HW, and OSSEO-ASARE K. Assessment of data from Three Electrochemical Instruments for Evaluation of Reinforcement Corrosion Rates in Concrete Bridge Components. *Corrosion* 1995; **51**: 602–609.
58. WENZLICK, J. An Electrochemical Method for Detecting Ongoing Corrosion of Steel in a Concrete Structure with CP Applied. SHRP 4003. 1993. Washington, DC, 1991, Strategic Highway Research Program, National Research Council.

Chapter 4

Corrosion Monitoring

4.1. What Is Corrosion Monitoring?
4.2. The Role of Corrosion Monitoring
4.3. Corrosion Monitoring System Considerations
    4.3.1. What Is the Monitoring Objective?
    4.3.2. Corrosion Monitoring Locations
        4.3.2.1. Process Industry
        4.3.2.2. Oil and Gas Production Gathering Lines
    4.3.3. Probe Design and Selection
        4.3.3.1. Flush Mounted Electrode Design
        4.3.3.2. Protruding Electrode Design
        4.3.3.3. Probes to Suit the Application
            4.3.3.3.1. Stress Corrosion Cracking Probe
            4.3.3.3.2. Corrosion in Hydrocarbon Environments
            4.3.3.3.3. Coupled Multielectrode Array Systems and Sensors
            4.3.3.3.4. Atmospheric Corrosion Monitoring
    4.3.4. Location of Monitoring Hardware
    4.3.5. Sensitivity and Response Time
    4.3.6. Data Communication and Analysis Requirements
    4.3.7. Define System Reliability
    4.3.8. Availability and Cost
    4.3.9. In a Nutshell
4.4. Corrosion Monitoring Techniques
    4.4.1. Direct Intrusive Techniques
        4.4.1.1. Physical Techniques
            4.4.1.1.1. Mass Loss Coupons
                4.4.1.1.1.1. Length of exposure and limitations
                4.4.1.1.1.2. Uniform corrosion coupons
                4.4.1.1.1.3. Galvanic-corrosion coupons
                4.4.1.1.1.4. Crevice-corrosion coupons
                4.4.1.1.1.5. Stress-corrosion cracking coupons
                4.4.1.1.1.6. Heat-transfer coupons
                4.4.1.1.1.7. Welded coupons

Corrosion Inspection and Monitoring, by Pierre R. Roberge
Copyright © 2007 John Wiley & Sons, Inc.

Chapter 4 Corrosion Monitoring

- 4.4.1.1.1.8. Sensitized metal
- 4.4.1.1.1.9. Coupon cleaning and evaluation
- 4.4.1.1.2. Electrical Resistance
- 4.4.1.1.3. Inductive Resistance Probes
- 4.4.1.2. Electrochemical Techniques
  - 4.4.1.2.1. Linear Polarization Resistance (LPR)
  - 4.4.1.2.2. Zero Resistance Ammetry
  - 4.4.1.2.3. Potentiodynamic–Galvanodynamic Polarization
  - 4.4.1.2.4. Electrochemical Impedance Spectroscopy
  - 4.4.1.2.5. Harmonic Distortion Analysis
  - 4.4.1.2.6. Electrochemical Noise (EN) Analysis
    - 4.4.1.2.6.1. Electrochemical noise analysis
    - 4.4.1.2.6.2. Probe design and signal analysis
- 4.4.2. Direct Nonintrusive Techniques
  - 4.4.2.1. Thin-Layer Activation and Gamma Radiography
  - 4.4.2.2. Field Signature Method
  - 4.4.2.3. Acoustic Emission
- 4.4.3. Indirect On-Line Techniques
  - 4.4.3.1. Hydrogen Monitoring
  - 4.4.3.2. Corrosion Potential
  - 4.4.3.3. On-Line Water Chemistry Analyses
    - 4.4.3.3.1. pH
    - 4.4.3.3.2. Conductivity
    - 4.4.3.3.3. Dissolved Oxygen
    - 4.4.3.3.4. Oxidation–Reduction (Redox) Potential
  - 4.4.3.4. Process Variables
    - 4.4.3.4.1. Flow Regime
    - 4.4.3.4.2. Pressure
    - 4.4.3.4.3. Temperature
    - 4.4.3.4.4. Dewpoint
  - 4.4.3.5. Fouling
- 4.4.4. Indirect Off-Line Measurement Techniques
  - 4.4.4.1. Off-Line Water Chemistry Parameters
    - 4.4.4.1.1. Alkalinity
    - 4.4.4.1.2. Metal Ion Analysis
    - 4.4.4.1.3. Concentration of Dissolved Solids
    - 4.4.4.1.4. Gas Analysis
    - 4.4.4.1.5. Residual Oxidant
    - 4.4.4.1.6. Microbiological Analysis
  - 4.4.4.2. Residual Inhibitor
  - 4.4.4.3. Chemical Analysis of Process Samples
    - 4.4.4.3.1. Sulfur Content
    - 4.4.4.3.2. Total Acid Number
    - 4.4.4.3.3. Nitrogen Content
    - 4.4.4.3.4. Salt Content of Crude Oil

4.5. Monitoring Microbiologically Influenced Corrosion
    4.5.1. Planktonic Organisms
    4.5.2. Sessile Organisms
    4.5.3. Sampling
    4.5.4. Biological Assessment
        4.5.4.1. Direct Inspection
        4.5.4.2. Growth Assays
        4.5.4.3. Activity Assays
            4.5.4.3.1. Whole Cell
            4.5.4.3.2. Enzyme-Based Assays
            4.5.4.3.3. Metabolites
            4.5.4.3.4. Cell Components
                4.5.4.3.4.1. Fatty acid profiles
                4.5.4.3.4.2. Nucleic acid-based methods
        4.5.4.4. Detailed Coupon Examinations
    4.5.5. Monitoring MIC Effects
        4.5.5.1. Deposition Accumulation Monitors
        4.5.5.2. Electrochemical Methods
4.6. Monitoring Pipeline CP Systems
    4.6.1. Close Interval Potential Surveys
    4.6.2. Pearson Survey
    4.6.3. Direct Current Voltage Gradient Surveys
    4.6.4. Corrosion Coupons
4.7. Atmospheric Corrosion Monitoring
    4.7.1. Relative Humidity and Time of Wetness
    4.7.2. Pollutants
    4.7.3. Airborne Particles (Chlorides)
    4.7.4. Atmospheric Corrosivity
References

4.1. WHAT IS CORROSION MONITORING?

Corrosion monitoring refers to corrosion measurements performed under industrial or practical operating conditions. In its simplest form, corrosion monitoring may be described as acquiring data on the rate of material degradation. However, such data are generally of limited use and need to be converted into useful information for inclusion in a corrosion management program. This requirement has led to the evolution of corrosion monitoring tools toward real-time data acquisition, process control tools, knowledge-based systems, and smart structures. Corrosion monitoring is more complex than the monitoring of most other process parameters because

- There are a number of different types of corrosion (see Chapter 2 for more details.
- Corrosion may be uniform over an area or concentrated in very small areas (pitting).

- General corrosion rates may vary substantially, even over relatively short distances.
- There is no single measurement technique that will detect all of these various conditions.

Before embarking in a corrosion monitoring program, it is therefore helpful to review historical data and consider the types of corrosion problems that need to be investigated. It is also advisable to use several complementary techniques rather than rely on a single monitoring method.

Real-time monitoring of pipelines, vessels, and other static equipment enables a near instantaneous appraisal of the corrosivity of produced and transported fluids. If the corrosion activity increases as a result of process nonconformities, the corrosion information can be viewed alongside process variables such that cause-and-effect can be determined and rapid action can be taken to stifle the progress of any associated problem. The same approach can be used to demonstrate the effectiveness of remedial actions or preventive treatments.

Some modern corrosion monitoring technologies are particularly apt at revealing the highly time-dependent nature of corrosion processes. The integration of these corrosion monitoring technologies in existing systems can thus provide early warnings of costly corrosion damage.

An extensive range of corrosion monitoring techniques and systems has evolved, particularly in the last two decades, for detecting, measuring, and predicting corrosion damage. The development of efficient corrosion monitoring techniques and user-friendly software have permitted us to field new techniques that were until recently perceived to be mere laboratory curiosities. Noteworthy catalysts to the growth of the corrosion monitoring market has been the expansion of oil and gas production under extremely challenging operational conditions (e.g., the North Sea), cost pressures brought about by global competition, and the public demand for higher safety standards. In several sectors, such as oil and gas production, sophisticated corrosion monitoring systems have gained successful track records and credibility, while in other sectors their application has made only a limited progress.

4.2. THE ROLE OF CORROSION MONITORING

Correct and effective corrosion monitoring strategies should be used as a proactive tool to assist with operating a plant or any other system more effectively, thereby prolonging its life and gaining optimum throughput. Fundamentally, four strategies are available to an organization in its dealings with corrosion:

1. Ignoring it until a failure occurs.
2. Inspection, repairs and maintenance at scheduled intervals.
3. Using corrosion prevention systems (inhibitors, coatings, resistant materials etc.).

4. Applying corrosion control selectively, when and where it is actually needed.

The first strategy represents corrective maintenance practices, whereby repairs and component replacement are only initiated after a failure has occurred. Corrosion monitoring is completely ignored in this reactive philosophy. Obviously, this practice is unsuitable for safety critical systems and in general is inefficient from maintenance cost considerations, especially during life extension of aging engineering systems.

The second strategy is one of preventive maintenance. The inspection and maintenance intervals and methodologies are designed to prevent corrosion failures, while achieving 'reasonable' system usage. Corrosion monitoring can assist in optimizing these maintenance and inspection schedules. In the absence of information from a corrosion monitoring program, such schedules may be set too conservatively with excessive downtime and associated cost penalties.

Alternatively, if set too infrequently, the inspection and maintenance intervals may represent an excessive corrosion risk with associated cost and safety consequences (cf. Fig. 3.10). Furthermore, without input from corrosion monitoring information, preventive inspection and maintenance intervals will be of the "routine" variety, without accounting for the time dependence of critical corrosion variables. In the oil and gas industry, for example, the corrosivity at a well head can fluctuate significantly over the lifetime of the production system, between benign to highly corrosive. In oil refining plants, the corrosivity can vary with time, depending on the grade or hydrogen sulfide content of crude that is processed.

The application of corrosion prevention systems is obviously crucial in most corrosion control programs. However, without corrosion monitoring information, their application may be excessive and overly costly. For example, a certain inhibitor dosage level on a particular system may combat corrosion damage, but real-time corrosion monitoring might reveal that a lower dosage would actually suffice. Ideally, the inhibitor feed rate should be continuously adjusted based on real-time corrosion monitoring information.

In an ideal corrosion control program, inspection and maintenance would be applied only where and when they are actually needed. In principle, the information obtained from corrosion monitoring systems can be of great assistance in reaching this goal. However, it is sometimes difficult for a corrosion engineer to get management's commitment to investing funds in such initiatives. The importance of corrosion monitoring in industrial plants and other engineering systems should be presented as an investment to achieve some of the following goals:

- Improved safety.
- Reduced downtime.
- Production of early warnings before costly serious damage sets in.
- Reduced maintenance costs.
- Reduced pollution and contamination risks.

- Longer intervals between scheduled maintenance.
- Reduced operating costs.
- Life extension.

Experience has shown that the potential cost savings resulting from the implementation of corrosion monitoring programs generally increase with the sophistication level (and cost) of the monitoring system.

4.3. CORROSION MONITORING SYSTEM CONSIDERATIONS

Corrosion monitoring systems vary significantly in complexity, from simple coupon exposures or handheld data loggers (Fig. 4.1) to fully integrated plant process surveillance units with remote data access and data management capabilities.

Corrosion sensors (probes) are an essential element of all corrosion monitoring systems. The nature of the sensors depends on the various individual techniques used for monitoring, but often a corrosion sensor can be viewed as an instrumented coupon. A single high pressure access fitting for insertion of a retrievable corrosion probe can be used to accommodate most types of retrievable probes. With specialized tools (and brave specialist operating crews!) sensor insertion and withdrawal can be possible under pressurized operating conditions.

The signal emanating from a corrosion sensor usually has to be processed and analyzed. Examples of signal processing include filtering, averaging, and unit conversions. Furthermore, in some corrosion sensing techniques, the sensor surface has to be perturbed by an input signal to generate a corrosion signal output. In older systems, electronic sensor leads were usually employed for these purposes and to relay the sensor signals to a signal-processing unit. Advances in microelectronics have facilitated the sensor signal conditioning and processing by

Fig. 4.1 Field corrosion monitoring using electrochemical noise recorded with a handheld data logger. (Courtesy of Kingston Technical Software.)

4.3. Corrosion Monitoring System Considerations

the introduction of microchips that have become an integral component of sensor units (1, 2). Wireless data communication with such sensing units is also a product of the microelectronic revolution. Figure 4.2a and b illustrates a recent wireless monitoring system developed to provide an early warning that a protective coating is degrading (3). Such a direct measurement contrasts with corrosivity sensors that monitor indirectly the problem by monitoring ambient environments.

Irrespective of the sensor details, a data acquisition system is required for on-line and real-time corrosion monitoring. On several plants, the data acquisition system is housed in mobile laboratories, which can be made intrinsically safe. A computer system often performs a combined role of data acquisition, data processing and information management system. In data processing, a process is initiated to transform corrosion monitoring data (low intrinsic value) into process relevant information (higher intrinsic value). Complementary data from other relevant sources, such as process parameter logging and inspection reports, can be acquired together with the data from corrosion sensors, for use as input to a management information system.

Numerous real-time corrosion monitoring programs in diverse branches of industry have revealed that the severity of corrosion damage is rarely uniform with time. Rather, serious corrosion damage is usually sustained in time frames where operational parameters have suffered upsets. These undesirable operational windows can only be identified with the real-time monitoring approach.

In general, it can be said that no individual technique alone is suitable for monitoring corrosion under complex industrial conditions. Therefore, a multitechnique approach is often preferred. In many cases, this approach does not require a higher number of sensors, but rather only an increased number of sensor elements for a given probe and access fitting.

Another important consideration is that, irrespective of the technique, most instrumented sensors only provide semiquantitative corrosion damage information. It is thus sensible to correlate monitoring data from these sensors to long-term coupon exposure programs and actual plant damage. Unfortunately, nonspecialists may put too much faith in the numerical corrosion rate displayed by a commercial corrosion monitoring device. An example of a widely available technique is linear polarization resistance (LPR), which many commercial monitoring systems use to measure corrosion rates, commonly displayed as mm/year or milli-inches/year (mpy).[1] Such systems are used extensively in industry for monitoring the effectiveness of water treatment additives and various other applications. However, from fundamental considerations, the derived LPR corrosion rate is only valid if the following assumptions are met, which is rarely the case in actual operational conditions:

- There is only one simple anodic reaction.
- There is only one simple cathodic reaction.

---

[1] Table 2.2 provides the conversion factors between commonly used corrosion rate units for all metals and Table 2.3 describes these conversion factors adapted to iron or steel (Fe) for which $n = 2$, $M = 55.85 \text{ g mol}^{-1}$ and $d = 7.88 \text{ g cm}^{-3}$.

196 Chapter 4 Corrosion Monitoring

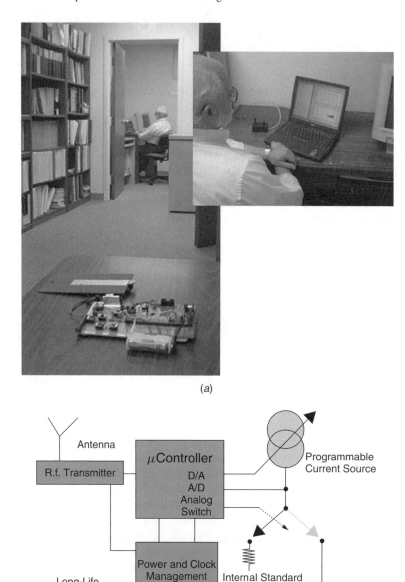

Fig. 4.2. (*a*) Test and demonstration of breadboard coating health monitor, including wireless communication. Inset shows transceiver unit attached to laptop computer. (*b*) Block diagram of the electronics. (Courtesy of Guy D. Davis, DACCO SCI, Inc).

- The anodic and cathodic Tafel constants are known and invariant with time.
- The corrosion reactions proceed by a simple charge-transfer mechanism under activation control, which essentially implies that the corroding surface is clean without corrosion product build-up, scale deposits, or solids settled out of solution;
- Corrosion proceeds in a uniform manner (whereas the vast majority of industrial corrosion problems are related to localized attack).
- The solution resistance is negligible (some instruments make a solution resistance compensation but this is not necessarily accurate).
- The corrosion potential has reached a steady state value.

The following sections describe in more specific details the various elements that should be considered when deciding on a corrosion monitoring program.

4.3.1. What Is the Monitoring Objective?

The first and most essential step of a corrosion monitoring program is to define the monitoring objective, a step often forgotten. If corrosion monitoring is done for corrosion control the purpose is to assure that asset life is not jeopardized by too many high corrosion rate events. The main objective of corrosion monitoring is in this case to limit the "corrosion events", without completely using the corrosion allowance of a system before the end of its design life. The main factors that govern the design of a monitoring system in this case are (4) available corrosion allowance; uncontrolled corrosion rates; event rates; corrosion rate detection sensitivity and response rate; and required service life.

If corrosion monitoring is used for corrosion control, it is thus essential that the corrosion mechanism and the corrosion rates during times of noncompliance be relatively well known. Corrosion monitoring can also be used in a corrosion control context to establish the corrosion mechanism of a system or optimize the corrosion control, for example, testing the efficiency of corrosion inhibitors, adjusting the corrosion inhibitor injection rate, or studying corrosion mechanisms. Such measurements can either be carried out on instrumented sections of the actual system or in a side stream.

Corrosion monitoring could also be needed in a broader context of integrity management to ensure that the operating envelope of a system is not exceeded. The time horizon of ongoing integrity activities can be much shorter than the plant lifetime, by virtue of the definition of integrity. This can be satisfied by ensuring that integrity is maintained up to the next inspection date and reassessed for another period.

4.3.2. Corrosion Monitoring Locations

An important decision in setting up a corrosion monitoring system is the selection of the monitoring points or sensor locations. As only a finite number of points can be considered for obvious economical reasons, it is usually desirable to monitor the "worst-case" conditions, at points where corrosion damage is expected to be most severe. Often, such locations can be identified by reasoning with basic corrosion principles, analysis of in-service failure records and in consultation with operational personnel. For example, the most corrosive conditions in water tanks are usually found at the water–air interface. Corrosion sensors could be attached to a floating platform to maintain these conditions independently of water level changes in order to monitor corrosion under these conditions.

In any monitoring situation, the ideal probe placement is most often not possible. Invariably, the flush or protruding electrode probes require to be placed in the most corrosive environment (e.g., at a 6 o'clock position in a pipeline, at the bottom of a vessel, or at a solution accumulation point in a separator tower). Almost equally invariably, the available location is entirely at odds with these requirements. Although certain steps can be taken to provide the application with a modified design (e.g., a protruding electrode probe installed at the 12 o'clock position with a long body to extend to the aqueous phase), this is without question a compromise on the part of the technology provider (5).

It is obviously imperative that corrosion sensors be positioned to reflect the state of the actual component or system being monitored. If this requirement is not met, all subsequent signal processing or data analysis is negatively impacted and the value of information greatly diminished or even rendered worthless. For example, if turbulence is induced locally around a protruding corrosion sensor mounted in a pipeline, the sensor will in all likelihood give a very poor indication of the risk of localized corrosion damage to the pipeline wall. In this particular case, a flush mounted sensor should be used instead if the goal is to monitor localized corrosion (Fig. 4.3).

In practice, the choice of monitoring points is also dictated by the existence of suitable access points, especially in pressurized systems. It is usually preferable to use existing access points, such as flanges for sensor installations. If it is difficult to install a suitable sensor in a given location, additional by-pass lines with customized sensors and access fittings may represent a practical alternative. One advantage of a bypass is the opportunity of experimenting with local conditions of highly corrosive regimes in a controlled manner, without affecting the actual operating plant.

4.3.2.1. Process Industry

The following example illustrates how critical sensor locations have been identified for a distillation column (6). The feed point, overhead product receiver, and bottom product line represent locations of temperature extremes and also points where products with different degrees of volatility concentrate. In many cases however, the highest corrosivity is encountered at an intermediate height in the column where the most corrosive species concentrate. Initially, therefore, several monitoring points would be required in such a column (Fig. 4.4). As

4.3. Corrosion Monitoring System Considerations

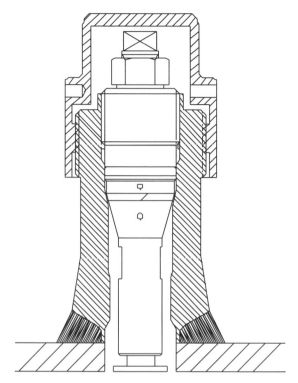

Fig. 4.3 Flush mounted corrosion sensor in an access fitting. (Courtesy of Metal Samples Company, www.metalsamples.com.)

monitoring progresses and data from these points become available, the number of monitoring points could be optimized further.

4.3.2.2. Oil and Gas Production Gathering Lines

Corrosion processes within oil and gas production gathering lines and process piping are usually monitored by the use of metal loss coupons or electrical resistance probes inserted into process fluids through access fittings (7). Once exposed to the process fluids for periods ranging from 90 days to 1 year, depending on the system and the corrosivity of the fluids, the coupons are removed and cleaned. A comparison of their initial and final weights is used to determine the general corrosion rate, based on the assumption of uniform corrosion throughout the exposure period. Corrosion pitting rates can also be determined on the same coupons by measuring the depth of the deepest pit.

In the same environment, electrical resistance (ER) and LPR probes can yield near real-time measurement of the corrosion rate within the process systems. These techniques provide corrosion rate measurements, such that short-term events that affect the rates can be easily identified. For example, the flowback of an acid stimulation treatment can be detected, provided the data readings are sufficiently frequent. Fortunately, remote data collectors and computers greatly reduce the time needed to record and analyze corrosion probe data.

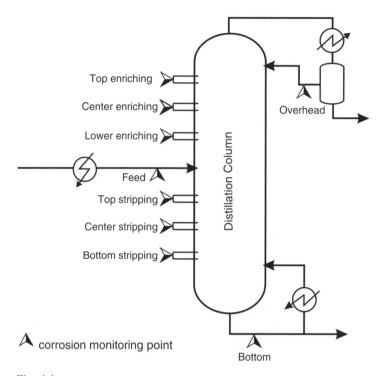

Fig. 4.4. Corrosion monitoring points in a distillation column.

Coupon monitoring and real-time corrosion monitoring techniques complement each other as they focus on different time intervals and should be integral parts of any comprehensive corrosion monitoring system.

The most important consideration when selecting corrosion monitoring locations within crude oil or wet gas production systems is to find locations near the end of the pipeline where the corrosion coupon or probe will be immersed in any produced water. This placement is typically at the 6 o'clock position on horizontal sections of pipeline because produced water is heavier than crude oil or gas condensates. In Fig. 4.5a, the monitoring probe installed at 6 o'clock is in an ideal position to sense corrosion processes.

Figure 4.5b illustrates a gas production line that contains a small quantity of condensate and produced water. The water is swept along the bottom of the pipe for horizontal runs. (Only at high fluid velocity is water swept along the circumference of the pipe.) Hence, monitoring locations on the side of the pipeline (3 or 9 o'clock) as shown in Fig. 4.5b cannot accurately measure the corrosion rates associated with the aqueous phase at the bottom of the pipeline.

Unfortunately, many pipeline and facility designers have installed monitoring locations on the sides of the pipelines rather than the bottom. Although the side locations may provide easier access for coupon crews, these coupons or probes cannot provide accurate data unless the pipelines are essentially full of water.

4.3. Corrosion Monitoring System Considerations

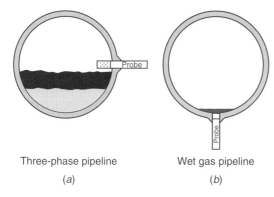

Fig. 4.5. (*a*) Probe at 9 o'clock. The intrusive probe is incorrectly positioned and cannot monitor corrosion associated with the water phase. This probe would yield invalid results because it is in the gas or oil phase. (*b*) Probe at 6 o'clock. The flush-mounted probe is best positioned to monitor corrosion, even when there are small volumes of produced water, as in wet gas or some three-phase pipelines.

Hence, operator convenience must not come at the expense of obtaining valid results when selecting coupon or probe monitoring locations.

It is best to design and install the coupon or probe access fittings during the initial fabrication and installation of pipeline systems rather than retrofitting. Although "hot taps" can be conducted safely, they require additional cost, preparation, and integrity checks.

Another consideration is the quantity of water within the pipeline. This determines whether intrusive or flush mounted coupons or probes are the better choice. For example, if an intrusive probe is installed within a wet gas pipeline, the sensing element extends well above the bottom pipe wall and cannot give valid readings. If, however, the pipeline contains three-phase production with a large water cut, inserting an intrusive probe to position the sensing element within the produced water might work. However, a flush mounted probe is generally the better choice because it is less dependent on the quantity of produced water. For pipelines carrying water, linear polarization probes may be used. If hydrocarbons are present, however, they will coat the sensing element and increase the noise/signal ratio. Thus, electrical resistance probes are recommended whenever hydrocarbons are present.

When natural gas is produced, small quantities of water and condensates may typically be produced. Quantities vary, but volumes are usually 160 L of water and up to 16,000 L of hydrocarbon condensate per 30,000 standard cubic meters. The produced water, being heavier than the gas or condensates, is generally swept along the bottom of the pipelines by the flow of gas. The water may accumulate in dead legs or low spots along the pipelines until the volumes are such that the gas will be able to push the fluids further down the pipelines.

Intrusive probes should be located where they can remain in place for extended periods, rather than having to be removed periodically to support pigging and other routine operations. Thus, an intrusive probe should be installed upstream of any pig launcher and downstream of any pig receiver so as not to block the path of the pig. Where this is not possible, flush-mounted probes should be used.

4.3.3. Probe Design and Selection

The probe design is another important consideration in corrosion monitoring since the probe element interfaces directly with the process environment and must be both suitable for the installation location and enable to make representative corrosion measurements. All too often, the quality and relevance of the corrosion data measured can be severely compromised by inappropriate probe design. In this context, knowledge of the probe surface condition is particularly crucial during the initial design and obviously remains important for the duration of the exposure period.

Other factors to consider relate to surface roughness, residual stresses, corrosion products, surface deposits, preexisting corrosion damage, and temperature that can all have an important influence on corrosion damage and need to be taken into account for making representative probes. By considering these factors, it can be desirable to manufacture corrosion sensors from a precorroded material that has experienced actual operational conditions. Heating and cooling may also be applied to corrosion sensors, using special devices, for their surface conditions to reflect certain plant operating domains. Sensor designs, such as spool pieces in pipes and heat exchanger tubes, flanged sections of candidate materials or test paddles bolted to agitators, also represent efforts to make these sensors representative of actual operational conditions.

The choice of a specific monitoring probe should also be based on the anticipated corrosion rates within the system, as well as on the required sensitivity. When conducting a short-term corrosion test, probes with high sensitivity are desired. For long-term monitoring, however, a thicker probe element with a longer measurement lifetime may be desired.

For rapid flow conditions or if there are concerns related to suspended solids, the sensing elements should be protected with a velocity shield. An ER probe can also be used to measure an "erosion rate" associated with production of sand or other solids. For this purpose, a non-corroding metal element should be selected (7).

4.3.3.1. Flush Mounted Electrode Design

The flush-mounted electrode design is most appropriate for use in applications, such as oil and gas flowlines where pigging operations are necessary. While the design is suited to this application operational need, it greatly limits the electrode exposed surface area and the accuracy of the measurements, particularly in low conductivity environments or with low sensitivity instrumentation. As with most measurement processes there is a trade-off between available area for measurement and the opportunity to actually measure corrosion events of low statistical probability (5).

Great care must also be taken in the manufacture of this type of probe as the opportunity exists to artificially create unwanted physical phenomena, such as crevices. A crevice created between the outer circumference of an electrode and the surrounding insulating material could provide a focal point for localized corrosion activity and introduce a significant error in the measured data. This effect would be reduced by using larger surface area electrodes.

While this type of probe can be supplied with electrodes of sufficient material to last the lifetime of the component into which it is installed, the monitoring program undoubtedly will benefit from the ability to replace the probe on a regular basis for visual inspection and to confirm the measured data.

4.3.3.2. Protruding Electrode Design

The protruding electrode design has more broad-ranging applications than the flush mounted design. A major benefit of this design is the possibility to use replaceable electrodes, as this provides a cost-effective solution. The possibility of crevice corrosion is also less important as the exposed length of an electrode increases the ratio of exposed surface area to the region where a crevice might occur, that is, the circumference of the electrode. However, this design relies on the complete exposure of the electrode surface to the corrosive environment, and so issues may arise in situations where the flow regime within the monitored system becomes turbulent or if the water cut is reduced significantly during operation (5).

4.3.3.3. Probes to Suit the Application

Each corrosion monitoring application has its specific needs and requirements. The following sections describe a few probe designs that have been developed for specific degradation mechanisms and environments.

4.3.3.3.1. Stress Corrosion Cracking Probe Corrosion probes have been developed that enable the working electrode of a three-electrode arrangement to be prestressed to yield stress in order to match the operating condition of a pipe or vessel. In the following example, corrosion monitoring and control of the double-shell tanks (DSTs) at Hanford had historically been provided through a waste chemistry sampling and analysis program. In this program, waste tank corrosion was inferred by comparing waste chemistry samples taken periodically from the DSTs with the results from a series of laboratory tests done on tank steels immersed in a wide range of normal and off-normal waste chemistries (8).

This method has been effective, but is expensive, time consuming, and does not yield real-time data. The Hanford Site near Richland, Washington has 177 underground waste tanks that store ~253 million liters of radioactive waste from 50 years of plutonium production. In 1996, the Department of Energy Tanks Focus Area launched an effort to improve Hanford's DST corrosion monitoring strategy and to help address questions concerning the remaining useful life of these tanks. Several new methods of on-line localized corrosion monitoring were evaluated. The electrochemical noise (EN) technique was selected for further study based on numerous reports that showed this technique to be the most appropriate for monitoring and identifying the onset of localized corrosion.

204   Chapter 4   Corrosion Monitoring

Based on a series of studies, a three-channel prototype field probe was designed, constructed and deployed in August 1996. Following the demonstration of the prototype for approximately a year, a longer more advanced eight-channel system was designed and installed in September 1997. Figure 4.6 shows the installation of this system. Unlike the previous prototype, the in-tank probe on this system reached from tank top to tank bottom exposing two channels of EN electrodes in the sludge at tank bottom, four channels in the tank supernate, and two channels in the tank vapor space. Four additional systems of similar design have been installed into other DSTs.

Like most EN based corrosion-monitoring systems, the active Hanford systems monitor EN on channels composed of three nominally identical electrodes immersed in the tank waste. Each system is composed of an in-tank probe and ex-tank data collection hardware. The in-tank probe is fabricated from a $\sim$17-m long piece of 2.5-cm diameter stainless steel tubing. Eight three-electrode channels are distributed along the probe body. Electrodes are fabricated from UNS K02400 steel that has been heat treated to match the tank wall heat treatment. Four channels on each probe are formed from sets of bullet-shaped electrodes (25 cm^2 electrode^{-1}). Four channels are formed from sets of thick-walled C-rings (44 cm^2 electrode^{-1}). Figure 4.7 shows two channels on the most recent probe. The unstressed bullet-shaped electrodes are used for pitting and uniform corrosion detection. The working electrode on each C-ring channel is notched, pre-cracked and stressed to yield prior to installation to facilitate the monitoring of SCC should tank chemistry conditions change to allow the onset of cracking. The other two C-rings on each C-ring channel are not stressed to match the operating conditions of the vessel. Bullet and C-ring channels alternate up the length of the probe. Current DST waste levels in monitored tanks immerses three channels of bullet-shaped electrodes and three channels of C-ring electrodes.

Fig. 4.6   Installation of first full-scale probe into a double-shell tank (DST) at the Hanford site. (Courtesy of HiLine Engineering & Fabrication.)

4.3. Corrosion Monitoring System Considerations 205

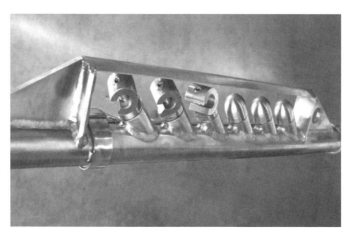

Fig. 4.7 Detail of the Hanford site 25 -cm^2 bullet and 47 -cm^2 C-ring channel electrodes. (Courtesy of Glenn Hedgemon, HiLine Engineering & Fabrication.)

In this way, the working electrode is allowed to behave in a manner most representative of the material in-service, thus providing corrosion information reflecting the real-life situation of the plant equipment. The exposure of single or multiple corrosion probes can enable informed decisions to be made regarding the choice of a material or of a stress relief process.

4.3.3.3.2. Corrosion in Hydrocarbon Environments In hydrocarbon environments there must be an electrolytically conductive phase present, which is generally provided by an aqueous phase or by polar solvents, for corrosion to occur. Examples may be flowlines in oil and gas applications or pipelines in chemical process environments where it may not be straightforward to introduce a complex probe system.

A circumferential spool probe has been developed specially for this application to allow a maximum contact of the electrodes with the process environment in "true" flowing conditions along flowlines or pipelines, especially in multiphase oil and gas applications.

The basic ring principle of these sensors allows elements to be made by "salami" slicing a pipe and reassembling pairs of the resulting rings, separated by insulation, to remake a pipe section capable of retaining line pressure. Each electrically isolated ring is measured using pick up wires attached to the outer face. If wires are attached at equally spaced intervals around the ring the instrumentation can be configured to measure the overall metal loss and the loss in the segment between each pick-up point Fig. 4.8.

One of each pair of rings is kept from contact with the process stream by means of a high integrity, thin-film, ceramic coating. The coated ring acts as the reference to the exposed sample ring.

A number of pairs of rings may be used enabling the study of different materials including weldment and HAZ material if preferential weld corrosion

206 Chapter 4 Corrosion Monitoring

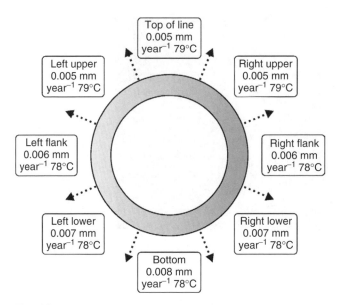

Fig. 4.8 The principle of ring pair corrosion monitoring. (Courtesy of Cormon Ltd.)

is an issue. In addition to the temperature data from the elements, it is a simple matter to include pressure measurement in the device. Together these may add considerably to the understanding of the behavior of the fluid in the line.

Standard rings have the same wall thickness as the original line so no problem should arise with element life in relation to the service life of the line. The thicker wall rings do have a lower speed of response that may not always meet the requirement, especially if real-time adjustment to chemical treatment is proposed. In this case, a combination of two concentric rings may be used to provide a fast response from a thin element inside a thicker supporting ring.

The potentially limited life of the element is balanced by the ability to maintain low corrosion rates through active control, thereby extending the life of the asset and the sensor. The spool sensor is, therefore, a very versatile measurement tool for looking at the character and degree of corrosion of various materials in conditions that are a true representation of line flow. It is capable of being finely tuned to perform the required task with great precision.

The sensor housing uses a double pressure barrier principle. The sensor rings, spacers, and isolators are held in compression by a clamp arrangement producing an inner-pressure tight cylinder. This cylinder is mounted in the outer housing using a pair of elastomer seal rings to complete the primary containment. The outer housing is a pressure tight assembly in its own right, sealed using flanges, ring type joints, and a spacer ring. Electronics, housings, power consumption, and telemetry are similar in most respects to those for intrusive probes. Figure 4.9 illustrates a monitoring system being deployed with a pipeline as it is submerged in sea.

Fig. 4.9 A ring pair corrosion monitoring system being deployed at sea during the installation of a submerged pipeline. (Courtesy of Cormon Ltd.)

4.3.3.3.3. Coupled Multielectrode Array Systems and Sensors The use of multielectrode array systems (CMAS) for corrosion monitoring is relatively new. The advantages of using multiple electrodes include the ability to obtain greater statistical sampling of current fluctuations, a greater ratio of cathode-to-anode areas that enhances the growth of localized corrosion once initiated, and, depending on the design, the ability to estimate the pit penetration rate and to obtain macroscopic spatial distribution of localized corrosion (9).

Figure 4.10 shows the principle of the CMAS in which a resistor is positioned between each electrode and the common coupling point. Electrons from a corroding or a relatively more corroding electrode flow through the resistor connected to the electrode and produce a small potential drop usually of the order of a few microvolts. This potential drop is measured by the high-resolution voltage-measuring instrument and used to derive the current of each electrode. The CMAS probes can be made in several configurations and sizes, depending on the applications. Figure 4.11 shows some of the typical probes that were reported for real-time corrosion monitoring.

The data from these CMAS probes are the large number of current values measured at a given time interval from all the electrodes. In a CMAS probe system, these data are reduced to a single parameter so that the probe can be conveniently used for real-time and online monitoring purposes. The most anodic current has been used as a one-parameter signal for the CMAS probes. Because the anodic electrodes in a CMAS probe simulate the anodic sites on a metal surface, the most anodic current may be considered as the corrosion current from the most corroding site on the metal.

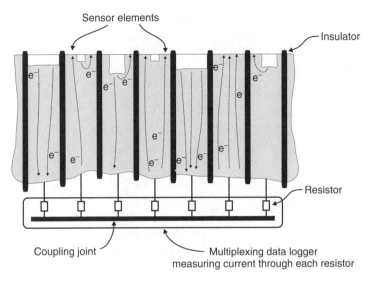

Fig. 4.10 Multielectrode array multiplexed current and/or potential measurement. (Adapted from Ref. 52.)

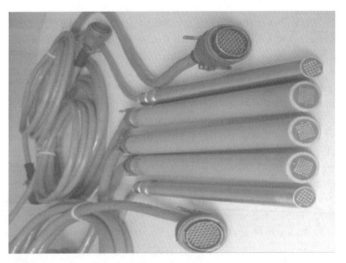

Fig. 4.11 Typical CMAS probes used for real-time corrosion monitoring. (Courtesy of Corr Instruments, LLC.)

The value based on three times of the standard deviation of currents is another way to represent the corrosion current from the most corroding site on the metal. Because the number of electrodes in a CMAS probe is always limited and usually far fewer than the number of corroding sites on the surface of a metal coupon, the value based on the statistical parameter, such as three times the standard deviation of current, was considered to be more appropriate than the single value of the most anodic current. The standard deviation value may be from the anodic currents or from both the anodic and cathodic currents.

In a less corrosive environment or with a more corrosion-resistant alloy, the most anodic electrode may not be fully covered by anodic sites until the electrode is fully corroded. Therefore, the most anodic electrode may still have cathodic sites available, and the electrons from the anodic sites may flow internally to the cathodic sites within the same electrode. The total anodic corrosion current, I_{corr}, and the measured anodic current, I_a^{ex} may be related by Eq. (4.1).

$$I_a^{ex} = \varepsilon I_{corr} \tag{4.1}$$

where ε is a current distribution factor that represents the fraction of electrons resulting from corrosion that flows through the external circuit. The value of ε may vary between 0 and 1, depending on parameters such as surface heterogeneities on the metal, the environment, the electrode size, and the number of sensing electrodes. If an electrode is severely corroded and significantly more anodic than the other electrodes in the probe, the ε value for this corroding electrode would be close to 1, and the measured external current would be equal to the localized corrosion current.

Because the electrode surface area is usually between 1 and 0.03 mm^2, which is ~2 to 4 orders of magnitude less than that of a typical LPR probe or a typical electrochemical noise (EN) probe, the prediction of penetration rate or localized corrosion rate by assuming uniform corrosion on the small electrode is realistic in most applications. CMAS probes have been used for monitoring localized corrosion of a variety of metals and alloys in the following environments and conditions:

- Deposits of sulfate-reducing bacteria.
- Deposits of salt in air.
- High-pressure simulated natural gas systems.
- H$_2$S systems.
- Oil–water mixtures.
- Cathodically protected systems.
- Cooling water.
- Simulated crevices in seawater.
- Salt-saturated aqueous solutions.
- Concentrated chloride solutions.
- Concrete.
- Soil.
- Low-conductivity drinking water.
- Process streams of chemical plants at elevated temperatures.
- Coatings.

4.3.3.3.4. Atmospheric Corrosion Monitoring For atmospheric corrosion monitoring, the required electrolyte for a probe to generate an electrochemical signal may either be a fine mist or a discontinuous film made by condensing humid air. The high resistivity of such an electrolyte means that the probe electrodes must be close while still electrically insulated to function properly. In addition, a further complication is the extremely low electrochemical activity normally produced by atmospheric corrosion, which means that the following requirements must be met to produce useful results:

- High sensitivity of the measuring instrumentation.
- Minimal IR drop throughout the monitoring system.
- Relatively large electrodes to maximize the opportunity to record corrosion signals.

Figure 4.12 shows schematically a probe configuration to achieve such measurements in a low conductivity condensing environment for monitoring aircraft corrosion. One such corrosion surveillance system was installed on an unpressurized transport aircraft. Electrochemical probes in the form of closely spaced probe elements were manufactured from an uncoated aluminum alloy. All but one of the probes were located inside the aircraft, in the areas that were most prone to corrosion attack and difficult to access. Another probe was located outside the aircraft, in its wheel bay (10). In flights from inland to marine atmospheres, a distinct increase in corrosivity was recorded by potential noise surveillance signals during the landing phase in the marine environment (Fig. 4.13). However, the strongest localized corrosion signals were recorded at ground level in a humid environment (Fig. 4.14).

Another example of an atmospheric corrosion probe is shown in Fig. 4.15. This sensor was fabricated using microcircuit technology in which a thin polyamide film was electroplated with two different metals (gold and cadmium) in a pattern to maximize the galvanic current produced in the presence of an even moderately corrosive environment. With this sensor the galvanic current produced by the thin-film bimetallic elements is integrated with a coulometer as

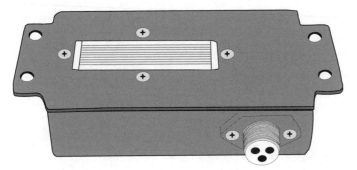

Fig. 4.12. Electrochemical probe manufactured from an uncoated aluminum alloy in the form of closely spaced elements.

4.3. Corrosion Monitoring System Considerations

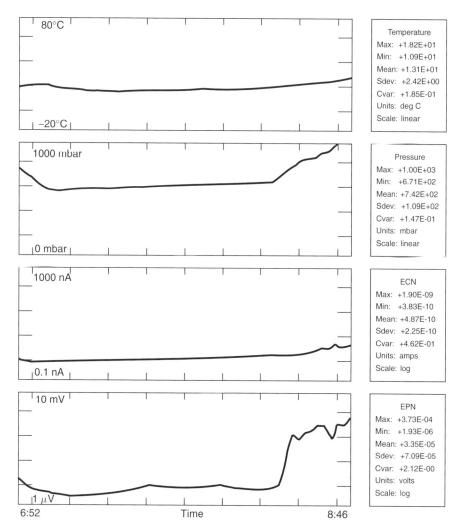

Fig. 4.13. Temperature, pressure, and electrochemical signals as a function of time during a flight to a marine environment in South Africa.

a function of time. The data are then stored in a memory chip for future download when queried through a radio frequency (rf) data-gathering transceiver (DGT) or transponder to a laptop computer.

4.3.4. Location of Monitoring Hardware

Many industrial plants have intrinsic safety requirements that impose important restrictions on corrosion monitoring systems. To ensure flexibility on large plants, some organizations have adopted the strategy of using a "mobile" corrosion

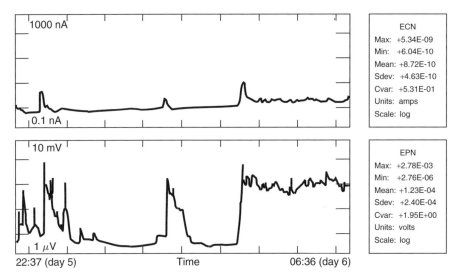

Fig. 4.14. Electrochemical signals as a function of time in a marine environment in South Africa.

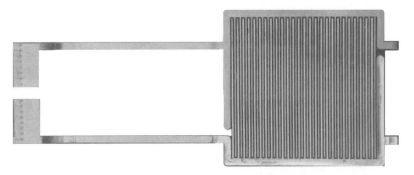

Fig. 4.15 Thin-film sensor for atmospheric corrosion monitoring. (Courtesy of Kingston Technical Software.)

monitoring laboratory that meets their safety regulations. Such a laboratory housing the corrosion monitoring instrumentation can be conveniently moved to different locations as required in order to overcome the problems associated with excessive lengths of sensor leads. Such an arrangement additionally provides a protective environment for measuring equipment and data storage hardware, which could otherwise be damaged in corrosive atmospheres.

Mobile laboratories may also be used for corrosion measurements on treated water circuits. A schematic of the water handling facility to carry out microbiologically influenced corrosion (MIC) field tests is shown as an example in Fig. 4.16 (11). This facility was used for the growth and monitoring of naturally

4.3. Corrosion Monitoring System Considerations 213

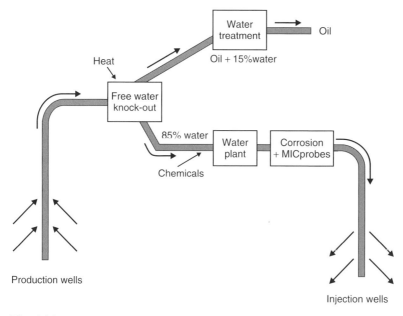

Fig. 4.16. Schematic of the water handling facility at the MIC field testing facility.

Fig. 4.17 Oil recovery microbial test lines to evaluate biocide programs. (Courtesy Kingston Technical Software.)

forming biofilms in five separate slip-streams. Each line accommodated 40 sample coupons and was equipped with an individual injector system that allowed testing of the effect of biocides on active biofilms in a fully equipped trailer (Fig. 4.17).

4.3.5. Sensitivity and Response Time

The usefulness of a corrosion monitoring system strongly depends on how well it can deliver warnings of unwanted corrosion conditions. For measuring techniques this translates into two closely related properties:

1. The sensitivity to detect a change in corrosion rate.
2. The time it requires to detect such a change, (i.e., the response time).

By virtue of the measuring principle of many systems, sensitivity and response time have an inverse relation to each other. In order to compare corrosion monitoring systems on the basis of their sensitivity, it is important to distinguish between the accuracy of the measurement and the sensitivity to measure a change in the corrosion rate. The sensitivity to measure corrosion rate is the combined result of measurement accuracy and the elapsed time (4).

Sensitivity (S) and response time (R) of a certain technique are closely related, and can be conveniently displayed in a single graph, as shown in Fig. 4.18. It is most convenient to display such S–R curves on a log–log graph. The example in this figure requires a 1–7 days response time while requiring sensitivity in the order of 1–20 mm/year^{-1} corrosion rate measurement. The S–R-curve of a specific technique lies below the window, hence will satisfy each requirement for this application. In Table 4.1, typical applications are given with their characteristics. These applications are depicted in Fig. 4.19 for their respective S–R windows.

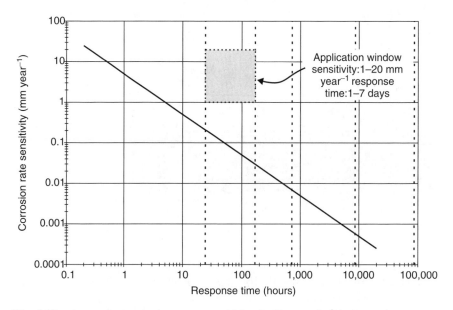

Fig. 4.18. A corrosion monitoring system sensitivity (1–20 mm year^{-1})/response time (1–7 days) application window for a given system performance threshold (solid line).

4.3. Corrosion Monitoring System Considerations

Table 4.1. Sensitivity and Response Time of Typical Corrosion Monitoring Applications

Application	Sensitivity range (mm year^{-1})	Response time	System characteristics
Corrosion tests	0.1–100	1 h–5 days	Continuous
Inhibition control	0.1–20	0.5–2 days	Continuous optimization
Corrosion control (upsets)	1–100	1 h–2 days	Continuous monitoring (upsets)
Corrosion control performance demonstration	1–10	1 week–1 month	Continuous/interval measurement
Inspection planning	0.2–10	1 month–0.5 year	Interval
Inspection	1–20	3 months–10 year	Interval

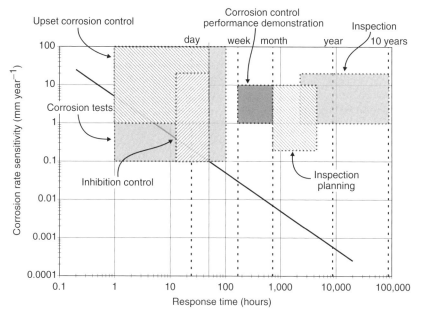

Fig. 4.19. Application windows depicted in the S–R plot. The size range of the application windows is given in Table 4.1.

A monitoring system can also be limited to measure accurately over long time scales because of inherent instability of the monitoring system. This might be the result of deterioration of the sensor, or drift in the recording instrumentation. For systems that have to measure with high sensitivity and over a long interval it is recommended to perform routine verifications and calibrations of the monitoring equipment.

4.3.6. Data Communication and Analysis Requirements

It is important, at the onset of a corrosion monitoring program, to define the full data communication chain from signaling unacceptable corrosion to the implementation of a remedial action. The times for each step in the chain should be in balance, that is, it is obviously not very useful to invest in a system with a response time of 1 day if it requires weeks to process the information and/or months to implement follow-up remedial measures. The following individuals may be involved in the communication process (4):

1. Process plant operator, to collect data.
2. Corrosion monitoring specialist (corrosion or inspection engineer), to process data.
3. Corrosion engineer, to assess the information and determine follow up.
4. Operations or maintenance engineer, to plan and implement remedial action.

The response time from "sensor-to-desk" for the steps 1–3 determine the actual response time obtained from a corrosion monitoring system. For a highly critical monitoring task, the data might go directly to the party responsible for remedial action, for example, to the control room, for action by an operator.

The perceived importance of the monitoring system and strategy has to be mirrored by commitment of all individuals involved in integrity management, that is, the asset holder, usually operations, but also maintenance and inspection staff, corrosion engineering, production chemists, and frequently the chemical treating contractor. It is essential that the approach is agreed and implemented by a team that includes these individuals, who together decide not only how corrosion should be controlled, but also how the corrosion monitoring should be implemented.

4.3.7. Define System Reliability

Note that if the monitoring system is an active operational component, a high demand is put on its reliability. For example, large economic losses could be incurred if the monitoring system used to control the inhibitor concentration of a system with a high uninhibited corrosion rate would fail. For such an application, a quantitative reliability study of the monitoring system may be warranted (4).

If a monitoring system has a passive role, for example, if it is used to verify that corrosion conditions have been within expected limits, a more qualitative analysis of possible failures might be adequate. It may well be that there are other sources of information to estimate corrosion progress besides monitoring, or it may be necessary to have more than one system in place to warn if extreme degradation is possible.

4.3.8. Availability and Cost

The final choice of the corrosion monitoring system will be influenced by availability of the tools and their cost. Some methods and in particular the installation and implementation of the overall corrosion monitoring system can be quite costly. In view of the implications of excessive corrosion and loss of integrity, the choice should be based on the total cost of ownership of the equipment (4).

4.3.9. In a Nutshell

An effective corrosion monitoring system or device should exhibit the following characteristics (12):

- **User Friendly:** The monitoring system must be simple to install, simple to use, and simple to interpret by system operators. At least some interpretation functions must be sufficiently developed so that the system can be interfaced to alarms and controllers for chemical treatment additions or on-line cleaning systems.
- **Rugged:** The monitoring system must be able to withstand the normal use and abuse if it is deployed in an industrial environment.
- **Sensitive:** The monitoring element or probe must be sensitive to the onset of a corrosion problem and provide a definitive indication in real time that may be used as a process control variable or to evaluate the effectiveness of a control measure.
- **Accurate:** False positives and negatives or any indications caused by interferences from effects, such as flow, erosion, and fouling, can be detrimental in many ways. Erroneous readings may seriously affect the credibility and straightforward usefulness of a corrosion monitoring program.
- **Maintainable:** Probes are expected to foul in service. A minimum time between servicing operations of several months to several years may be required for most applications. Periodic servicing and calibration should be simple and easy to perform.
- **Cost Effective:** The cost of the monitoring system must be significantly less than the cost of the downtime that is avoided or the treatment costs that are saved. The speed and accuracy of the technique are also factors in the cost effectiveness of the monitoring system.

4.4. CORROSION MONITORING TECHNIQUES

Assessment of corrosion in field conditions is complex due to the wide variety of applications, process conditions, and fluid phases that exist in industrial systems.

As discussed in Chapter 3, the expectations of a corrosion monitoring program will vary greatly between organizations that have well-established proactive corrosion management programs and other organizations where corrosion damage is simply a nuisance. Many of the possible corrosion monitoring and inspection techniques available have been recently organized by a group of experts and interested users in different categories as shown in Tables 4.2 and 4.3 (13).

In the report produced by this group, a direct technique is one that measures parameters directly affected by the corrosion processes while an indirect technique provides data on parameters that either affect, or are affected by the corrosivity of the environment or by the products of the corrosion processes. Additionally, a technique can be described as being intrusive if it requires access

Table 4.2. Direct Corrosion Measurement Techniques

Intrusive techniques

Physical techniques
  Mass-loss coupons
  Electrical resistance (ER)
  Visual inspection

Electrochemical dc techniques
  Linear polarization resistance (LPR)
  Zero-resistance ammeter (ZRA) between dissimilar alloy electrodes: galvanic
  Zero-resistance ammeter (ZRA) between the same alloy electrodes
  Potentiodynamic–galvanodynamic polarization
  Electrochemical noise (ECN)

Electrochemical ac techniques
  Electrochemical impedance spectroscopy (EIS)
  Harmonic distortion analysis.

Nonintrusive techniques

Physical techniques for metal loss
  Ultrasonics
  Magnetic flux leakage (MFL)
  Electromagnetic: eddy current
  Electromagnetic: remote field technique (RFT)
  Radiography
  Surface activation and gamma radiometry
  Electrical field mapping

Physical techniques for crack detection and propagation
  Acoustic emission
  Ultrasonics (flaw detection)
  Ultrasonics (flaw sizing)

Table 4.3. Indirect Corrosion Measurement Techniques

On-line techniques

Corrosion products
  Hydrogen monitoring

Electrochemical techniques
  Corrosion Potential (Ecorr)

Water chemistry parameters
  pH
  Conductivity
  Dissolved oxygen
  Oxidation reduction (redox) Potential

Fluid detection
  Flow regime
  Flow velocity

Process parameters
  Pressure
  Temperature
  Dewpoint

Deposition monitoring
  Fouling

External monitoring
  Thermography

Off-line techniques

Water chemistry parameters
  Alkalinity
  Metal ion analysis (iron, copper, nickel, zinc, manganese)
  Concentration of dissolved solids
  Gas analysis (hydrogen, H_2S, other dissolved gases)
  Residual oxidant (halogen, halides, and redox potential)
  Microbiological analysis (sulfide ion analysis)

Residual inhibitor
  Filming corrosion inhibitors
  Reactant corrosion inhibitors

Chemical analysis of process samples
  Total acid number
  Sulfur content
  Nitrogen content
  Salt content in crude oil

through a pipe or vessel wall in order to make the measurements. Most commonly intrusive techniques make use of some form of probe or test specimen, which include flush mounted probe designs. Some indirect techniques can serve to monitor various parameters on-line in real-time while others provide information off-line after samples collected from process streams or other operational locations are further analyzed following an established method.

The following sections will generally follow the organization used in that report with the notable exceptions of visual inspection and the direct nonintrusive techniques that are usually described as either nondestructive evaluation (NDE), nondestructive testing (NDT), or nondestructive inspection (NDI) techniques, the subject of Chapter 5.

4.4.1. Direct Intrusive Techniques

4.4.1.1. Physical Techniques

Physical corrosion monitoring techniques determine corrosion damage by measuring changes in the geometry of exposed coupons or test specimens. There are many properties of a test specimen that may change to some degree as a result of corrosion, such as its mass, electrical resistance, magnetic flux, reflectivity, stiffness, or or any other mechanical properties. When the physical property is measured by electronic means, the test specimen can remain *in situ* and frequent readings are possible. However, frequent readings are less easy to implement if the test specimen needs to be removed from the process to have its physical property measured.

4.4.1.1.1. Mass Loss Coupons Mass loss coupons are usually designed to monitor the damage rate occurring on existing equipment, to evaluate alternative materials of construction, and sometimes to determine the effects of process conditions that cannot be reproduced in the laboratory. This simple and low cost method of corrosion monitoring consists in exposing small specimens in the environment of interest for a specific period of time and subsequently removing them for weight loss and more detailed examination. Even though the principle is relatively simple, numerous potential pitfalls exist, which can be avoided by following the recommendations of a comprehensive standard guide, such as ASTM G-4 (14).

Various means of introducing coupons of materials of construction into operating equipment have been devised. For equipment operating under considerable pressure, special attachments are available from corrosion equipment suppliers. A single high-pressure access fitting for insertion of a retrievable corrosion probe (Fig. 4.20) can be used to accommodate most types of retrievable probes (Fig. 4.21). As mentioned earlier, specialized tools exist for sensor insertion and withdrawal under pressurized operating conditions (Fig. 4.22).

Mass loss coupon testing provides several specific advantages over laboratory coupon testing. A large number of materials can be exposed simultaneously and

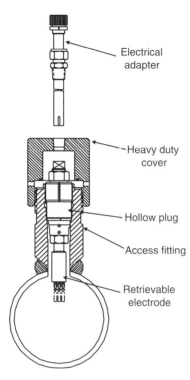

Fig. 4.20 High-pressure access fitting for insertion of a retrievable corrosion probe. (Courtesy of Metal Samples Company, www.metalsamples.com.)

ranked in actual process streams with actual process conditions. A slip-in rack, for example, can be inserted into and removed from equipment with a retractable coupon holder full (Fig. 4.23). A rod-shaped coupon holder is contained in the retraction chamber, which is flanged to a full-port gate valve. The other end of the retraction chamber contains a packing gland through which the coupon holder can pass. Coupons mounted on the rod in the extended position are drawn into the retraction chamber. The chamber is bolted to the gate valve, which is then opened to allow the coupons to be slid into the process stream. The sequence is reversed to remove the test coupons. It is essential that the rod or handle be equipped with a restraining chain or other device to prevent the blowout of the specimen holder when retracting the specimens in a system under pressure.

Because many coupons can be exposed simultaneously, they can be tested in duplicate or triplicate (to measure scatter), and be fabricated to simulate such conditions as welding, residual stresses, or crevices. These variations can provide the engineer with increased confidence in selecting materials for new equipment, maintenance, or repair (15).

Coupons can be designed to detect such phenomena as crevice corrosion, pitting, and dealloying corrosion. The environment of interest can be the full process flow at a location where the conditions are deemed to be suitably severe to give a meaningful representation. Alternatively, coupons can be exposed in a side stream that can be isolated from the main process stream. The design of

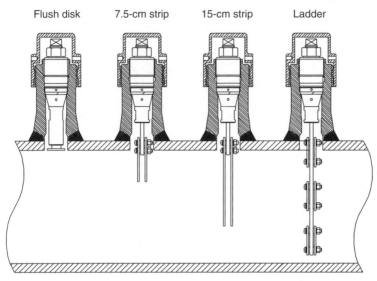

Fig. 4.21 A single high-pressure access fitting can be fitted with different types of retrievable corrosion probes. (Courtesy of Metal Samples Company, www.metalsamples.com.)

the coupon should match the objective of the test, that is, simple flat sheets for general corrosion or pitting, welded coupons for local corrosion in weldments, stressed or precracked test specimens for SCC problems (13).

4.4.1.1.1.1. Length of exposure and limitations In general, the length of exposure is typically as long as possible to allow for initiation of localized corrosion and adequate evaluation of service conditions. A minimum exposure of 3 months has normally been used for evaluation of pitting and crevice corrosion. For general corrosion, ASTM G-31 describes a minimum test duration in hours that is ~50 divided by the expected corrosion rate in millimeter per year (16).

For example, when the corrosion rate is estimated to be ~0.05 mm year^{-1}, the exposure period should be at least 1000 h. This can vary according to whether the coupons are exposed in well-controlled situations, such as laboratory tests or in the field. The use of coupons is, however, subject to the following limitations (13, 15):

- The technique only determines average rate of metal loss over the period of exposure.
- Coupon testing cannot be used to detect rapid changes in the corrosivity of a process.
- Localized corrosion cannot be guaranteed to initiate before the coupons are removed even with extended test durations.
- Reinsertion of a used coupon is generally not recommended.

4.4. Corrosion Monitoring Techniques 223

Fig. 4.22 Retrieval tool for removing corrosion probes under pressure. (Courtesy of Metal Samples Company, www.metalsamples.com.)

- Short exposure periods normally yield unrepresentative average rates of metal loss. This is often the result of higher metal loss during initial acclimation to the process environment.

- Corrosion rate calculated from coupons may not reflect the corrosion of the plant equipment due to factors, such as multiphase flow, where the aqueous phase is much more corrosive than the organic or vapor phase, or turbulence from mixers, elbows, valves, pumps, and other items that accelerate the corrosion in a specific location in the equipment removed from where coupons were exposed.

- Procedures for mass-loss coupon analysis can be quite labor intensive.

- Another limitation of coupon testing is the simulation of erosion–corrosion and heat-transfer effects. Careful placement of the coupons in the process equipment can slightly offset these weaknesses.

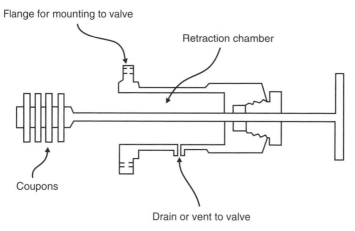

Fig. 4.23. Illustration of retractable coupon holder (15).

- Coupons may be misleading in situations where the corrosion rate varies significantly over time because of unrealized process factors.

Contamination of the process fluid can also be an issue for corrosion monitoring for particular industries (e.g., food processing, medical, and electronic equipment manufacturing industries).

4.4.1.1.1.2. Uniform corrosion coupons The most common coupon shape for the evaluation of uniform corrosion is rectangular because most alloys are available in plate or sheet form. Other shapes are used when there are restrictions on available product forms or when a specific material condition is required. Coupon identification must be legible and permanent. The simplest, and preferred, method of identification is stencil stamping (15).

Coupon finish represents a significant contribution to the overall cost. The least expensive finish that is consistent with the monitoring requirements should be selected. For example, an inexpensive surface finish is acceptable where carbon steel coupons are used routinely to monitor additions of inhibitor in water treatment programs. This may be achieved by punching or shearing, followed by glass bead blasting. On the other hand, when it is necessary to rank alloys in a process environment, the coupons must be finished with ground or machined parallel edges and sanded faces. Ideally, the surface finish of the coupon should match the finish of the equipment. However, this is hardly achievable for several reasons including the aging and scaling of real surfaces when exposed to process conditions.

Coupons that are cut by punching or shearing have cold-worked edges. The effects of cold work extend back from the cut edge a distance equal to the material thickness. These affected areas can be removed by grinding or machining. Cold-worked edges may affect the corrosion rate in some cases, and the residual stresses it induces may cause SCC in some materials. Extensive edge preparation can be a major contributor to the cost of a coupon.

4.4.1.1.1.3. Galvanic-corrosion coupons Pairs of test coupons can be coupled electrically to study the effects of galvanic corrosion. The relative areas exposed usually vary from 1:1 to 10:1 or greater. A major concern with electrically coupled coupons is maintaining electrical continuity for the entire exposure period. Corrosion product films can wedge mechanically joined coupons apart, thereby eliminating the electrical contact and any galvanic corrosion effect. ASTM G-71 provides valuable information on galvanic corrosion testing in both field and laboratory environments (15, 17).

With metals that can become embrittled by hydrogen absorption (e.g., titanium, zirconium, tantalum, and hardenable steels) the cathodic or protected member of the galvanic couple may be subject to the greater damage. However, the typical mass loss measurements would not reveal such damage.

4.4.1.1.1.4. Crevice-corrosion coupons Equipment crevices are quite common in complex systems. These crevice sites can easily trigger the onset of localized corrosion in even benign process environments. Many metals perform differently in crevices as opposed to unshielded areas. The various techniques that can be used for crevice corrosion testing include rubber bands, spot-welded lap joints, and wire wrapped around threaded bolts. Each crevice test creates a particular crevice geometry between specific materials and has a particular anode/cathode area ratio.

The two most widely used crevice geometries in field coupon testing employ insulating spacers to separate and electrically insulate the coupons. Spacers are usually either flat washers or multiple-crevice washers. Either type of spacer can be made of materials ranging from hard ceramics to soft thermoplastic resins. ASTM G-78 and G-48 provide valuable information on crevice corrosion specimen design and should be consulted before attempting a crevice corrosion test (18, 19). In this ASTM G-78 test, washers make a number of contact sites on either side of the specimens (Fig. 4.24 and 4.25*a* and *b*). The number of sites showing attack in a given time can be related to the resistance of a material to initiation of localized corrosion, and the average or maximum depth of attack can be related to the rate of propagation.

4.4.1.1.1.5. Stress-corrosion cracking coupons Typical sources of sustained tensile stress that cause SCC of equipment in service are the residual stresses resulting from forming and welding operations and the assembly stresses associated with interference fitted parts, especially in the case of tapered, threaded connections. Therefore, the most suitable coupons for plant tests are the self-stressed bending and residual-stress specimens. Convenient coupons are the cup impression, U-bend (20), C-ring (21), tuning fork, and welded panel (22). All these methods of stressing coupons produce a decreasing load as the cracks form and begin to propagate. Therefore, complete fracture is seldom observed, and careful examination is required to detect cracking.

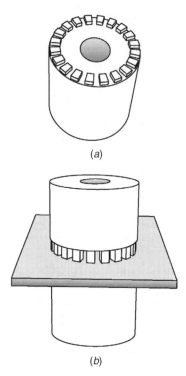

Fig. 4.24. A schematic representation of the washer (*a*) and a washer assembly (*b*) for conducting an ASTM G 78 crevice susceptibility test.

4.4.1.1.1.6. Heat-transfer coupons For heat-transfer effects, specially designed coupons are required that simulate effects, such as those found in heating elements or condenser tubes. Coupons range in design from thermowell-shaped devices to sample tubes in a test heat exchanger. Thermowell-shaped devices are heated or cooled on the inside and project into the process stream (14). Heat-transfer tests can also be conducted in the laboratory. In this environment, the coupon forms part of the wall of the test vessel, and can therefore be heated or cooled from one side. Because of the cost involved, heat-transfer coupon tests are usually carried out on only one (or perhaps two) alloys that have been selected from a larger group (15).

4.4.1.1.1.7. Welded coupons Because welding is a principal method of fabricating equipment, the use of welded coupons is often desirable. Aside from the effects of residual stresses, the primary concern is the behavior of the weld bead and the heat-affected zone (HAZ). In some alloys, the HAZ becomes sensitized to severe intergranular (sometimes called knife-line) attack, and in certain other alloys, the HAZ is anodic to the parent metal. When possible, it is more realistic to remove welded coupons from production-sized weldments than to weld the small coupons (15).

Because the thermal conditions in both the weld metal and the HAZ are a function of the number of weld passes, the metal thickness, the weld position, and

4.4. Corrosion Monitoring Techniques 227

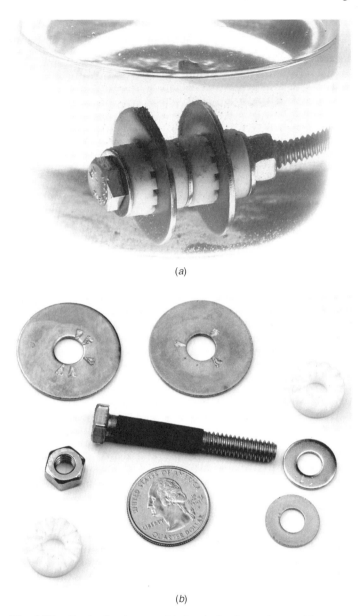

Fig. 4.25. Crevice forming assembly in a beaker immersion test (*a*); components of the assembly and S30400 washer coupons after thirty days in a solution containing 4% NaCl and 8% FeCl$_3$ (*b*).

the welding technique, it is not usually considered good practice to use welded coupons to assess the possibility of sensitization from welding. It is generally better to carry out a sensitizing heat treatment on an unwelded coupon before it is tested and then look for evidence of intergranular corrosion or cracking.

4.4.1.1.1.8. Sensitized metal
Sensitization is a metallurgical change that occurs when certain noble alloys (e.g., austenitic and ferritic stainless steels, nickel base alloys) are heated under specific conditions. This may result in the precipitation of carbides or other intermetallic phases at grain boundaries that may reduce the corrosion resistance. Any heat-inducing process may cause sensitization, a process that is time and temperature dependent. There is a specific temperature range over which each particular alloy sensitizes rapidly.

Welding is the most common cause of sensitization. However, welded coupons may not exhibit sensitization because they may have been given insufficient weld passes when compared to actual process equipment. As a result, they spend insufficient time in the sensitizing temperature range, and susceptibility to intergranular corrosion may not be detected.

An appropriate sensitizing heat treatment guarantees that any corrosion susceptibility induced by welding or heat treatment is detected. The optimal temperature and time ranges for sensitization vary for different alloys. For example, 30 min at 650°C is usually sufficient to sensitize S31600 stainless steel. Some of the corrosion-resistant aluminum magnesium alloys containing 3–6% Mg, (e.g., 5XXX series) are also subject to sensitization when heated at temperatures in the range of 65–175°C (23).

4.4.1.1.1.9. Coupon cleaning and evaluation
The test coupons should be cleaned as soon as possible after removal from a test. The procedures for cleaning and weighing, which depend on the test material and the extent of corrosion, are described in ASTM standard G-1 (24). Examination of corrosion coupons after cleaning and weighing should reveal the forms of corrosion that may be expected in equipment made of the coupon material. Coupons are examined with the unaided eye, and then at increasing magnifications, up to 30–50X, with a binocular microscope. A scanning electron microscope (SEM) has often proved to be an extremely useful tool for detecting superficial localized defects (15).

In some cases, coupons must be bent and/or sectioned and metallographically examined to reveal certain types of corrosion damage. There are special localized corrosion effects that may not only jeopardize the determination of realistic corrosion rates, but also signal other serious types of behavior. Once the coupons have been cleaned thoroughly through repetitive cleaning processes, the corrosion or penetration rate can be estimated from a mass loss plot (Fig 4.26) (24, 25). The rate is estimated by Eq. (4.2).

$$R = \frac{K(m_1 - m_2)}{A(t_1 - t_2)\rho} \tag{4.2}$$

where R is the penetration (corrosion) rate (mm/year^{-1}); A is the exposed area (cm^2); m_1 and m_2 are the initial and final masses (g), with m_2 being the intercept made by extrapolating line BC to the y axis in Fig. 4.26; t_1 and t_2 are the starting and ending times (h); ρ is the density (g cm^3); and K is a constant for unit conversion.

One question that arises when estimating the reproducibility of immersion test results is the amount of uncertainty that each measurement of the observables

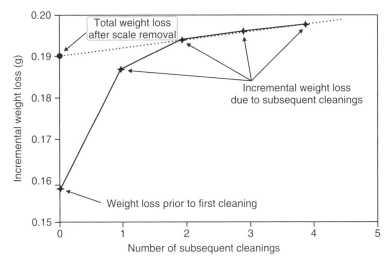

Fig. 4.26. Cleaning procedure of corrosion coupons for weight loss determination yields scale weight, total corrosion weight loss, and error due to cleaning procedure (24, 25).

(e.g., time, mass loss, and area) contributes to the total uncertainty in Eq. (4.2). This error defines the minimum uncertainty in the penetration rate that is possible during a given experiment. Such minimum uncertainty would be possible when (26):

- There is no localized corrosion.
- The penetration is uniform across the coupon surface.
- The projected and actual surface areas are the same.
- Weights are unaffected by corrosion product removal.
- Areas have not changed during the exposure.
- The penetration rate is independent of time.

Accurate weight determination is essential to minimize the uncertainty. The balance should have an accuracy of at least 0.1 mg and weighing each coupon at least three times to obtain an average would decrease the uncertainty somewhat.

Assuming that the environment remains unchanged, the longer the coupons are exposed to the environment the smaller will be the error. As discussed earlier, adhering to the ASTM Standard G-31 recommendation is a good rule-of-thumb to minimize the error. However, the ability to control the environment may limit the test duration so that the long test times needed for accurate measurement of low corrosion rates may not be achievable.

4.4.1.1.2. Electrical Resistance Acceptance of the electrical resistance (ER) corrosion monitoring method grew quickly after the correlation with corrosion rates was established in the 1950s (27, 28). The principle of the widely used ER technique is quite simple, that is, the electrical resistance of a sensing element increases as its cross sectional area is reduced by corrosion damage. The electrical resistance of a metal or alloy element is given by Eq. (4.3):

$$R = r\frac{L}{A} \tag{4.3}$$

where L is the probe element length (cm); A is the cross-sectional area (cm^2); and r is the specific resistance of the probe metal (Ω cm).

Reduction or metal loss in the element cross section A due to corrosion will be accompanied by a proportionate increase in the element electrical resistance (R). Since temperature influences the electrical resistance of the probe element, ER sensors usually measure the resistance of a corroding sensor element relative to that of an identical shielded element (Fig. 4.27). Commercial sensor elements are in the form of plates, tubes, plates or wires (Fig. 4.28).

Reducing the thickness of the sensor elements can increase the sensitivity of these sensors. However, improved sensitivity involves a trade-off against reduced sensor lifetime. The ER probe manufacturers provide guidelines showing this trade-off for different sensor geometries (Fig. 4.29). These probes usually have a useful life up to the point where their original thickness has been halved with the exception of wire sensors. For ER wire sensors the lifetime is lower, corresponding to a quarter original thickness loss.

It is obvious that erroneous ER corrosion results will be obtained if conductive corrosion products or surface deposits form on the sensing element. Iron sulfide formed in sour oil–gas systems or in microbial corrosion and carbonaceous deposits in atmospheric corrosion are relevant examples. The same restriction also applies to electrically conductive environments, (e.g., molten salts or liquid metals).

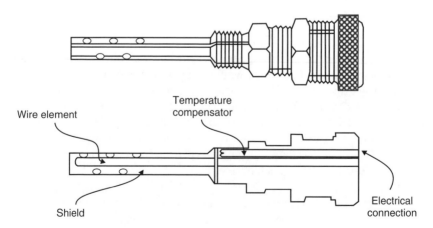

Fig. 4.27. Illustration of an electrical resistance probe (15).

4.4. Corrosion Monitoring Techniques 231

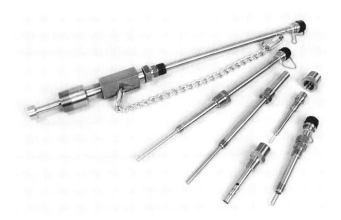

Fig. 4.28 Commercial sensor elements to carry out ER measurements. (Courtesy of Metal Samples Company, www.metalsamples.com.)

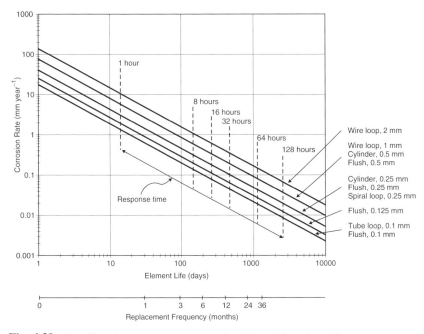

Fig. 4.29 The ER probe element selection guide. (Adapted from Metal Samples.)

To obtain the corrosion rate, a series of measurements are made over a period of time, and the results are plotted as a function of exposure time. The corrosion rate can be determined from the slope of the resulting plot (29).

There are several advantages to the ER corrosion monitoring method. Because probes are relatively small, they can be installed easily and the system wired directly to a control room location or to a portable resistance bridge at the probe location. For systems that are wired directly to control rooms, a computer system can be used to obtain the data and to transform the results in corrosion rate values. On the other hand, it is time consuming and sometimes impossible to take measurements at the probe site with a portable bridge. The temperature-compensation device reacts slowly, and it can be a source of error if the temperature varies when the measurement is taken (15).

Corrosion rate measurements obtained in short periods of time can also be inaccurate because the method measures only the remaining metal, which produces significant errors by estimating small differences between large numbers. Measurement resolution is typically 1 part in 1000 of the total measuring range of the probe, and the probe range is typically from 0.05 to 0.64 mm. Measured resistance changes are small so that thermal, stress, or electrical noise can affect the signal, necessitating hardware and software filtering.

The ER results provide a good measure of metal loss by general corrosion. However, the probes are less sensitive to effects of localized attack, which increase the element resistance on only a small area of the element, except near the end of probe life on loop element probes, where the localized attack completely corrodes through the element, increasing its resistance to infinity. Special probes have been prepared for sensitivity to crevice corrosion by creating multiple crevices on the measurement element, such as beads on wire loop probes.

4.4.1.1.3. Inductive Resistance Probes Following advances in electronics, signal processing, and measuring techniques, this new metal loss monitoring technology is a derivative of ER corrosion sensing. This corrosion monitoring technology has been developed to combine very high resolution measurement with long probe life and the capability of intrinsically safe operation in hydrocarbon process plant environments (30).

The thickness reduction of a sensing element is measured by changes in the inductive resistance of a coil embedded in the sensor (Fig. 4.30). Sensing elements with high magnetic permeability intensify the magnetic field around the coil; therefore thickness changes affect the inductive resistance of the coil. Sensitivity has been claimed to be several orders of magnitude higher than with comparable ER probes.

The measurements are virtually unaffected by other process variables, such as temperature, hydrostatic pressure, impact loading (slugging), or flow regimes. The system is also immune to extraneous industrial noise, specifically electromagnetic induction, and thermally induced electromotive force voltages. The inductive resistance sensor elements have very high geometric and physical symmetry providing sensor surfaces with identical metallurgy and microstructures.

4.4.1.2. Electrochemical Techniques

In view of the electrochemical nature of corrosion, it is not surprising that measurements of the electrical properties of the metal solution interface are so

4.4. Corrosion Monitoring Techniques 233

(a)

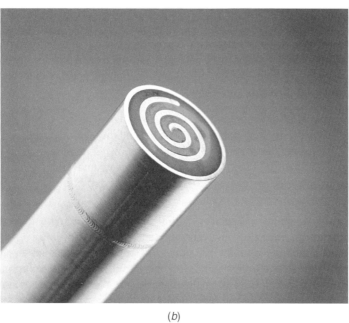

(b)

Fig. 4.30 Subsea inductive resistance probe (*a*); probe element (*b*); spool instrument package for subsea (*c*). (Courtesy of Cormon Ltd.)

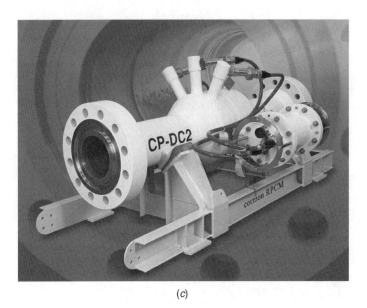

(c)

Fig. 4.30. (*continued*).

extensively used across a wide spectrum of corrosion science and engineering activities, from fundamental studies to monitoring and control in service. Electrochemical monitoring methods involve the determination of specific interface properties that can be divided into three broad categories:

- **Corrosion potential measurements:** The potential at a corroding surface arises from the mutual polarization of the anodic and cathodic half-reactions constituting the overall corrosion reaction. Corrosion potential is intrinsically the most readily observable parameter and understanding its relation to the thermodynamics of a system can provide very useful information on the state of a system;

- **Reaction rate as current density:** Partial anodic and cathodic current densities cannot be measured directly unless they are purposefully separated into a bimetallic couple. By polarizing a metal immersed in an aqueous environment, it is possible, with the use of simple assumptions and models of the underlying electrochemical behavior, to estimate net currents for both the anodic and cathodic polarizations from which a corrosion current density can be deduced;

- **Surface impedance:** A corroding interface can also be modeled for all its impedance characteristics, therefore revealing subtle mechanisms not visible by other means. Electrochemical impedance spectroscopy (EIS) is now well established as a powerful technique for investigating corrosion processes and other electrochemical systems.

4.4. Corrosion Monitoring Techniques

Corrosion potential or current produced by naturally occurring or externally imposed conditions can be measured with a variety of electrochemical techniques. Conversions of the measurements into corrosion rates or other meaningful data use equations or algorithms that are specific to each technique. The conversion of the measured current density into a corrosion rate, for example, can be made by using Faraday's law (Tables 4.4 and 4.5) and other empirically determined constants and factors specific to the system. Some techniques use analysis of the interface with direct current (dc) methods. Other techniques use alternating current (ac) methods to provide further characterization of the corrosion interface and the conductivity of the process fluids.

Limits of operation for field work are more serious than those experienced in a laboratory environment, mostly for reasons of practical probe geometry. For example, capillary salt bridges (e.g., Luggin capillary) commonly used in laboratory setups to reduce the interference of solution resistance are definitively too delicate or cumbersome for field use (13).

Table 4.4. Conversion between Current, Mass Loss, and Penetration Rates for All Metals[a,b]

	mA cm^{-2}	mm year^{-1}	mpy	g m^{-2} day^{-1}
mA cm^{-2}	1	3.28 M nday^{-1}	129 M nday^{-1}	8.95 M n^{-1}
mm year^{-1}	0.306 nday M^{-1}	1	39.4	2.74 day
mpy	0.00777 nday M^{-1}	0.0254	1	0.0694 day
g m^{-2} day^{-1}	0.112 n M^{-1}	0.365/day	14.4/day	1

[a] mpy = milliinch per year; n = number of electrons freed by the corrosion reaction; M = atomic mass; d = density.
[b] Note: The table should be read from left to right, that is, 1 mA cm^{-2} = (3.28 M nday^{-1})mm year^{-1} = (129 M nday^{-1}) mpy = (8.95 M n^{-1}) g m^{-2} day^{-1}.

Table 4.5. Conversion between Current, Mass Loss, and Penetration Rates for Steel[a]

	mA cm^{-2}	mm y^{-1}	mpy	g m^{-2} day^{-1}
mA cm^{-2}	1	11.6	456	249
mm year^{-1}	0.0863	1	39.4	21.6
mpyear	0.00219	0.0254	1	0.547
g m^{-2} day^{-1}	0.00401	0.0463	1.83	1

[a] Note: The table should be read from left to right, that is, 1 mA cm^{-2} = 11.6 mm year^{-1} = 456 mpy = 249 g m^{-2} day^{-1}.

The widespread use of electrochemical polarization techniques, dc or ac, does not mean that they are without complications. The main complications or obstacles in performing polarization measurements can be summarized in the following categories:

- **Effect of scan rate:** the rate at which a potential or current is scanned may have a significant effect on the resulting polarization (31). The goal is for the polarization scan rate to be slow enough to minimize surface capacitance charging, otherwise some of the current being generated may serve to charge the surface capacitance with the net result that the measured current can be greater than the current actually generated by the corrosion reactions alone.
- **Effect of solution resistance:** the distance between the reference electrode and the working electrode is purposely minimized in most measurements to limit the effect of the solution resistance. In solutions that have extremely high resistivity (e.g., concrete, soils, and organic solutions), this effect can be an extremely significant.
- **Changing surface conditions:** corrosion reactions take place on the metallic surface exposed to the environment and that surface can be modified by changing process conditions. This can have a strong effect on the polarization curves (31).
- **Determination of pitting potential:** in analyzing polarization curves the presence of a hysteresis loop between the forward and reverse scans often indicates that localized corrosion (e.g., pitting or crevice corrosion) is in progress. This observation has led to the creation of special potentiodynamic techniques to reveal the severity of localized corrosion problems. However, it can be a serious hindrance for monitoring uniform corrosion.

4.4.1.2.1. Linear Polarization Resistance (LPR) In this popular electrochemical technique, a small potential perturbation (typically 10–20 or even 30 mV) is applied to the sensor electrode of interest and the resulting current is measured. The ratio of the potential to current perturbations is known as the polarization resistance, which, according to Eq. (4.4), is inversely proportional to the uniform corrosion rate. More specifically, the polarization resistance of a metal is defined as the slope of the potential-current density ($\Delta E/\Delta i$) curve at the free corrosion potential (Fig. 4.31), yielding the polarization resistance R_p that can be itself related to the corrosion current (i_{corr}) with the help of the following Stern–Geary Eqs. (32–34):

$$R_p = \frac{B}{i_{corr}} = \frac{(\Delta E)}{(\Delta i)}_{\Delta E \to 0} \qquad (4.4)$$

where R_p is the polarization resistance[2]; i_{corr} is the uniform corrosion current; and B is an empirical polarization resistance constant that can be related to the

---

[2]The accuracy of the technique can be improved by measuring the solution resistance independently and subtracting it from the apparent polarization resistance value.

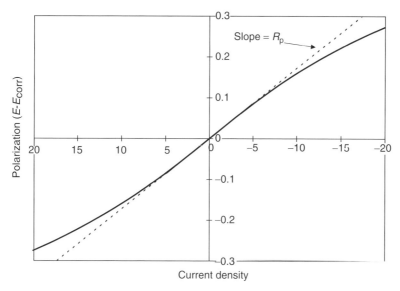

Fig. 4.31. Hypothetical linear polarization plot.

anodic (b_a) and cathodic (b_c) Tafel slopes with Eq. (4.5).

$$B = \frac{b_a \cdot b_c}{2.3(b_a + b_c)} \quad (4.5)$$

The Tafel slopes required to perform these calculations can be either determined empirically from polarization plots such as shown in Fig. 4.32 or obtained from the literature (34). Tafel slopes can also be determined by other techniques, such as curve fitting of polarization resistance curves (i.e., nonlinear polarization resistance), by potentiodynamic polarization scans, or by harmonic distortion analysis.

In a plant situation, it is necessary to use a probe as one of those shown in Fig. 4.33 such that it enters the vessel in the area where the corrosion rate is desired (Fig. 4.34). An electronic power supply polarizes the specimen from the corrosion potential. The resulting current is recorded as a measure of the corrosion rate. Several commercially available probes and analyzing systems can be directly interfaced with remote computer data-acquisition systems. Alarms can also be used to signal plant operators when high corrosion rates are experienced (15, 29).

The LPR probes are typically a two or three electrode configuration with either flush or projecting electrodes. With a three electrode system, corrosion measurements are made on the test electrode. Because these measurements take only a few minutes, the need for a stable reference electrode is minimized. For field monitoring, the reference electrode is typically made of stainless steel or of the same alloy as that being monitored on the test electrode. The auxiliary electrode is also normally of the alloy being monitored. The proximity of the reference electrode to the test electrode governs the degree to which compensation for solution resistance is effective. With a two-electrode system, the corrosion

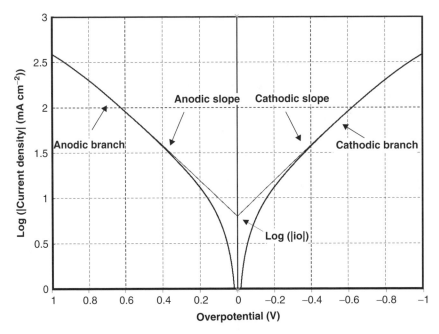

Fig. 4.32. E of a plot of η against log $|i|$ or Tafel plot showing the exchange current density can be obtained with the intercept.

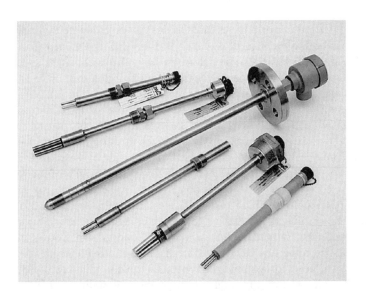

Fig. 4.33 Commercial sensor elements to carry out linear polarization resistance (LPR) measurements. (Courtesy of Metal Samples Company, www.metalsamples.com.)

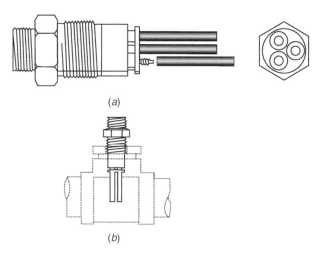

Fig. 4.34. Typical linear polarization resistance probe (*a*) and probe in pipe tee (*b*) (15).

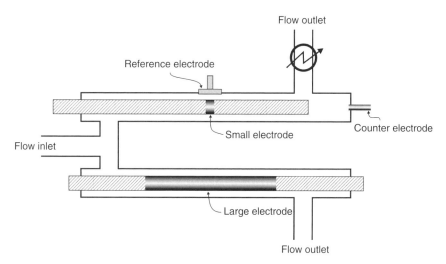

Fig. 4.35. Schematic of differential flow cell with a fast flow electrode (FE), a slow flow electrode (SE), a reference electrode (RE), and an inert counter electrode (CE) (35).

measurement is an average of the rate for both electrodes. Both electrodes would then be made of the alloy being monitored (13).

A combination of LPR and zero resistance ammeter (ZRA) measurements has been used in a special study to evaluate the rate of localized corrosion in a flowing environment by placing a large-area electrode in fast flow conditions and one small electrode in slow flow conditions in a side-stream differential flow cell schematically shown in Fig. 4.35 (35, 36). When the large area electrode and the small electrode were connected together through a ZRA, the large electrode became a cathode and the small electrode an anode, due to differential aeration forcing the small electrode to experience preferential corrosion. As immersion time increased, the small electrode became covered with deposits and its corrosion rate a good representation of an underdeposit or localized corrosion situation.

There are several limitations to corrosion monitoring with LPR. The corroding environment must be an electrolyte with reasonably low resistivity. High-resistivity electrolytes produce erroneously low corrosion rates. This need for a sufficiently conductive medium precludes the use of this technique for many applications in oil and gas, refinery, chemical, and other low-conductivity applications.

Another problem with the technique is that the vessel or pipe wall must be penetrated, and this involves concerns regarding leaks, personnel safety, and other problems. The ability to use direct wiring from the probe location to a remote control room is desirable, but the installation of these wiring systems is costly. In addition, LPR measurements do not provide information on localized corrosion, such as pitting and SCC and since the corrosion-rate values obtained with LPR are at best approximate, the method is best suited for use during periods when substantial corrosion-rate changes may occur (15, 29).

4.4.1.2.2. Zero Resistance Ammetry With this electrochemical technique galvanic currents between dissimilar electrode materials are measured with a zero resistance ammeter.[3] The design of dissimilarities between sensor elements may be made to target a feature of interest in the system being monitored (e.g., different compositions, heat treatments, stress levels, or surface conditions). The technique may also be applied to nominally identical electrodes in order to reveal changes occurring in the corrosive environment and thus serve as an indicator of changing corrosion rates.

The main principle of the technique is that differences in the electrochemical behavior of two electrodes exposed to a process stream give rise to differences in the redox potential at these electrodes. Once the two electrodes are externally electrically connected, the more noble electrode becomes predominantly cathodic, while the more active electrode becomes predominantly anodic and sacrificial. When the anodic reaction is relatively stable the galvanic current monitors the response of the cathodic reaction to the process stream conditions. When the cathodic reaction is stable, it monitors the response of the anodic reaction to process fluctuations (13).

This technique has been found particularly useful to study depolarization effects of the cathode of a galvanic pair of electrodes to obtain feedback of low levels of dissolved gases, particularly oxygen, or the presence of bacteria, which depolarize the cathode of the galvanic pair and increase the coupling current. When used for detection of low levels of oxygen, other dissolved gases may

---

[3]A zero resistance ammeter is a current to voltage converter that produces a voltage output proportional to the current flowing between its to input terminals while imposing a 'zero' voltage drop to the external circuit.

interfere. Calibration against a dissolved oxygen meter is usually required if quantitative values are needed.

The ZRA method can provide a quantitative measure if the number of influencing factors is limited and preferably verifiable through other means. For monitoring the effect of dissolved gases, the conversion of the signal to a gas concentration level is not accurate. When other factors play a role as well, for example, during the formation of biofilm compounds or in the presence of inhibitors, the method cannot really provide a quantitative indication of any kind.

Additionally, results from galvanic probes do not always reflect the actual galvanic corrosion rates, because galvanic corrosion depends on the relative areas and specific geometries of the components, which can easily vary between a probe design and the actual plant components being monitored. The ZRA method cannot distinguish between either activation of the anodic or the cathodic reaction. For example, an increase in the measured current can result from cathodic activation by increased dissolved oxygen, from anodic activation by increased bacterial activity, or by a combination of these. Separate analysis is sometimes performed if it is necessary to distinguish between these electrochemical components.

4.4.1.2.3. Potentiodynamic–Galvanodynamic Polarization In this technique, a three electrode corrosion probe is used to polarize the electrode that serves as the sensing element. The current or potential response is measured as the potential (potentiodynamic) or current (galvanodynamic) is shifted away from the free corrosion potential. The basic difference with the LPR technique is that the applied polarization can be of several hundred millivolts (Fig. 4.36). While the technique is used quite commonly in a laboratory environment, it is used only occasionally in the field mostly to estimate the anodic and cathodic Tafel

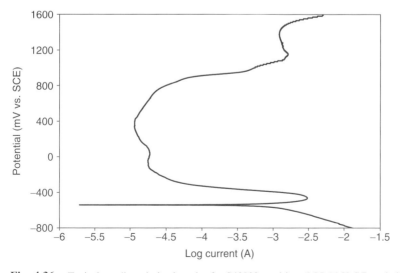

Fig. 4.36. Typical anodic polarization plot for S43000 steel in a 0.05 M H_2SO_4 solution.

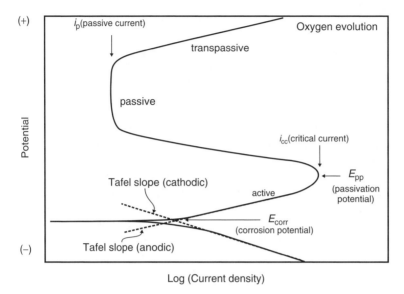

Fig. 4.37. Generalized polarization diagram showing the various potential regions of a passivable metal and the Tafel extrapolation lines.

slopes for systems on which the basic corrosion rate theory is based (Fig. 4.37). The formation of passive films and the onset of pitting corrosion can also be identified at characteristic potentials, which can assist in assessing the overall corrosion risk.

Potentiodynamic polarization is generally used in aqueous systems while galvanodynamic polarization has been used in systems containing oil, so that current density is controlled. One of the most common uses of the technique is to determine whether crevice corrosion or pitting is a problem and whether general corrosion can be estimated by another technique. It is also often used to estimate the relative susceptibility of various materials to localized corrosion in the process stream (13).

Ideally, each point on the current–potential graph should be made by allowing a polarization time of tens of seconds to several minutes to permit the complete charging of the double-layer capacitance associated with the metal–fluid interface. Such long-time intervals also permit the electrode surface to oxidize or reduce all of the surface deposits and to polarize to the applied potential. In practice, the applied current or applied potential, whichever is the controlling variable, is changed continuously in an analog form at some preset rate (potentiodynamic or galvanodynamic) or in a digital form of small discrete steps at some preset rate (potential staircase or galvanic staircase).

In either case, the faster the rate of change of the applied signal, the greater is the lag of the measured signal behind the true steady-state values desired. This compromise between a practical and reasonable time to complete a scan and the degree of lag from measured-to-ideal response is a key question with this technique, particularly so in the field where the tested environment is continually changing and cannot usually be controlled as it can be in a laboratory.

A potentiodynamic polarization variant is cyclic voltammetry which involves sweeping the potential in a positive direction until a predetermined value of current or potential is reached, then the scan is reversed toward more negative values until the original value of potential is reached. In some cases, this scan is done repeatedly to determine changes in the current–potential curve produced with scanning.

Another variation of potentiodynamic polarization is the potentiostaircase method. This refers to a technique for polarizing an electrode in a series of potential steps where the time spent at each potential is constant, while the current is often allowed to stabilize prior to changing the potential to the next step. The step increase may be small, in which case, the technique resembles a potentiodynamic curve (31).

Electrochemical potentiodynamic reactivation (EPR) is another polarization method that evaluates the degree of sensitization of stainless steels, such as S30400 and S30403 steels. This method uses a potentiodynamic sweep over a range of potentials from passive to active (called reactivation).

However, probably the most popular variant is the cyclic polarization test. This test is often used to evaluate the pitting susceptibility of a material. The potential is swept in a single cycle or slightly less than one cycle usually starting the scan at the corrosion potential. The voltage is first increased in the anodic or noble direction (forward scan). The voltage scan direction is reversed at some chosen current or voltage toward the cathodic or active direction (backward or reverse scan) and terminated at another chosen voltage. The presence of a hysteresis between the currents measured in the forward and backward scans is believed to indicate pitting, while the size of the hysteresis loop itself is often related to the amount of pitting.

This technique has been especially useful to assess localized corrosion for passivating alloys such as S31600 stainless steel, nickel-based alloys containing chromium, and other alloys, such as titanium and zirconium. Though the generation of the polarization scan is simple, its interpretation can be difficult (37).

In the following example, the polarization scans were generated after 1 and 4 days of exposure to a chemical product maintained at 49°C. The goal of these tests was to examine if S31600 steel could be used for short-term storage of a 50% commercial organic acid solution (aminotrimethylene phosphonic acid) in water. A small amount of chloride ion (1%) was also potentially present in this acidic chemical.

In this example, the potential scan rate was 0.5 mV s^{-1} and the scan direction was reversed at 0.1 mA cm^{-2}. Coupon immersion tests were run in the same environment for 840 h. The S31600 steel specimens were exposed to the liquid, at the vapor–liquid interface, and in the vapor phase. The reason for the three exposures was that in most storage situations, the containment vessel would be exposed to a vapor–liquid interface and a vapor phase at least part of the time.

Corrosion in these regions can be very different from liquid exposures. The specimens were also fitted with artificial crevice formers.

Figure 4.38 shows the polarization scan generated after 1 day and Fig. 4.39 shows the polarization scan generated after 4 days of exposure. The important parameters considered were the position of the "anodic-to-cathodic" transition relative to the corrosion potential, the existence of the repassivation potential and its value relative to the corrosion potential, the existence of the pitting potential and its value relative to the corrosion potential, and the hysteresis (positive or negative). The interpretation of the results is summarized in Table 4.6.

The presence of the negative hysteresis would typically suggest that localized corrosion may be a problem depending on the value of the corrosion potential relative to the characteristic potentials present in these polarization plots. After the first day of exposure, pitting was not expected to be a problem because the pitting potential was far away from the corrosion potential. The currents generated were much higher than those normally associated with S31600 steel in a passive state. These observations suggested that there was a risk of initiation of corrosion, particularly in localized areas where the pH can decrease drastically (37).

After 4 days, the risk of localized corrosion increased. At this time, the repassivation potential and the potential of the change from anodic-to-cathodic current were equal to the corrosion potential. The pitting potential was only ~0.1 V more noble than the corrosion potential and the hysteresis still negative. The risk of pitting had increased to become a concern.

Coupon immersion tests confirmed the long-term predictions. Slight attack was found under the artificial crevice formers in the complete liquid exposure.

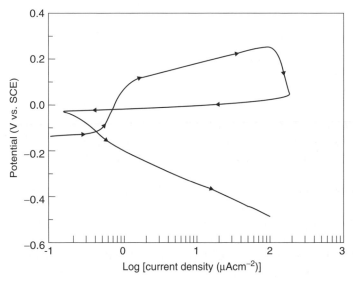

Fig. 4.38. Polarization scan for S31600 steel in 50% aminotrimethylene phosphonic acid after one day of exposure (the arrow indicates scanning direction).

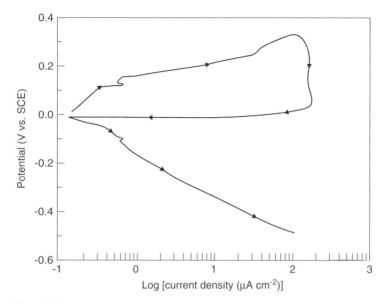

Fig. 4.39. Polarization scan for S31600 steel in 50% aminotrimethylene phosphonic acid after four days of exposure (the arrow indicates scanning direction).

Table 4.6. Features and Values Used To Interpret Figs. 4.38 and 4.39

Feature	Value in Fig. 4.38	Value in Fig. 4.39
Repassivation potential–corrosion potential	0.12 V	0.0 v
Pitting potential–corrosion potential	0.22 V	0.12 V
Potential of anodic-to-cathodic transition–corrosion potential	0.12 V	0.0 V
Hysteresis	Negative	Negative
Active-to-passive transition	No	No

The practical conclusion of this in-service study was that, since localized corrosion often takes time to develop, a few days of exposure to this chemical product could be acceptable. However, it was recommended to avoid long-term exposure since both pitting and crevice corrosion would be expected for longer exposure periods.

While this example was an interesting study, the potentiodynamic techniques are generally not considered for on-line or real-time monitoring because the electrodes have to be replaced after only one or two test runs. The high anodic potentials used in the tests simply corrode the sensing elements and permanently changes their surface. Correcting the ohmic(IR) drop due to the solution resistance may also be particularly important with this technique because of the relatively high currents used compared with other electrochemical techniques. The IR induced error can be much smaller in highly conductive environments.

4.4.1.2.4. Electrochemical Impedance Spectroscopy With electrochemical impedance spectroscopy (EIS), the sensing element is polarized by the application of an alternating potential that in turn produces an alternating current response. For corrosion monitoring, the frequency range of the applied ac polarization is typically between 0.1 Hz and 100 kHz with a polarization level within 10 mV of the corrosion potential. Full frequency scans provide phase shift information that can be utilized in combination with equivalent circuit models to obtain useful information on the system complex interface.

Amongst the numerous equivalent circuits that have been proposed to model electrochemical interfaces only a few really apply to a freely corroding system. The circuit shown in Fig. 4.40a is the simplest equivalent circuit that can describe a metal–electrolyte interface. Its behavior is described by Eq. (4.6).

$$Z(\omega) = R_s + \frac{R_p}{1 + (j\omega R_p C_{dl})^\beta} \tag{4.6}$$

Where R_s is the solution resistance, R_p is the polarization resistance, ω is the ac polarization frequency, and C_{dl} is the double layer capacitance

The term Q in Fig. 4.40(a) describes the "leaky capacitor" behavior corresponding to the presence of a constant phase element (CPE) (38). Figure 4.40a also illustrates the complex plane presentation of this EIS model circuit in which $R_s = 10$ Ω, $R_p = 100$ kΩ, and Q is a function of $C_{dl} = 40$ μF and $n = 0.8$. Figure 4.40b shows how the same data would appear in a Bode plot format.

The high-frequency response is used to determine the component of solution resistance (R_s) included in the measurement. The polarization resistance (R_p) can then be determined by subtracting the R_s value from the low-frequency measurement. To convert the polarization resistance into a corrosion rate involves an empirical measurement of the Tafel slopes that have to be determined by other techniques, such as potentiodynamic polarization and harmonic distortion analysis, or again obtained from the literature (34).

The measurement cycle time depends on the frequency range used, especially the low frequencies. A single frequency cycle at 1 mHz, for example, takes 15 min. A high-to-low frequency scan going to such a low frequency would take >2 h. In order to make routine corrosion monitoring with EIS certain simplifications are needed to maximize the use of high frequency data and drastically shorten the measurement time. It is also important to simplify the data processing and analysis to make the technique user friendly for field corrosion monitoring. However, the need for a field instrument that can be easily deployed has always been an impediment for on-line corrosion monitoring with EIS.

In order to simplify the analysis of field EIS results, a method was developed that consists of finding the geometric center of an arc formed by three successive data points on a complex impedance diagram (Fig. 4.41) (11, 39). This technique was designed as an improvement over the two point method based on the comparison of high-and low-frequency data points for which the impedance would

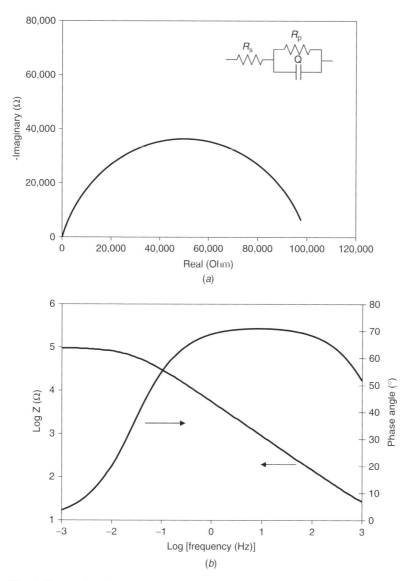

Fig. 4.40. (a) Complex plane and RC model circuit with simulated data, where $R_s = 10\ \Omega$, $R_p = 100\ k\Omega$ and Q decomposes into $C_{dl} = 40\ \mu F$ and $n = 0.8$. (b) Bode plot corresponding to the same simulated data.

be proportional to the R_s at the high frequency point and the summation of R_s and R_p at the low frequency point (40). In real-world situations, one difficult assumption to satisfy with the two point method is that data points should contain negligible imaginary components (i.e., 0 phase shift), a condition usually hard to achieve in a meaningful manner at low measuring frequencies.

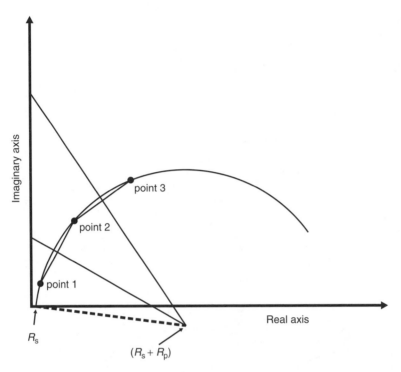

Fig. 4.41. Schematic of the extrapolation method to obtain the polarization resistance from EIS data.

The three point analysis technique was further developed by permuting the data points involved in the projection of centers in order to obtain a population of projected centers. This improvement has permitted to automate the data analysis while providing some information concerning the adherence of the results with the RC behavior described by Eq. (4.6), what is assumed for the evaluation of the parameters associated with uniform corrosion. The technique was extensively tested in a laboratory environment before being applied in field tests with the same laboratory equipment (11).

Very recently, a full-spectrum relatively low cost EIS corrosion monitoring system has been developed, which is wireless, small (5-cm diameter, 1.2-cm height), requires nominally 10-mW power during its 200-s measurement period, and has an electronic identifier, which allows for a single data logger to monitor multiple devices in the same general vicinity (Fig. 4.2a and b) (3).

The wireless EIS sensor determines the impedance at 15–20 independent frequencies, by measuring amplitude and phase at each frequency (Fig. 4.42). It computes corrosion rate, conductivity and coating impedance, and transmits the result wirelessly to a data logger. The miniature and wireless features make it suitable for embedding in concrete or placing in hidden and inaccessible locations,

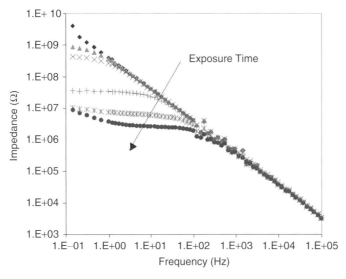

Fig. 4.42 Magnitude of impedance of a coating versus frequency. Low-frequency impedance values show good correlation with long-term exposure behavior. (Courtesy of Guy D. Davis, DACCO SCI, Inc.)

for example, in HVAC systems. Its minimal power consuming aspect lends itself useful for long-term monitoring of coating integrity.

The miniature EIS system has been tested in various environments, namely, concrete, water, and under coatings (Fig. 4.43a and b). In addition, it was also tested against a simple resistor capacitor circuit and validated with two different commercial instruments. The miniature EIS system successfully monitored the corrosivity of water contaminated with chloride, concrete in contact with water containing salt, and health and integrity of coatings on metals. These results correlated well with data obtained using conventional bench-top EIS and LPR instruments. However, unlike the conventional instruments, the wireless EIS sensor is small, requires very little power to operate and can be hermetically sealed. Therefore, it can be embedded in structures and immersed in liquids without much concern about drawing wires and cables.

4.4.1.2.5. Harmonic Distortion Analysis With this technique, a low-frequency sinusoidal potential is applied to a three-electrode measurement system, and the resulting current is measured. As the corrosion process is nonlinear in nature, a potential perturbation by one or more sine waves should generate responses at more frequencies than the frequencies of the applied signal. Current responses can be measured at zero, harmonic, and intermodulation frequencies.

Measuring the dc at frequency "zero" is called the Faraday rectification (FR) technique. The FR technique can be used for corrosion rate measurements if at least one of the Tafel parameters is known. The corrosion rate and both Tafel

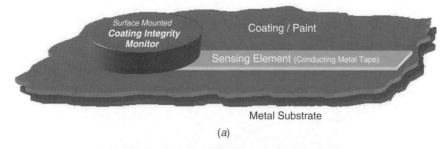

(a)

(b)

Fig. 4.43 (a) Schematic of the Coating Health Monitor (CHM) with a tape sensing element mounted on a coated metal and (b) actual sensor tape electrode and electronics housing mounted on the frame of a commercial vehicle. (Courtesy of Guy D. Davis, DACCO SCI, Inc.)

parameters can be obtained with one measurement by analyzing the harmonic frequencies. The speed with which the Tafel slopes can be determined (typically <1 min) is a particular attraction of the technique (41).

With harmonic distortion analysis (HDA), a single, low-frequency and low-distortion, sinusoidal voltage is applied to the corroding interface. As a quality check, three different frequencies are used to verify the repeatability of the technique. The amplitude is in the range of 10–30-mV peak to peak. The frequency used is typically 0.1–10 Hz. Other frequencies can be used depending on the corrosion process.

The theoretical analysis for the computation of Tafel slopes and corrosion current makes no assumptions about solution resistance effects and measurements cannot be performed unless the system is free of significant electrical noise in the frequency range of the applied measurement potential, or one of its harmonic frequencies (13).

In principle, analyzing the primary frequency and the harmonics makes it possible to extract all the information to calculate Tafel slopes and corrosion rate. In practice, this has been used in a very limited number of applications to date. A serious limitation of the existing nonlinear techniques is that their application has been restricted mainly to activation controlled corrosion processes, for example, corrosion rate measurements in acid media with and without inhibitors.

Besides FR, variations of harmonic distortion analysis developed have been called nonlinear EIS (NLEIS), harmonic analysis (HA), harmonic impedance spectroscopy (HIS), and electrochemical frequency modulation (EFM). These techniques all analyze nonlinear response of a corrosion system after perturbation with a mono -or biharmonic signal, and on this basis allow extraction of the required kinetic parameters of the corrosion process. Except for the FR method, all other HAD techniques allow the simultaneous determination of corrosion current and both Tafel coefficients (42).

4.4.1.2.6. Electrochemical Noise (EN) Analysis Fluctuations of potential or current of a corroding metallic specimen are a well known and easily observable phenomenon. Electrochemical noise analysis (ENA) as a corrosion tool has increased steadily since Iverson's paper in 1968 (43) and the number of industrial applications of electrochemical noise monitoring has grown significantly in recent years. The study of EN has been found to be uniquely appropriate for monitoring the onset of events leading to localized corrosion and understanding the chronology of the initial events typical of this type of corrosion.

The EN technique differs in many ways from the other electrochemical techniques described so far. One important difference is that ENA does not require that the sensing element be polarized in order to generate a signal. However, it is also possible to measure current noise under an applied potential, or measure potential noise under an applied current. The potential and current between freely corroding electrodes (in many cases <1 μV and <1 nA) are measured with sensitive instrumentation. A measurement frequency of 1 Hz is usually appropriate to provide meaningful data. For simultaneous measurement of electrochemical potential and current noise, a three-electrode sensor is required. In field corrosion monitoring, the three sensor elements are usually made of the same material.

While the measurement of electrochemical noise is relatively straightforward, the data analysis can be complex and inconclusive. Even if ENA was first applied in field corrosion monitoring in the late 1960s, an understanding of the method of analysis is still evolving, partly because the technique has been used to look at several types of corrosion. The relationships between potential and current noise are inherently complex to analyze quantitatively because the naturally occurring fluctuations do not have controlled frequencies as are applied, for example, in EIS. For these reasons, much of the investigations with EN have been centered on frequency analysis of the data. There are still varying conclusions about the accuracy and effectiveness of the technique in its own right (13).

4.4.1.2.6.1. Electrochemical noise analysis The analysis of EN, begun a few decades ago, has only been recently introduced as a credible corrosion monitoring technique. In this context, the pioneering work of Eden et al. has been instrumental in introducing the idea of a corrosion cell with two working electrodes (WE), where both current and voltage fluctuations can be measured (44). The remaining question was how to establish the data interpretation on a firm basis. Because EN measurements on a single corroding electrode are not sufficient to evaluate corrosion rates, most applications in the field are based on the use of cells with two identical electrodes (same material, same size, same surface preparation), connected through a zero-resistance ammeter (ZRA) so as to have both working electrodes set at a common corrosion potential (45).

There are many cases, particularly in field applications where the use of low-noise reference electrodes commonly used in laboratories would be impractical. In these cases, one can use a third electrode made of material similar to the other two working electrodes. Obviously, such a reference electrode would contribute to the noise of the system. It turns out, however, that in such an arrangement the noise impedance Z_n is equal to a fraction of the total noise impedance (Z), that is, $\sqrt{3}|Z|$, so that a simple correction is sufficient to correct the problem. However, a more serious concern is that the noise signals depend on the three electrodes having the same impedance and contributing the same noise. As every corrosion worker knows, initially identical electrodes tend to diverge in behavior with time. Experience has shown that this is particularly troublesome in the case of localized corrosion for which it could introduce significant errors difficult to correct.

There are basically three categories of ENA: visual examination, sequence-independent methods that treat the collection of voltage or current values without regard to their position in the sequence of readings (moments, mean, variance, standard deviation, skewness, and kurtosis), and those that take the sequence into account (autocorrelation, power spectra, fractal analysis, stochastic process analysis) (46].

Visual examination of the time record trace can give indications as to the type of corrosion processes that are occurring. The following example illustrates how a simple examination of EN measurements could reveal the corrosivity of various points of an industrial gas scrubbing system where highly corrosive thin-film electrolytes may form (47). These conditions arise when gas streams are cooled to a temperature below the dewpoint. The resulting thin electrolyte layer (moisture) is often highly concentrated in corrosive species.

The corrosion probe used in this example is illustrated in Fig. 4.44 and 4.45. A retractable probe with flexible depth was selected, in order to mount the sensor

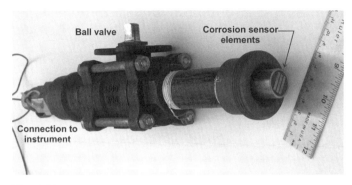

Fig. 4.44 Corrosion sensor and access fitting used for thin-film corrosion monitoring. (Courtesy Kingston Technical Software.)

Fig. 4.45 Close-up of corrosion sensing elements used for thin-film corrosion monitoring. (Courtesy Kingston Technical Software.)

surface flush with the internal scrubber wall surface. The close spacing of the carbon steel sensor elements was designed to work with a discontinuous thin surface electrolyte film. This corrosion sensor was connected to a handheld multichannel data recorder by shielded multi-strand cabling Fig. 4.1. As the ducting of the gas scrubbing tower was heavily insulated, no special precautions were taken to cool the corrosion sensor surface.

Potential noise and current signals recorded during the first hour of exposure at the conical base of the gas scrubbing tower are presented in Fig. 4.46. According to the operational history of the plant, condensate had a tendency to

254 Chapter 4 Corrosion Monitoring

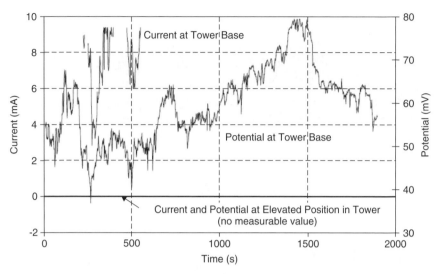

Fig. 4.46. Potential and current noise records at two locations in a gas scrubbing tower.

accumulate at this location where highly corrosive conditions had been noted. The high levels of potential noise and current noise in Fig. 4.46 are indicative of a massive pitting attack that is consistent with the operational experience. Note that the current noise is actually off-scale for most of the monitoring period, in excess of 10 mA. The high corrosivity indicated by the electrochemical noise data from this sensor location was confirmed by direct evidence of severe pitting attack on the sensor elements, revealed by scanning electron microscopy (Fig. 4.47). In contrast, both current and voltage signals remained relatively constantly small at a position higher up in the tower, where the sensor surface remained mostly dry.

An improvement over this simple analysis is commonly practiced in industry by tracking the width of the potential and current signals as an indication of corrosion activity in the system being monitored. Figure 4.48 illustrates how the decrease in the current band obtained with a monitoring system was interpreted as a reduction in general corrosion activity in a debutanizer overhead piping where the interaction between operational changes and the corrosion mechanism were being investigated.

Most ENA techniques have been developed to relate sometimes subtle features in the noise data records to changes in the corrosivity of a system and particularly on how these changes affect localized corrosion. Three techniques have also been proposed for obtaining the polarization resistance as a measure of uniform corrosion rate if the Tafel constants of the system are known or measured.

The noise resistance (R_n), a sequence-independent method, is the ratio of the standard deviation of potential to the standard deviation of current. The impedance resistance (R_{imp}), a sequence-dependent method, is the ratio of the square root of the potential power to the square root of the current power in

4.4. Corrosion Monitoring Techniques   255

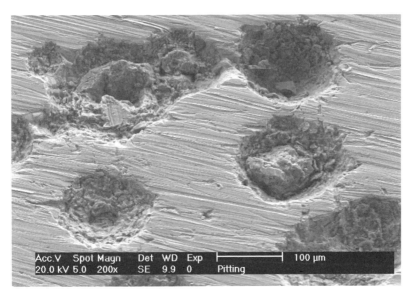

Fig. 4.47. Scanning electron microscope image of a sensor element surface after exposure at the base of the scrubbing tower clearly showing corrosion pits.

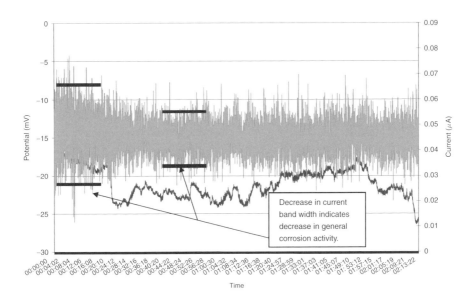

Fig. 4.48 Electrochemical current noise (large band) and potential noise (lower signal) in a debutanizer overhead piping obtained with the Concerto VT noise system. (Courtesy of CAPCIS Ltd.)

the low frequency limit. The potential and current power densities are calculated from power spectrum distribution (PSD) techniques, such as fast Fourier transform (FFT) or the maximum entropy method (MEM) (48). Finally, the EN transient resistance (R_{tran}) is the ratio of the amplitude of potential to current for single transients (49). The parameter R_{tran} provides the most exact corrosion resistance value, but the data analysis is also the most complicated since the technique hinges on finding transients amenable to a clear interpretation.

While these techniques have been extensively used in controlled laboratory experiments, the high variability of field environments combined with the availability of well established methods such as ER, LPR, or the use of coupons have created the impression that ENA is not an appropriate option for obtaining corrosion rates when uniform corrosion is the prevalent corrosion mechanism.

In contrast, localized corrosion processes, such as pitting and SCC, have characteristic transients in the time record traces that may help to distinguish between various possible types of corrosion. The modern success of ENA is in great part based on this unique property since more traditional on-line techniques are quite ineffective in this area. This success is also due to the fact that most forms of localized corrosion are much more troublesome in the management and operation of many engineering systems than uniform corrosion.

The effect of localized corrosion on EN signals tends to create a deviation from a normal distribution (Poisson distribution) as may be determined from the skewness and kurtosis of the EN signals. If techniques, such as linear polarization resistance (LPR) and harmonic distortion analysis (HDA), are also used, then independent derivation of the general corrosion current (I_{corr}) is possible. The ratio of the current noise (standard deviation of the current) to the general corrosion current may then be used as an improved indicator of localized corrosion activity (50).

Skewness and kurtosis are measures of the probability density of the noise signals in relation to the mean value. Skewness provides an indication of the symmetry of the distribution. A value of zero implies that the distribution is symmetrical whereas a positive skewness implies that there is a tail in the positive direction and a negative skewness a tail in the negative direction. Kurtosis is a measure of the shape of the distribution compared with the normal distribution. A kurtosis of zero implies that the distribution has a shape similar to that of the normal distribution. A positive kurtosis implies a narrower distribution, whereas a negative kurtosis implies a flatter distribution. It was found experimentally that the kurtosis of a signal generated by uniform corrosion was approximately 1.5–2 while a kurtosis value >5 would be indicative of localized corrosion (51).

The localization parameter is the ratio of the standard deviation of the current, σ_I, to the root-mean-square (rms) current (48). However, the rms current is related to the mean coupling current, which, over an infinite period of time with two similar electrodes is zero. This gives the localization parameter the possibility of having the unfortunate value of infinity. The pitting factor is a modification of the localization parameter (10). The idea is to use a measure of i_{corr} instead of rms current. If one assumes that R_n is representative of the polarization resistance then the pitting factor is equivalent to the ratio of the standard deviation of potential, σ_V, to the Stern–Geary constant (B).

Figures 4.49 and 4.50 illustrate how the corrosion pitting indication and associated system instability were tracked during the debutanizer overhead piping evaluation project mentioned earlier.

The shot-noise charge is a measure of the average charge of a transient based on shot-noise analysis (48). This is the ratio of $\sigma_I \sigma_V / bB$, where b is the bandwidth of measurement (s^{-1}). The bandwidth is the frequency range from essentially zero to the Nyquist frequency, $1/(2\Delta t)$, where Δt is the sampling period.

Certain characteristics of localized corrosion have been interpreted by analyzing EN data in the frequency domain. The slope of the linear region of the PSD plot of current or potential versus frequency has been interpreted as being related to the shape, amplitude and frequency of elementary transients (48). The expected slopes can be in the range of 0 to -4 with zero for white noise processes, -0.5 for diffusion controlled processes, -1 for processes having a Gaussian distribution and less than -2 for processes with fluctuations (51). However, the exact value of this slope corresponding to localized corrosion behavior may differ significantly between systems and environments.

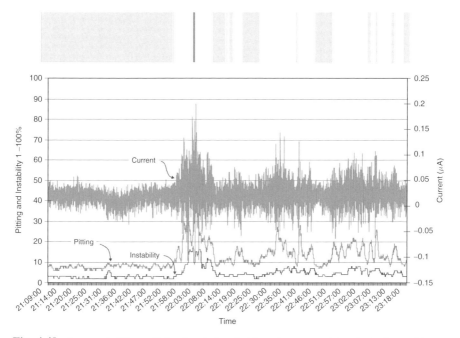

Fig. 4.49 Increased pitting activity at end of period from large current transients observed in a debutanizer overhead piping obtained with the Concerto VT noise system. (Courtesy of CAPCIS Ltd.)

258 Chapter 4 Corrosion Monitoring

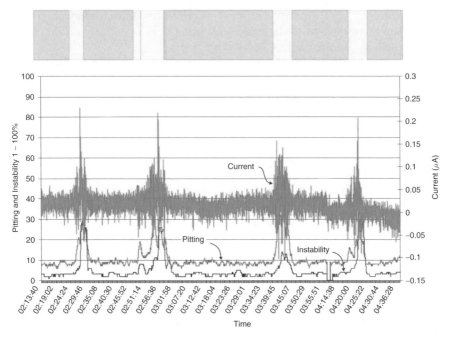

Fig. 4.50 Moderate pitting events on fairly regular intervals observed in a debutanizer overhead piping with the Concerto VT noise system. (Courtesy of CAPCIS Ltd.)

4.4.1.2.6.2. Probe design and signal analysis An appreciable level of effort has been directed towards developing electrode arrangements for monitoring localized corrosion with EN, in particular pitting, such as the use of both very small (≤ 1 mm^2) or very large (≥ 100 cm^2) electrodes, as well as electrode arrays. A great variety of instrumentation approaches have also been used (51):

- Three identical electrodes, configured as two coupled via a ZRA and one pseudoreference, with simultaneous (Fig. 4.51) or sequential (Fig. 4.52) measurements of current and potential.

- Three identical electrodes configured in potentiostatic mode (working, reference and auxiliary) with the reference potential set at 0 V (Figure 4.53), with simultaneous measurement of current and potential. This particular arrangement is useful in that the potentiostatic control may be used to perform other controlled polarization measurements such as LPR or HDA.

- Multiple electrode arrays of the test material using scanning techniques to measure current and/or potential between individual or combinations of elements in order to identify locally anodic behavior on one or more of the electrodes (Fig. 4.10) (52).

- Single electrode probe consisting of a working electrode with current (Fig. 4.54) or potential (Fig. 4.55) measured with respect to the structure

4.4. Corrosion Monitoring Techniques 259

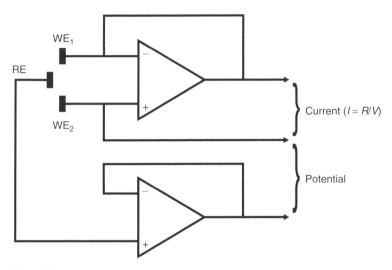

Fig. 4.51 Simultaneous current and potential monitoring using three identical electrodes. (Adapted from Ref. 51.)

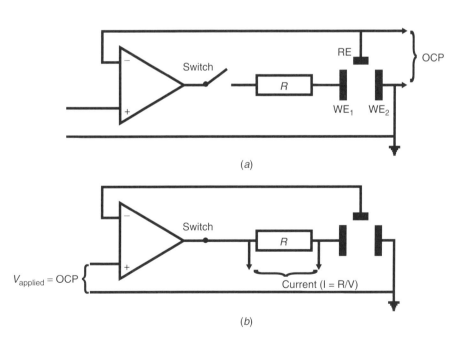

Fig. 4.52 Sequential current and potential logging of three electrodes: (*a*) the open circuit potential of WE_2, and (*b*) switched-on mode of operation. (Adapted from Ref. 51.)

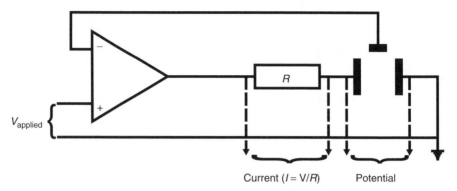

Fig. 4.53 Three identical electrodes simultaneous current and potential measurement. (Adapted from Ref. 51.)

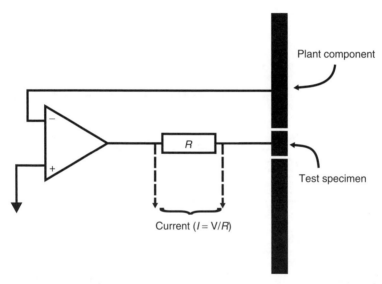

Fig. 4.54 Single electrode measurement of current with respect to plant. (Adapted from Ref. 51.)

of interest. As with the three electrode arrangement this may also be used to evaluate polarization response.

Higher levels of electrochemical corrosion activity are generally associated with higher EN levels. Certain electrochemical phenomena, such as the breakdown of passivity during pit initiation show up distinct noise "signatures", which can be exploited for corrosion monitoring purposes. Pit initiation and growth, for example, can be detected with EN measurements long before it becomes evident by visual examination.

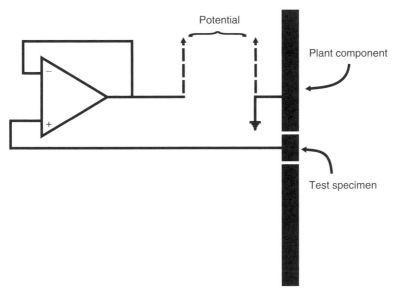

Fig. 4.55 Single electrode measurement of potential with respect to plant. (Adapted from Ref. 51.)

The analysis of EN results obtained between two working electrodes (WEs) with an imposed or naturally developing asymmetry can be carried out by considering Eq. (4.7), which indicates that the noise impedance of a cell (Z_n) depends on the impedances of the two WEs, as well as their noise levels, represented by the power density spectra (Ψi_1 and Ψi_2) obtained by performing the analysis of noise signals with either FFT or with the maximum entropy method (MEM).

$$Z_n(f) = \sqrt{\frac{\Psi_V(f)}{\Psi_I(f)}} = |Z_1(f) Z_2(f)| \sqrt{\frac{\Psi_{i_1}(f) + \Psi_{i_2}(f)}{|Z_1(f)|^2 \Psi_{i_1}(f) + |Z_2(f)|^2 \Psi_{i_2}(f)}}$$
(4.7)

For the simplest case of two WEs with the same impedance ($Z_1 = Z_2$) the noise impedance is equal to the modulus of the electrode impedance $|Z(f)|$. This result is valid independently of the origin of the noise signals (localized or uniform corrosion, bubble evolution due to the cathodic reaction) and the shape of the impedance plot, even if the noise levels of the two electrodes are different. In such case, noise measurements are equivalent to impedance measurements for which the external signal perturbation has been replaced by the internal noise generated by the corrosion processes (45).

However, when the two WEs do not have the same impedance, the noise impedance analysis requires a more cautious interpretation. Depending on the source of the current noise the measured impedance may either be that of a quiet cathode or that of a quiet anode while in intermediate cases the results of the measurement of Z_n would show a mixed behavior and be more difficult to interpret. For example, if hydrogen bubbles are evolving on the cathode while the anode undergoes generalized corrosion, the noise of the cathode is orders of magnitude larger than that of the anode, so that Z_n becomes equal to the impedance modulus of the anode, $|Z_a|$. In these conditions, while the time records appear to show only the cathodic processes, the impedance measured is that of the anode, using the noise of the cathode as input signal.

An opposite case would be a cell where the anode is undergoing pitting, while the cathodic reaction is the reduction of dissolved oxygen or an imposed galvanic situation. Since the anodic noise is preponderant, Eq. (4.7) shows that Z_n is equal to the impedance modulus of the cathode, $|Z_c|$. The anodic noise is the internal signal source utilized for the measurement of the impedance of the cathode.

However, this special arrangement can produce current noise signals valid on their own (53, 54). The design of asymmetric corrosion probes opens new possibilities for the evaluation of the susceptibility of a system to localized problems difficult to instrument otherwise: galvanic cells (inhibitors, presence of oxygen), crevice corrosion, stress corrosion cracking, etc. This is surely a very promising area of research since no other field technique has close to the sensitivity and flexibility that ENA could have for the detection of such problems.

4.4.2. Direct Nonintrusive Techniques

By definition, direct nonintrusive measurement techniques do not require access to the inside of the system being monitored and so avoid the usual risks associated with on-line insertion and retrieval operations of intrusive devices. An important advantage of these techniques over intrusive corrosion monitoring techniques is that the actual plant material is monitored instead of a possibly different test coupon material. Depending on the specific technique and implementation, the area of the plant material monitored is generally larger than on an inserted specimen.

As mentioned earlier, most of the techniques belonging to this category in the NACE International report (13) are described in Chapter 5 since they are usually described as NDE, NDT, or NDI techniques.

4.4.2.1. Thin-Layer Activation and Gamma Radiography

With thin-layer activation (TLA) and gamma radiography, a technique developed from the field of nuclear science, a small section of material is exposed to a high energy beam of charged particles to produce a radioactive surface layer. For example, a proton beam may be used to produce the radioactive isotope Co-56 within a steel surface. This isotope decays to Fe-56 with the emission of gamma radiation. The concentration of radioactive species is sufficiently low that metallurgical properties of the monitored component are essentially unchanged. The radioactive effects utilized are much lower than conventional radiography and so the health risk concerns. The low levels of radiation present involve modest handling procedures and product quality, other than for human consumables, is not compromised.

Changes in gamma radiation emitted from the surface layer are measured with a separate detector to study the rate of material removed from the surface. The radioactive surfaces can be produced directly on components (nonintrusive) or on separate sensors (intrusive). The measured gamma radiation depends on the quantity of the original radioactive tracer present and the natural decay of the isotope with time (related to its half-life). The technique has been applied to measure wear, erosion, and corrosion when the corrosion product is removed from the surface. Different components in a system can be irradiated with different isotopes to allow simultaneous measurements of these components. Activation area can be small (<1 mm^2), which allows monitoring from specific metallurgical areas such as welds and weld heat-affected zones.

The gamma radiation can penetrate typically up to 5 cm of steel with acceptable signal attenuation, so that on-site external monitoring of internally generated activated layers is possible without any physical interconnection between internal and external surfaces. To compensate for the natural isotope decay and the natural background levels, three measurements are used for each reading, namely the activated component, the reference sample of the same isotope (for natural decay), and the background radiation (for naturally occurring radiation in the system and atmosphere) (13). However, these techniques measure strictly material loss and do not distinguish between erosion and corrosion effects. Additionally, corrosion products still adherent to the material surface are not measured as material loss.

4.4.2.2. Field Signature Method

Field signature method (FSM) is a nonintrusive technique that can monitor corrosion of a pipe wall directly. The original development of this technique was largely directed at oil and gas production. Typical applications involve pin attachment to the external surface of a pipeline, which is normally protected against corrosion, to monitor corrosion damage to the inside of the pipe wall. The technique measures corrosion damage over several meters of an actual structure and can be used for locations where access for traditional intrusive probes is difficult, and for high temperatures ($>150°$C) where the application of probes and ultrasonic testing (UT) is limited. The technique is useful in difficult access areas because generally no access is required after the initial installation. There are no consumable parts in this system except for the pipe spool itself that may be consumed when used in high corrosion rate conditions.

In FSM, a current is induced in the monitored section of interest, typically 3–10 m apart for a large pipe or a few centimeters for small pipes, and the resulting voltage distribution is measured to detect corrosion damage (Fig. 4.56). An array of pins is attached by stud welding, gluing, or spring-loading to make electrical connections externally over the structure.

From a corrosion perspective, the location and type of corrosion to be expected is critical in designing a FSM monitoring matrix. Up to 84 pin pairs can

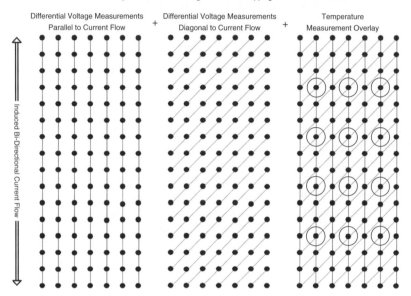

Fig. 4.56 Schematic of the electric field monitoring sensor array for high-resolution mapping of isolated pit defects with FSM. (Courtesy of Eric Kubian, PinPoint Corrosion Monitoring Inc.)

be typically utilized for one location. If monitoring for localized attack, the pin pairs are set up closer together, whereas for uniform corrosion the pin pairs can be spread out to cover a much larger area. A general rule for covering both types of corrosion is to position the pin pairs at three times the pipe wall thickness (55).

The voltage readings are monitored and compared looking for any nonuniformity that could be due to cracking or pitting in the monitored section (Fig. 4.57 and 4.58). Comparison of the average voltage drop with that of a reference element is used to monitor general metal loss. The area monitored is dependent on the pin spacing with a resolution inversely proportional to pin spacing and wall thickness (13). When a reference element is used, the sensitivity to general metal loss is approximately one part per thousand of wall thickness. The FSM monitoring unit has a lower sensitivity when not permanently attached because of the need to relocate the contact pins accurately for voltage measurements.

Once the locations vulnerable to the corrosion mechanism are determined, the process temperature needs to be considered. Two types of FSM systems are

4.4. Corrosion Monitoring Techniques   265

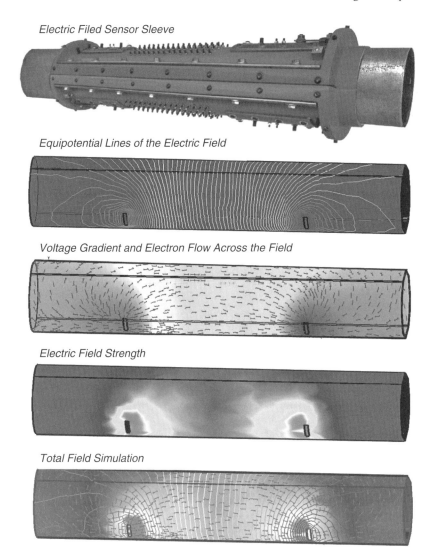

Fig. 4.57 Electric field sensor sleeve and various representations of the electric behavior between two sensor points when monitoring with FSM. (Courtesy of Eric Kubian, PinPoint Corrosion Monitoring Inc.)

available. One is a high-temperature system and the other a low temperature system, with the cut off being ~150°C due to hardware requirements. The high-temperature system requires a further junction box located outside the insulation. However, in theory the FSM technique has no temperature limits (55). Following the installation of the FSM system, attention must be given to data interpretation, which can be quite complex. This technique does not distinguish between internal and external flaws or general material loss.

266 Chapter 4 Corrosion Monitoring

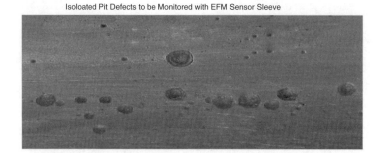

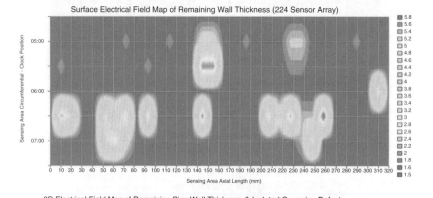

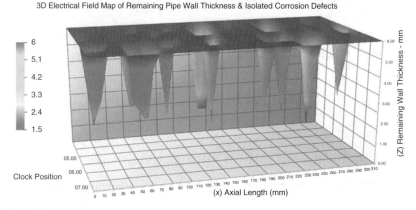

Fig. 4.58 Example of FSM results plotted as a three-dimensional (3D) map obtained on a pitted pipe using a 224 sensor array. (Courtesy of Eric Kubian, PinPoint Corrosion Monitoring Inc.)

4.4.2.3. Acoustic Emission

Acoustic emission (AE) is based on measuring acoustic sound waves that are emitted during the growth of microscopic defects, such as stress corrosion cracks. The sensor elements are thus essentially very sensitive microphones, which are strategically positioned on structures. The sound waves are generated from mechanical stresses generated during pressure or temperature changes. Background noise effects have to be taken into consideration and can be particularly troublesome in on-line measurements. The technique produces large amounts of data and requires relatively sophisticated filtering and analysis.

It is the flaw growth or even plastic deformation that generates the sound waves that are detected by the acoustic sensors. Secondary emissions from crack fretting or breaking of corrosion pockets can also be employed for the detection of flaws. The technique is essentially qualitative in identifying areas with flaws that may be further investigated with other nondestructive techniques, (e.g., ultrasonics) (13).

Testing is most commonly done off-line by applying pressure changes, or during shutdown as the temperature is changing, to minimize the background noise effects. On-line testing is also done, but the range of detection is reduced due to background noise. Frequency and gain of the sensors are modified according to the procedure to adjust filtering. Off-line tests are short-term tests to monitor for integrity whereas on-line tests are used to track the operational conditions that promote flaw propagation. A frequency range of 150–175 kHz is used for initial wide coverage in off-line applications. For on-line applications in which flaw noise is the targeted problem, the frequency used is ~1 MHz, which reduces the range of coverage to ~0.50 m.

Triangulation can be used to estimate the location of growing flaws. Acoustic energy measured at the sensor depends on the magnitude and distance of the flaw from the sensor. In off-line tests, overpressure only detects flaws that have sufficient energy during the monitoring period to be on the verge of propagation. It does not detect flaws that are not actively growing under the applied stress excursions, even though these flaws may be potential weak points or actively growing under a more severe process excursion.

4.4.3. Indirect On-Line Techniques

A great variety of techniques have been developed to measure the indirect changes, either in the environment or in the metallic component of interest, that can occur during corrosion processes or that can increase the rate at which these processes are occurring. This variety is reflected in the number of disciplines (e.g., metallurgy, physics, biology, chemistry, nuclear science) involved in the development of these tools designed to track and assess the effects and factors associated with corrosion damage.

4.4.3.1. Hydrogen Monitoring

The principle of the hydrogen probe relies on the fact that one of the cathodic reaction products in nonoxidizing acidic systems is hydrogen which, in its atomic form, can diffuse through the thickness of vessel or pipe walls and recombine into hydrogen molecules at the exterior surface. This "uptake" of hydrogen occurs when the recombination of hydrogen atoms and subsequent release as molecular hydrogen in the environment is inhibited by the presence of some chemical species or poisons (e.g., cyanides, arsenic, antimony, selenium, or sulfur compounds). The generation of atomic hydrogen can be used for corrosion monitoring purposes in both intrusive and nonintrusive ways. In the latter, hydrogen monitoring sensors are attached to the outside walls of vessels and piping. It is the diffusion of atomic hydrogen through the metallic substrate that is of most concern, as it can lead to problems such as hydrogen induced cracking.

Hydrogen monitoring is highly applicable to the oil refining and petrochemical industries with hydrocarbon process streams. The presence of hydrogen sulfide in these industries promotes the uptake of hydrogen into plant components. Several types of hydrogen probe have been developed to monitor hydrogen flux (Fig. 4.59) (56). These probes are based on one of the following three principles:

1. **Hydrogen pressure (or vacuum) probe:** One version of this probe technology is intrusive and consists in the insertion of a steel tube or cylinder that includes an inner cavity. The pressure in the inner cavity of the tube or cylinder is measured with a pressure gauge. Another version is nonintrusive and consists of a patch-type device, in which a patch or foil is welded or otherwise sealed to the outside of the pipe or vessel to create a cavity. In some implementations, the pressure range from zero absolute to atmospheric pressure, that is, the vacuum region. Other types use positive pressure above atmospheric pressure. Hydrogen passing through the wall of the tube or cylinder is detected as an increase in pressure in the cavity as a function of time. The higher the hydrogen flux, the greater the rate of pressure increase with time (13).

2. **Electrochemical hydrogen patch probe:** This nonintrusive device is attached to the outer surface of the pipe or vessel to be monitored. The patch probe itself consists of a small electrochemical cell containing an appropriate electrolyte in contact with the component to be monitored (Fig. 4.60). Typically, this cell uses the nickel electrode from a nickel

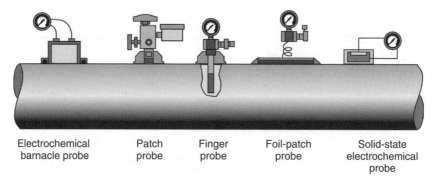

Fig. 4.59 Various hydrogen probes, which are inserted into a vessel or pipe, or are external. (Adapted from Ref. 56.)

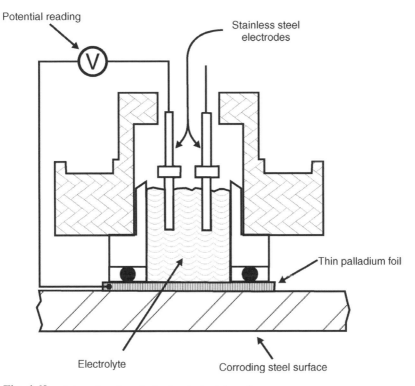

Fig. 4.60. Schematic of an electrochemical patch probe.

cadmium battery with a caustic electrolyte put directly on the pipe. Often the electrolyte is isolated from the outer surface of the pipe or vessel by a thin palladium foil. This foil is also used as an electrode in the detection circuit. The cell is operated at a potential that oxidizes hydrogen as it enters the cell. The current used to maintain this potential is proportional to the hydrogen flux into the cell (13).

3. **Hydrogen fuel cell probe:** This other nonintrusive technique is also an electrochemical device that consists of a small fuel cell. As shown in Fig. 4.61, the cell contains a solid electrolyte membrane, a cathode material, and an anode which consists of a surface catalyzed with palladium. Hydrogen entering the cell is reacted on the activated palladium surface while the cathode material reacts with ambient oxygen (air) causing a current flow in an external circuit between the cathode and anode. This current is directly proportional to the hydrogen flux through the palladium membrane (57).

When the relative distribution of atomic hydrogen in a material is constant, the permeating fraction of hydrogen can give relative corrosion rate information. However, the technique is used more commonly to detect high flow rates of the

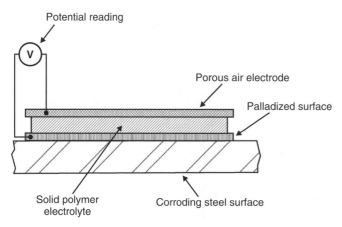

Fig. 4.61. Schematic of the hydrogen detector using a solid electrolyte proton exchange membrane fuel cell (57).

hydrogen passing through the steel that could lead to hydrogen blistering and hydrogen-induced cracking (HIC).

Hydrogen evolution is only one of a few possible cathodic reactions that can occur concurrently with the actual purely anodic corrosion reaction. This technique is not considered suitable to indicate corrosion rates in conditions or environments when any of the other cathodic reactions (e.g., reduction of dissolved oxygen) would be dominant. Currently there is no known absolute correlation between hydrogen diffusion rates and corrosion rates, cracking, or blistering without specific analysis of a piece of the actual steel affected.

4.4.3.2. Corrosion Potential

Measurement of the corrosion potential (E_{corr}) is a relatively straightforward intrusive technique that is widely used in industry for monitoring reinforcing steel corrosion in concrete and structures, such as buried pipelines under cathodic or anodic protection. Changes in corrosion potential can also give an indication of active–passive behavior in stainless steel. However, this technique only indicates a corrosion risk and does not measure corrosion or pitting rates.

The parameter E_{corr} (also known as rest potential, open-circuit potential, or freely corroding potential) is measured relative to a reference electrode, which is characterized by a stable half-cell potential. Either a reference electrode has to be introduced into the corrosive medium for these measurements, or an electrical connection has to be established to a structure in conjunction with an external reference electrode in contact with the structure with a wet electrolyte.

The success of corrosion-potential measurements depends on the long-term stability of the reference electrode. Such electrodes have been developed for continuous application for measuring corrosion potentials. However, conditions of temperature, pressure, electrolyte composition, pH, and other possible variables can limit the usage of these electrodes for corrosion-monitoring service. Special reference electrodes are needed for operating temperatures above about the boiling point of water (100°C).

The use of E_{corr} measurements for in-service corrosion monitoring in process environments is not as widespread as the use of polarization resistance. However, this approach can be valuable in some cases, particularly where an alloy could show both active and passive corrosion behavior in a given process stream. For example, stainless steels can provide excellent service, as long as they remain passive. However, if an upset occurs that might introduce either chlorides or reducing agents into the process stream, these alloys could become active and exhibit excessive corrosion rates. Corrosion-potential measurements would indicate the development of active corrosion that could be confirmed by using LPR, for example (15).

4.4.3.3. On-Line Water Chemistry Analyses

Various types of chemical analyses can provide valuable information to a corrosion monitoring programs. The measurement of pH, conductivity, dissolved oxygen, metallic and other ion concentrations, water alkalinity, concentration of suspended solids, inhibitor concentrations, and scaling indexes ices all fall within this domain. Several of these measurements can be made on-line using appropriate sensors.

All of the following measurements are used particularly in water treatment, and in general apply only to aqueous environments. Many of these parameters are used in combination to generate various indices, such as the Langelier index,[4] in order to rate the scaling tendency and corrosivity of waters.

4.4.3.3.1. pH The pH[5] of an aqueous environment may be measured with a meter or calculated if certain parameters are established. Water itself dissociates to a small extent to produce equal quantities of H^+ and OH^- ions displayed in the equilibrium described in Eq. (4.8).

$$H_2O \rightleftarrows H^+ + OH^- \tag{4.8}$$

The term pH was derived from the manner in which the hydrogen ion concentration is calculated. It is the negative logarithm of the hydrogen ion (H^+) concentration as shown in Eq. (4.9).

$$pH = -\log_{10}(a_{H^+}) \tag{4.9}$$

where log is a base 10 logarithm and a_{H+} is the activity (related to concentration) of hydrogen ions.[6]

---

[4] See Appendix B for additional information on Langelier, Ryznar, and other scaling indices.
[5] The pH, originally defined by Danish biochemist Søren Peter Lauritz Sørensen in 1909, is a measure of the concentration of hydrogen ions.
[6] The *p* in Eq. (4.9) stands for the German word for "power", potenz, so pH is an abbreviation for power of hydrogen.

A higher pH means that there are fewer free hydrogen ions. A change of one pH unit reflects a 10-fold change in the concentrations of the hydrogen ion. For example, there are 10 times as many hydrogen ions available at pH 7 than at pH 8. Substances with a pH < 7 are considered to be acidic and substances with pH > 7 are considered to be basic or alkaline. Thus, a pH of 2 is very acidic and a pH of 12 very alkaline. In general, for steel, low pH (or high acidity) produces a particularly corrosive environment. For other alloys this can vary.

The pH is an important corrosion factor for two reasons. First, the H^+ ion can be reduced into hydrogen and thus participate to the overall corrosion processes as the cathodic reaction. Second, the pH influences the solubility of the products of the chemical corrosion reactions, particularly the passivation reactions involving oxides, sulfides, or carbonates. As a measure of alkalinity, pH is a very important component in scaling indices.[7]

Generally, pH measurements have been limited to a maximum temperature of boiling water and to a maximum pressure of ~2 MPa, although some special-purpose probes are available for up to 70 MPa.

The pH measurements can be affected by interfering ions, such as lithium, sodium, and potassium ions that can interact with the sensing glass membrane of a pH electrode. However, because lithium ions are normally not found in sample solutions and potassium ions cause little interference, the most significant interference is from sodium ions.

Additionally, fouling of the probe measurement element by hydrocarbons, for example, can reduce or even completely block the probe response to pH changes. Frequent maintenance might thus be required to ensure cleanliness and maintain calibration. Low-conductivity solutions can also be a problem. Even low-ionic-strength pH probes can have problems at conductivities less than 20 $\mu S\ cm^{-1}$. Some low-conductivity probes inject an electrolyte to correct for high solution resistance.

4.4.3.3.2. Conductivity In the present context, conductivity is the electrolytic current-carrying capacity of water, which is largely determined by the concentration of ions from dissolved salts and solids (58). When a salt dissociates, the resulting ions interact with surrounding solvent molecules or ions to form charged clusters known as solvated ions. These solvated ions can move through the solution under the influence of an externally applied electric field. Such motion of charge is known as ionic conduction. Ionic conductance, which for the bulk solution is the only conductance present, is the reciprocal of ionic resistance. The dependence upon the size and shape of the conductor can be corrected by using conductivity κ rather than conductance G, as expressed in Eq. (4.10) for the simple geometry shown in Fig. 4.62.

$$\kappa = \left(\frac{\ell}{A}\right)\frac{1}{R} = G\left(\frac{\ell}{A}\right) \qquad (4.10)$$

---

[7]See Appendix B for additional information on Langelier, Ryznar, and other scaling indices.

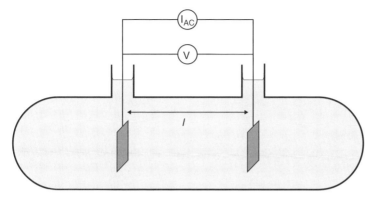

Fig. 4.62. Schematic of a conductivity cell containing an electrolyte and two inert electrodes of surface A parallel to each other and separated by distance ℓ.

where ℓ is the length of the conductor, that is, the gap separating the electrodes in Fig. 4.62; A is the cross-sectional area of each electrode, assuming that both electrodes have the same dimensions.

Because corrosion is an electrochemical process, an increase in conductivity is generally associated with an increase in corrosivity. However, this does not mean that zero electrolyte conductivity signifies that corrosion is stifled. Extra pure water such as used in nuclear reactors, for example, has proven to be quite corrosive unless small amounts of oxygen or other passivating agents are added to counter this aggressiveness.

Electrochemical reactions at the electrode interfaces can affect the readings and can be dependent on the measurement frequency used. Therefore a frequency of 1 kHz is commonly used as the measurement frequency to avoid this problem. Routine cleaning of the measurement cell has often been necessary to prevent false readings due to bridging of the electrodes, or fouling on the membrane.

4.4.3.3.3. Dissolved Oxygen   Dissolved oxygen refers to the amount of oxygen dissolved in a liquid, usually expressed as parts per billion (ppb) or parts per million (ppm) in the engineering world or in milligrams per liter (mg L^{-1}) in the chemical world. The solubility of oxygen depends on temperature, pressure, and molarity of the solution. Increased pressure increases oxygen solubility while an increase in temperature decreases its solubility (see Chapter 1 for additional details).

Dissolved oxygen can be measured by ion-selective electrodes, or for approximate low levels of oxygen a galvanic probe can be used (Fig. 4.63). The oxygen ion-selective electrode has a thin organic membrane covering a layer of electrolyte and two metallic electrodes. Oxygen diffuses through the membrane and is electrochemically reduced at the cathode. There is a fixed voltage between the cathode and anode so that only oxygen is reduced. The greater the oxygen partial pressure, the more oxygen diffuses through the membrane in a given time. The result is a current that is proportional to the oxygen in the sample (59).

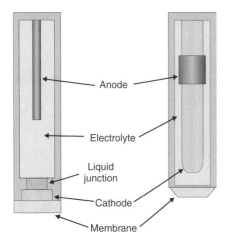

Fig. 4.63. Two different electrode chemical cell designs to measure dissolved oxygen.

Temperature sensors can be built into the probe to make corrections for the sample and membrane temperatures. The cathode current, sample and membrane temperatures, barometric pressure, and salinity information are used to calculate dissolved oxygen content of the sample in either concentration (ppb or ppm or mg L^{-1}) or percent saturation (% sat).

The affinity of oxygen for most structural metals is the cause of many corrosion phenomena. Oxygen is responsible for both corrosive attack and passivation. The approach to controlling corrosion of carbon steel in oil fields and in boilers in North America has generally been to remove all or most of the molecular oxygen by various combinations of physical and chemical means. Bringing oxygen levels <20 ppb or 20 µg L^{-1} has a significant effect on corrosion. With an oxygen scavenger, typical levels are <1 ppb or 1 µg L^{-1}. In some boilers that use high-purity deionized water, no oxygen scavenger is used but instead small amounts of oxygen or hydrogen peroxide are deliberately added to the feedwater to act as a passivating agent under carefully controlled boiler water chemistry conditions.

Generally, dissolved oxygen measurements have been limited to a maximum temperature of 65°C and a maximum pressure of 200–350 kPa. Consumption of oxygen by the probe may also lower the oxygen concentration near the probe, so that a flowing or stirred sample is generally recommended. The electrolyte is also depleted of its active ingredients as it reduces the oxygen and needs to be replaced at regular intervals of a few weeks or months depending on the design and use of the probe. Periodic calibration of the probes is therefore a recommended practice. Nonetheless, poisoning of the electrode can occur. This has precluded the use of the method in many industrial environments such as processes involving heavy hydrocarbon and water.

4.4.3.3.4. Oxidation–Reduction (Redox) Potential Oxidation reduction (redox) potential is the potential of a reversible oxidation–reduction electrode measured with respect to a reference electrode in a given electrolyte. The measurement system comprises a potentiometer (high-sensitivity and high-impedance voltmeter) connected to a reference electrode and a noble metal sensing electrode. The change in potential of the sensing electrode produced by the oxidizing or reducing species, is measured relative to the reference electrode. This voltage difference is then converted electronically to display on a meter or recorder for real-time on-line measurement.

This technique has also been used to detect the end point of oxidation or reduction reactions during a water treatment to control more closely the addition of oxidizing biocides, such as chlorine and bromine, which may have a significant effect on increasing the corrosion rates.

Redox potential is also a key measure to evaluate soil corrosivity where the redox potential is essentially related to the degree of aeration. A high redox potential indicates a high oxygen level. Low redox values may provide an indication that conditions are conducive to anaerobic microbiological activity. Sampling of soil will obviously lead to oxygen exposure and unstable redox potentials are thus likely to be measured in disturbed soil.

4.4.3.4. Process Variables

4.4.3.4.1. Flow Regime Corrosion usually occurs on the metal surface wetted by the aqueous phase. This wetting action and transport of corrosive agent to the surface can be controlled by the flow regime. In some flow regimes, such as slug flow, high surface shear stresses and extreme turbulence can make corrosion inhibition difficult and increase corrosion rates to several centimeters per year. At higher flow regimes, impingement, cavitation, and erosive conditions created by two-phase flows can be extremely corrosive.

In single-phase flows, the flow regime is typically described as laminar, transition, or turbulent flow as defined by the Reynolds number, which depends on the flow velocity, the pipe diameter, the fluid viscosity, and the fluid density. In multiphase flows the relative pattern of multiple phases is important as defined by mist flow, annular flow, or slug flow. In single-phase environments, the flow regime affects the mass transfer to the metal surface and the shear stresses on the metal surface. These have a direct impact on the corrosion rates at the metal surface (see Chapter 2 for a description of the effects of flow on corrosion).

In multiphase flows the flow regime varies with varying flow rates of each phase, elevation changes, or specific geometry of a line or conduit. Acoustic monitoring and on-line γ-radiography have been used to detect slug flow. Flow characteristics can also be used to provide predictions of corrosion rates based on modeling and other empirical data.

4.4.3.4.2. Pressure Pressure can affect the proportion of phases present in a vessel or pipe, or the composition of the process fluids. Different phases and constituents can produce quite different corrosive environments. For example, the partial pressure of CO_2 affects the amount of CO_2 dissolved in water, which in turn affects the corrosivity of the fluid due to the presence of carbonic acid. Similarly, the partial pressure of H_2S is a major determining factor in the susceptibility of various alloys to sulfide stress cracking (SSC).

Total pressure measurements can be made on-line, and are a real-time measurement. Determination of partial pressures utilizes knowledge of the composition of the process fluid, the temperature, and the total pressure. This involves sampling of the process fluid. In some gas–liquid systems, pressure affects gas solubility, but the relationships are complex. Consequently, the pressure is usually used only to analyze and predict which phases are present, rather than as an on-line measurement.

4.4.3.4.3. Temperature The temperature of a process fluid can have a direct effect on its corrosion dynamics. These temperature effects can be nonlinear. Low temperature can produce condensation of water or other corrosive liquids. High temperature increases chemical reaction rates and can change the composition of the process. In general, increased temperature increases the chemical reaction rates. The temperature can lead either to vaporization (a dry condition) or condensation (from a dry to a wet condition). Both these changes in state can affect corrosion.

4.4.3.4.4. Dewpoint Dewpoint is the temperature at which a liquid starts condensing from a vapor or gaseous phase. Dewpoint monitoring is important because the region of condensation of water and corrosive fluids in otherwise dry environments has a major impact on corrosion rates. This can be complex in some environments where multiple gases and condensable combinations are present. In atmospheric environments in which the condensing liquid is water, the dewpoint temperature is measured with a wet-bulb thermometer. In a process system, this may not be practical, and a system in which cooling of the flow to the point of condensation on an optical mirror has been used.

The direct effects of dewpoint conditions can be measured with mass-loss coupons or electrical resistance techniques. Electrochemical methods, such as electrochemical noise and multielectrode array systems (CMAS), have also been used in a similar way, either as a time-of-wetness indicator or for corrosion rate measurements, since these electrochemical techniques will function even when surfaces are only partially wetted. Additionally, forced cooling of the corrosion measurement surfaces can be used to generate the appropriate dewpoint on the measurement surfaces.

4.4.3.5. Fouling

Fouling is an accumulation of both organic and inorganic substances from a fluid stream onto the surfaces of the equipment through which the fluid is circulated as well as *in situ* corrosion products and inorganic deposits, such as hardness salts that occur on the metal surface. Fouling is a prime cause of underdeposit corrosion since it can produce highly anodic areas and give rise to hot spots in boilers, for example. Fouling may also restrict flow by causing an increase in pressure drop through the equipment or retard heat transfer by formation of an insulating deposit. Side-stream and in-line intrusive measurement techniques, together with visual inspection, have been used for determination of fouling. One of two techniques has been employed, that is, heat-transfer or pressure-drop monitoring.

4.4.4. Indirect Off-Line Measurement Techniques

A great variety of chemical analysis techniques have been used over the years to provide operators and providers of corrosion control services some estimates of the effects of corrosion on systems, on the corrosivity of given environments, or on the effective usage of chemical additives. The following sections describe the type of information commonly sought with these techniques (13).

4.4.4.1. Off-Line Water Chemistry Parameters

4.4.4.1.1. Alkalinity Alkalinity is considered to be an important measurement when handling waters because carbonate and hydroxide scales are common problems. The alkalinity of a water as determined by titration with standard acid solution to the methyl orange endpoint (pH ~4.5) is sometimes abbreviated as "M alkalinity". Total alkalinity includes many alkalinity components, such as hydroxides (OH^-), carbonates (CO_3^{2-}), and bicarbonates (HCO_3^-).

4.4.4.1.2. Metal Ion Analysis Some corrosion products are soluble and can be detected by metal ion analysis of the process stream for determining the amount of metal lost that has dissolved in the process stream or the amount that has been carried along in the process stream as corrosion product (e.g., iron, copper, nickel, zinc, manganese). The analysis can be done easily, inexpensively, and quickly in the field. The assumption that metal loss occurred over the total surface area or that the concentration of ions in solution is proportional to the corrosion rate is often incorrect, in which case the technique may only provide a trend indication. An increase in the metal ion concentration can indicate an increase in corrosion. However, a low metal ion concentration is not a guarantee of low corrosion, due to the possibility of localized corrosion, deposition of the metal ion due to temperature or pH changes, or a significant time delay before analysis.

Metal ion analysis is most useful when applied in closed systems and if the corrosion products are soluble or relate to particular concentrations of soluble species. In open systems, the relative changes in ion concentrations between locations can provide some useful information. The technique is generally not reliable in fluids containing hydrogen sulfide due to the precipitation of insoluble sulfides from soluble metallic ions, or in alkaline solutions because of precipitation of hydroxides.

Much care should be used in obtaining a representative sample of the aqueous phase, including the sampling point design, because the sampling point can accumulate corrosion products. Obtaining a representative sample from a process stream can also be difficult, because fluid velocity, temperature, and pressure can vary greatly with time unless specific precautions are taken to prevent this.

4.4.4.1.3. Concentration of Dissolved Solids Total dissolved solids (TDS) is the sum of minerals dissolved in water. It is therefore an important parameter for the determination of a scaling index.[8] Total dissolved solids is determined by evaporating the water from a preweighed sample. A calculated TDS value can be determined by adding the various cations and anions from an analysis.

4.4.4.1.4. Gas Analysis Gas analysis commonly includes hydrogen, H_2S, or other dissolved gases. This technique is generally done in a laboratory, but can also be done for certain gases in the field to determine potentially corrosive constituent gases, particularly acid gases, that become corrosive when hydrated, for example, when the temperature falls below the dewpoint. However, the technique is limited to very specific gaseous process situations and the equipment for analysis can be relatively expensive to acquire and maintain.

4.4.4.1.5. Residual Oxidant Ozone and halogens, specifically chlorine, chlorine dioxide, and bromine, are powerful oxidizing agents and are widely used to control microbiological fouling in aqueous systems. Residual halogen can directly oxidize the inhibitors used to protect against corrosion or fouling. Dissolved halogens in aqueous environments can be measured using redox potential or one of a variety of colorimetric techniques.

Halides are salts that come from halogen acids, typically HF, HCl, HBr, and HI. Halides have been directly implicated as a causative agent in SCC and have been indirectly linked to galvanic corrosion by increasing the conductivity of the environment. Halides can be measured using specific ion electrodes or with a colorimetric method.

4.4.4.1.6. Microbiological Analysis Microbiological influenced corrosion is a very pervasive factor in many corrosion situations. Due to the importance of the subject, a special section of this chapter is devoted entirely to the techniques and methods that have been developed to monitor microbes and bacteria that are the most susceptible to cause corrosion damage.

4.4.4.2. Residual Inhibitor

Measurement of corrosion inhibitor residuals in a system provides an indication of the concentrations of inhibitor at various locations of a system. When the reliability of an analytical test method is established and the minimum acceptable inhibitor concentration has been determined, measurement of inhibitor concentrations throughout a system can indicate whether adequate corrosion protection

---

[8]See Appendix B for additional information on Langelier, Ryznar, and other scaling indices.

is likely to be achieved at each sample location. Additional inhibitor loss due to reaction with the system hardware or reactions with the environment (e.g., absorption, neutralization, precipitation, adsorption on solids or corrosion products) can be detected and compensated for by dosage selection.

- **Filming corrosion inhibitors:** Filming corrosion inhibitors are chemical products used in low concentrations to adsorb on system surfaces and shield them from corrosive agents. Filming inhibitor residual measurements are usually made to ensure that an adequate supply of inhibitor has been introduced in the system to compensate for a reduction of the inhibitor concentration by adsorption and maintenance of inhibitor film;
- **Reactant corrosion inhibitors:** Reactant corrosion inhibitors are designed to react with potential corrosive agents in a system to neutralize their harmful effect and combine with constituents in a system to produce *in situ* corrosion inhibitors. Reactant corrosion inhibitor residual measurements are used to ensure that sufficient reactant inhibitors have been injected into the system to provide an excess of reactant. Measurement of residual reactant inhibitor can be taken after the point of introduction of oxidant (oxygen or air entry, e.g.) to ensure that sufficient inhibitor has been injected to perform the reaction while leaving some residual inhibitor.

Reactant corrosion inhibitors can be alkaline neutralization chemicals used to maintain a safe operating pH and, for example, adjust the pH at the point of condensation of acid gases in a distillation system. A typical parameter in determining the effectiveness of such an inhibitor is the ability to measure the pH at appropriate points in a system. Reactant inhibitors can also be used to react with or reduce oxidants in a system and thus remove them from the environment. This type of inhibitor or oxygen scavenger can include sulfites, bisulfites, and hydrazine for example.

4.4.4.3. Chemical Analysis of Process Samples

Chemical analysis of process samples taken at times of high and low corrosion rates can be useful in identifying the constituents that cause the high corrosion rates. From this information, the source of aggressive constituents can be found and corrected or modified. In petroleum production, crude oil and gas condensate samples are typically analyzed for organic nitrogen and acid. Sulfur, organic acid, nitrogen, and salt content are typically measured for refining purposes. These parameters are used in combination to predict the corrosivity of the oil.

In gas handling and gas processing operations, samples of produced and processed natural gas and gas–liquids are typically analyzed to determine hydrogen sulfide (H_2S), carbon dioxide (CO_2), water (H_2O), carbonyl sulfide (COS), carbon disulfide (CS_2), mercaptans (RSH), and/or oxygen (O_2) to predict and assess corrosion potential in producing wells, gas gathering systems, and gas processing operations.

In this monitoring technique, a representative sample of a process stream is taken and kept in a vessel that maintains the original condition of the sample for subsequent analysis. The sampling methods can be quite complex since the samples need to be maintained under the pressure conditions of the process stream. This can be particularly difficult for high-pressure process systems.

4.4.4.3.1. Sulfur Content Sulfur is the most abundant element in petroleum other than carbon and hydrogen. Corrosion of carbon steel may become extremely high ($\sim$100 mm year^{-1}!!) when the sulfur >0.2%. It can be present as elemental sulfur, dissolved hydrogen sulfide, mercaptans, sulfides, and polysulfides. The total sulfur content is generally analyzed according to ASTM D 4294. Halides and heavy metals interfere with this method and the capability of a sulfur compound to form H_2S during heating in the refinery process, rather than the total amount of the compound, is believed to correlate with corrosion in the plants.

4.4.4.3.2. Total Acid Number Acid content is generally expressed in terms of total acid number (TAN) or neutralization (neut) number. For the purpose of predicting corrosion in crude distillation units in refineries, the TAN threshold is believed to be $\sim$0.5 for whole crude and 2.0 for the cuts. In petroleum production, it was found that the corrosion rate is usually inversely proportional to the algebraic product of the nitrogen concentration and the TAN.

With this monitoring technique, an oil sample is dissolved in a mixture of toluene and isopropyl alcohol containing a small amount of water. The solution is then titrated with an alcoholic potassium hydroxide solution (KOH). Both ASTM D 664 and ASTM D 974 can be used to determine the TAN. However, ASTM D 664 yields numbers 30–80% higher than does ASTM D 974.

Inorganic acids, esters, phenolic compounds, sulfur compounds, lactones, resins, salts, and additives, such as inhibitors and detergents, sometimes interfere with the measurement. Universal Oil Products (UOP) Methods 565 and 587 give procedures to remove most of the interfering materials before performing the analysis for organic acids.

4.4.4.3.3. Nitrogen Content The total nitrogen is generally analyzed according to ASTM D 3228 for the assessment of the corrosivity of process feedstocks. High nitrogen content indicates corrosion-inhibitive properties of the crude oil or condensate in petroleum production. As already stated, the corrosion rate is inversely proportional to the algebraic product of the organic nitrogen concentration in weight percent and the TAN. In refining, however, when the organic nitrogen concentration exceeds 0.05%, cyanide and ammonia can form, collect in the aqueous phase, and corrode certain materials.

4.4.4.3.4. Salt Content of Crude Oil Salt (primarily sodium chloride with lesser amounts of calcium and magnesium chloride) is present in produced water, and this produced water can be dispersed, entrained, and/or emulsified in crude oil. Salt can cause corrosion of refinery equipment and piping by the formation of hydrochloric acid through hydrolysis and heating. Salt precipitates can form scale in heaters and heat exchangers and this can result in accelerated corrosion of equipment. Refineries generally limit salt content of crude oil for processing to 2.5–12 mg L^{-1}.

The salt content of crude oil can be determined using ASTM D 3230. The analytical procedure assumes that calcium and magnesium exist as chlorides and all the chloride is calculated as sodium chloride.

4.5. MONITORING MICROBIOLOGICALLY INFLUENCED CORROSION

An effective biocorrosion mitigation program needs to include corrosion monitoring as a periodic or continual means of assessing whether program goals are being achieved. This is particularly true in industrial water-handling systems with known susceptibility to biocorrosion (e.g., cooling water and injection water systems, heat exchangers, wastewater treatment facilities, storage tanks, piping systems, and all manner of power plants, including those based on fossil fuels, hydroelectric, and nuclear) (60). Table 4.7 lists potential problem areas by industry (61).

As mentioned in Chapter 1, the interaction of microbial metabolism and corrosion processes can produce localized attack at very high rates. Monitoring techniques that detect the presence of microbes, especially on metallic surfaces, can provide an early indication of incipient MIC or the potential for MIC. A number of methods for the detection of microorganisms, including specific types of organisms and estimates of their numbers and activity, have been developed (12).

The first biocorrosion monitoring systems were focused on assessing the number of microbes per unit volume of water sampled from the system. These data were combined with electrochemical corrosion measurements, using ER or LPR probes in addition to coupon weight loss measurements. The problem with this approach is that the number of free-floating planktonic organisms in the water does not correlate well with the organisms present in biofilms on the metal surface where the corrosion actually takes place. An effective monitoring scheme for controlling both biofouling and biocorrosion should include data gathering of as many of the following types of data as possible (60):

- Sessile bacterial counts of the organisms in the biofilm on the metal surface done by either conventional biological techniques or optical microscopy.
- Direct observation of the community structure of the biofilm. This can be done on metal coupons made from the same alloy used for the system. Several types of probe systems are commercially available for holding and inserting such coupons into the system. Examination of the biofilm has been done by SEM, epifluorescence optical microscopy, or confocal laser scanning microscopy.
- Identification of the microorganisms found in both the process water and on the metal surface.

Table 4.7. Where MIC Problems Are Most Likely To Occur

Industry/Application	Potential Problem Sites for MIC	Organisms Responsible
Pipelines-oil, gas, water, wastewater	Internal corrosion primarily at the bottom (6:00) position Dead ends and stagnant areas Low points in long-distance pipes Waste pipes-internal corrosion at the liquid/air interface Buried pipelines-on the exterior of the pipe, especially in wet clay environments under disbonded coating	Aerobic and anaerobic acid producers, SRB, manganese and iron-oxidizing bacteria, sulfur-oxidizing bacteria
Chemical process industry	Heat exchangers, condensers, and storage tanks-especially at the bottom where there is sludge build-up Water distribution systems (See also "Cooling Water Systems," "Fire Protection Systems," and "Pipelines" in this table)	Aerobic and anaerobic acid producers, SRB, manganese, and iron-oxidizing bacteria In oil storage tanks also methanogens, oil-hydrolyzing bacteria
Cooling water systems	Cooling towers Heat exchangers-in tubes and welded areas-on shell where water is on shell side Storage tanks-especially at the bottom where there is sludge build-up	Algae, fungi, and other microorganisms in cooling towers Slime-forming bacteria, aerobic and anaerobic bacteria, metal-oxidizing bacteria, and other microorganisms and invertebrates
Fire protection systems	Dead ends and stagnant areas	Anaerobic bacteria, including SRB
Docks, piers, oil platforms, and other aquatic structures	Just below the low-tide line Splash zone	SRB beneath barnacles, mussels, and other areas sequestered from oxygen
Pulp and paper	Rotating cylinder machines Whitewater clarifiers	Slime-forming bacteria and fungi on paper making machines Iron-oxidizing bacteria SRB in waste water
Power generation plants	Heat exchangers and condensers Firewater distribution systems Intakes	As above for heat exchangers and fire protection systems Under mussels and other fouling organisms on intakes
Desalination	Biofilm development on reverse osmosis membranes	Slime-forming bacteria

- Surface analysis to obtain chemical information on corrosion products and biofilms.
- Evaluation of the morphology of the corrosive attack on the metal surface after removal of biological and corrosion product deposits. Conventional macrophotography, as well as low-power stereomicroscopy, optical microscopy, metallography, and SEM may all be helpful in this regard.
- Electrochemical corrosion measurements.
- Water quality and redox potential measurements.
- Other types of information specific to each operational system, including duty cycle and downtime information, concentrations and timing for addition of biocides and other chemical inputs, local sources and nature of pollutants, and so on.

4.5.1. Planktonic Organisms

Depending on the type of industrial system, planktonic organisms may include, besides bacteria, unattached algae, diatoms, fungi, and other microorganisms present in a system bulk fluids. In most cases, it is planktonic bacteria that are the focus of monitoring for MIC using microbiological detection techniques, since system fluids are generally easier to sample than metallic surfaces. Unfortunately, the levels of planktonic bacteria present in the liquids are not necessarily indicative of MIC problems or their severity (12).

At best, the detection of viable planktonic bacteria may serve as an indicator that living microorganisms are present in a particular system, some of these organisms are capable of participating in the microbial attack. It is generally very important that additional monitoring methods be performed to confirm that actual corrosion due to microbial processes has occurred.

4.5.2. Sessile Organisms

Microorganisms that are attached to a surface are termed "sessile" organisms. These organisms are most often present as a consortium or community, collectively referred to as a biofilm. Complex assemblages of various species may occur within both planktonic and sessile microbial populations. The environmental conditions largely dictate whether the microorganisms will exist in a planktonic or sessile state.

Sessile microorganisms do not attach directly to the actual surface, but rather to a thin layer of organic matter adsorbed on the surface (Fig. 4.64, Stages 1 and 2). As microbes attach to and multiply, a biofilm composed of immobilized cells and their extracellular polymeric substances builds up on the surface.

The growing biofilm increasingly prevents the diffusion of dissolved gases and other nutrients coming from the bulk liquid. These changing conditions become inhospitable to some microorganisms at the base of the biofilm, and eventually many of these cells die. As the foundation of the biofilm weakens,

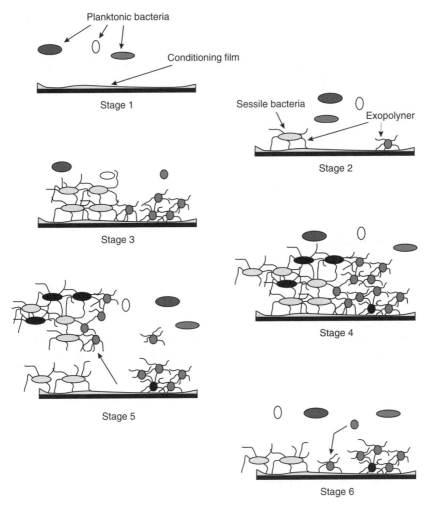

Fig. 4.64. Different stages of biofilm formation and growth. Stage 1: Conditioning film accumulates on submerged surface. Stage 2: Planktonic bacteria from the bulk water colonize the surface and begin a sessile existence by excreting exopolymer that anchors the cells to the surface. Stage 3: Different species of sessile bacteria replicate on the metal surface. Stage 4: Microcolonies of different species continue to grow and eventually establish close relationships with each other on the surface. The biofilm increases in thickness. Conditions at the base of the biofilm change. Stage 5: Portions of the biofilm slough away from the surface. Stage 6: The exposed areas of surface are recolonized by planktonic bacteria or sessile bacteria adjacent to the exposed areas (62).

shear stress due to adjacent fluid flow, for example, may cause sloughing of cell aggregations exposing the bare surface to the bulk fluid in localized areas (Fig. 4.64, Stage 5). The exposed areas are subsequently recolonized and new microorganisms and their exopolymers are woven into the fabric of the existing biofilm (Fig. 4.64, Stage 6). This phenomenon of biofilm instability occurs even when the physical conditions in the bulk liquid remain constant. Thus, biofilms are constantly in a state of flux (62).

Since MIC occurs directly on metal surfaces, the presence of sessile organisms is an important component to monitor. Monitoring sessile organisms either requires that the system be regularly opened for sampling or that accommodations be made in the system design to allow for regular collection or on-line tracking of attached organisms while the system continues to operate. It should be pointed out, however, that the presence of viable sessile organisms does not always translate into actual attack on the metal surfaces. Again, it is a good idea to use additional methods that directly determine the presence of active MIC.

4.5.3. Sampling

Samples for analysis can be obtained by scraping accessible surfaces. In open systems or on the outside of pipelines or other underground facilities, this can be done directly. Bull plugs, coupons or inspection ports can provide surface samples in low pressure water systems (63). More sophisticated devices are commercially available for use in pressurized systems (64). In these devices, coupons are held in an assembly that mounts on a standard pressure fitting.

If the biofilm developing on the coupon is to be representative of the behavior in a system it is important for the sampling coupons to be made of a material similar to the system material and that they be flush mounted in the wall of the system to have the same flow effects as those of the surrounding surface. While pressure fittings allow inserting coupons directly in process units, these fittings can be expensive. Pressure vessel codes and accessibility can also restrict possible locations. For these reasons, sidestream installations are often preferred.

Handling of field samples should be done carefully to avoid contamination with foreign matter including biological materials. Many different types of sterile sampling tools and containers are commercially available. Because many systems are anaerobic, proper sample handling and transport is essential to avoid exposure to oxygen from the air. One option is to analyze samples on the spot with special kits. Where transportation to a laboratory is required, Torbal jars or similar anaerobic containers can be used (65). In many cases, simply placing samples directly in a large volume of the process water in a completely filled screw cap container is adequate.

Processing in the lab should also be done anaerobically using special techniques or in anaerobic chambers designed for this purpose. Because viable organisms are involved, processing should be done quickly to avoid growth or death of cells stimulated or inhibited by changes in temperature, oxygen exposure or other factors (66).

4.5.4. Biological Assessment

Biological assays can be performed on liquid samples or on suspensions of solid deposits to identify and enumerate viable microorganisms, quantify metabolic or specific enzyme activity, or determine the concentration of key metabolites (67). Table 4.8 summarizes some of the methods that have been used to detect and describe microorganisms in terms of total cells present, viable cell numbers, and metabolic activity. Table 4.8 also identifies assays that can be used to establish the presence of biomass in a field sample, identify the organisms present, and assess the activity of enzymes, such as hydrogenase, that are thought to accelerate corrosion through cathodic depolarization. A more detailed description of these assays and other techniques is provided in the following sections.

4.5.4.1. Direct Inspection

Direct inspection is best suited for the enumeration of planktonic organisms suspended in relatively clean water. In liquid suspensions, cell densities $>10^7$ cells cm^{-3} may cause the sample to appear turbid. Quantitative enumerations using a phase contrast microscopy can be done quickly using a counting chamber that holds a known volume of fluid in a thin layer. Visualization of microorganisms can be enhanced by fluorescent dyes that cause cells to light up under ultraviolet radiation. The technique is usually performed in the laboratory (13).

Using a stain, such as acridine orange, cells separated by filtration from large aliquots of water can be visualized and counted on a 0.25-μm filter using the epifluorescent technique. Newer stains, such as fluorescein diacetate, 5-cyano-2,3-ditolyltetrazolium chloride, or p-iodonitrotetrazolium violet, indicate active metabolism by the formation of fluorescent products (66). Antibody fluorescence microscopy is similar to general fluorescent microscopy, except that the fluorescent dye used is bound to antibodies specific to SRBs. Only bacteria recognized by the antibodies fluoresce. Results can be analyzed within 2 h. The technique detects viable and nonviable bacteria, but it is limited to the type of SRB used in the manufacture of these antibodies (13)

Identification of organisms can be accomplished by use of antibodies generated as an immune response to the injection of microbial cells into an animal, typically a rabbit. These antibodies can be harvested and will selectively bind to the target organism in a field sample. A second antibody tagged with a fluorescent dye is then used to light up the rabbit antibody bound to the target cells. In effect, the staining procedure can selectively light up target organisms in a mixed population or in difficult soil, coating or oily emulsion samples (68).

Such techniques can provide insight into the location, growth rate, and activity of specific kinds of organisms in mixed biofilm populations. Antibodies that bind to specific cells can also be linked to enzymes that produce a color reaction in an enzyme-linked immunosorbent assay. The extent of color produced in solution can then be correlated with the number of target organisms present (69). While antibody based stains are excellent research tools, their high specificity means that they can only identify the target organisms. Other organisms potentially capable of causing problems are therefore missed.

Table 4.8. Inspection, Growth, and Activity Assays for Microbial Populations

Assay	Method	Comments
Microorganisms Cell numbers Specific organisms	*Microscopic examination:* Cell numbers are obtained with a Petroff–Hausser counting chamber. Fluorescent dyes can light up specific microbes.	Requires a microscope with high magnification, phase contrast, and ultraviolet fluorescence, as appropriate. Cell counts are straightforward, but particulates and fluorescent materials may interfere.
Viable cell counts SRB APB Other organisms	*Most probable number:* Sample is diluted into a series of tubes of specific growth medium. Cell numbers based on growth at various dilutions	Commercially available kits can be inoculated in the field, but growth takes days to weeks.
	Dip slides: A tab coated with growth medium is immersed, then incubated 2–5 days. Microbial colonies are counted visually.	Crude numbers of bacteria, yeast, or fungi can be estimated in contaminated crankcase oil or fuels, for example. Commercial kits require minimal cost and expertise.
Identification of microorganisms	*Fatty acid methyl ester analysis:* Methyl esters of fatty acids from field sample are analyzed by gas chromatograph.	Fatty acid composition fingerprints specific organisms. The technique is available commercially.
	Probes based on nucleic acid base sequences: Probes bind to DNA or RNA for specific proteins or organisms.	Probes require special expertise and lab facilities.
	Reverse sample genome probing: DNA from MIC microbes is spotted on a master filter. Labeled DNA from field samples will bind to that DNA, if the reference organism is present.	The assay answers the question, "Is this organism present?" It can also be used to track changes in a population after chemical treatment but it requires special expertise and equipment.

Table 4.8. (continued)

Assay	Method	Comments
Biomass	*Protein,lipopolysaccharide,nucleicacid analysis:* Established methods, widely available	Concentration of cell constituents correlates to level of organisms present in field sample.
Metabolic activity	*Adenosine triphosphate (ATP):* Fluorometer and supplies commercially available for field use	ATP level correlates well with microbial metabolism level. Its concentration reflects the level of activity in a sample.
Sulfate reduction	^{35}S *sulfate reduction:* Radioactively labeled sulfate is incubated in field sample. Hydrogen sulfide formed is liberated by acid addition, trapped on a zinc acetate wick, and measured by scintillation counter.	A specialized lab technique useful in discovering nutrient sources for MIC and in quickly screening biocide activity in samples taken from the target system
Enzyme activity	*Hydrogenase assay or sulfate reductase assay:* Measures enzyme activity in a field sample as a rate of reaction	Commercial kits are available. Hydrogenase activity may be related to MIC, while sulfate reductase assesses the presence of SRB.

4.5.4.2. Growth Assays

The most common way to assess microbial populations in water samples is through growth tests using commercially available growth media for groups of organisms most commonly associated with industrial problems. These are packaged in a convenient form suitable for use in the field. Serial dilutions of suspended samples are grown on solid agar or liquid media. Based on the growth observed for each dilution estimates of the most probable number (MPN) of viable cells present in a sample can be obtained (70). The test can show some results in a few days, but the usual incubation period for the test is 14–28 days (13).

However, despite the common use of these assays, only a small fraction of wild organisms actually grow in commonly available artificial media. Estimates for SRB in marine sediments, for example, indicate that only 1 in 1000 of the organisms present actually show up in standard growth tests (71).

4.5.4.3. Activity Assays

4.5.4.3.1. Whole Cell Approaches based on the conversion of a radioisotopically labeled substrate can be used to assess the potential activity of microbial populations in field samples. This technique is specific to SRBs. It depends on bacterial growth for detection, but it generates results in $\sim$2 days.

The sample is incubated with a known trace amount of radioactive-labeled sulfate. (SRBs reduce sulfate to sulfide.) After incubation, the reaction is terminated with acid to kill the cells and the radioactive sulfide is fixed in zinc acetate, which is sent to a laboratory for evaluation. This is a highly specialized technique, involving expensive laboratory equipment and the handling of radioactive substances (13).

The radiorespirometric method that can use field samples directly without the need to separate organisms is very sensitive. Selection of the radioactively labeled substrate is key to interpretation of the results, but the method can provide insights into factors limiting growth by comparing activity in native samples with supplemented test samples under various conditions. Oil degrading organisms, for example, can be assessed through the mineralization of ^{14}C labeled hydrocarbon to carbon dioxide.

Radioactive methods are not routinely used by field personnel. However, they have been particularly useful in a number of applications including biocide screening programs, identification of nutrient sources, and assessment of key metabolic processes in various corrosion situations (66).

4.5.4.3.2. Enzyme-Based Assays An increasingly popular approach is the use of commercial kits to assay the presence of enzymes associated with microorganisms suspected to cause problems. For example, kits are available for the sulfate reductase enzyme common to SRB associated with corrosion problems (72). This technique takes advantage of the fact that SRBs reduce sulfate to sulfide through the presence of an enzyme, APS-reductase, common to all SRBs. Measurement of the amount of APS-reductase in a sample gives an estimate of total numbers of SRBs present. The test does not require bacterial growth and the entire test takes 15–20 min (13).

Another example is the hydrogenase enzyme implicated in the acceleration of corrosion through rapid removal of cathodic hydrogen formed on the metal surface (73). The test analyzes for the hydrogenase enzyme that is produced by bacteria able to use hydrogen as an energy source. The test is usually performed on sessile samples. The sample is exposed to an enzyme-extracting solution, and the degree of hydrogen oxidation in an oxygen-free atmosphere is detected by the addition of a dye (13).

Performance of several of these kits has been assessed by field personnel in round robin tests. Correlation of activity assays and population estimates is variable. In general, these kits have a narrower range of application than growth-based assays making it important to select a kit with a range of response appropriate to the problem under consideration (74).

4.5.4.3.3. Metabolites Adenosine triphosphate (ATP) is present in all living cells, but it disappears rapidly on death. The measure of ATP may thus provide a measure of living material. The ATP can be measured using an enzymic reaction, which generates flashes of light that are detected by a photomultiplier (13). Commercial instruments are available that measure the release of light by the firefly luciferin–luciferase with ATP. The method is best suited to clean aerobic aqueous samples since suspended solids and chemical quenching can affect the results. Detection of metabolites, such as organic acids in deposits or gas compositions including methane or hydrogen sulfide by routine gas chromatography can also indicate biological involvement in industrial problems (66).

4.5.4.3.4. Cell Components Biomass can be generally quantified by assays for protein, lipopolysaccharide, or other common cell constituents, but the information gained is of limited value. An alternate approach is to use cell components to define the composition of microbial populations with the hope that the insight gained may allow future damaging situations to be recognized and managed. Fatty acid analysis and nucleic acid sequencing provide the basis for the most promising methods in this category.

4.5.4.3.4.1. Fatty acid profiles Analyzing fatty acid methyl esters derived from cellular lipids can fingerprint organisms rapidly and, provided pertinent profiles are known, organisms in industrial and environmental samples can be identified with confidence. The immediate impact of events, such as changes in operating conditions or the application of biocides, can be monitored by such analysis. Problem populations of certain organisms can also be identified in order to implement appropriate management responses in a timely fashion.

4.5.4.3.4.2. Nucleic acid-based methods Specific DNA probes have been developed to detect segments of genetic material coding for known enzymes. A gene probe developed to detect the hydrogenase enzyme that occurs broadly in

SRB from the genus *Desulfovibrio* has been tested on samples from an oilfield waterflood plagued with iron sulfide related corrosion problems. The enzyme was detected with this probe in 12 of 20 samples, suggesting that sulfate reducers that did not have this enzyme were also present in the operation (75). In principle, probes could be developed to detect potentially all sulfate-reducers. However, the operation of a battery of probes could be a daunting task where large numbers of field samples have to be analyzed.

To overcome this obstacle, the reverse sample genome probe (RSGP) was developed. With RSGP, the DNA from organisms previously isolated from field problems is spotted on a master filter following which the DNA isolated from field samples is labeled with either a radioactive or fluorescent indicator and exposed to the filter. Labeled DNA from the field sample sticks to the corresponding spot on the master filter when complementary strands of DNA are present. Organisms represented by the labeled spots are then known to be in the field sample (75).

4.5.4.4. Detailed Coupon Examinations

A great deal of information can be learned by careful, in-depth examination of corrosion coupon surfaces using commonly available analytical techniques. A wide variety of samplers for introducing metallic surfaces of interest into a system are available. A popular sampling device is shown in Fig. 4.65. Special handling of coupons after removal from the system being monitored is crucial

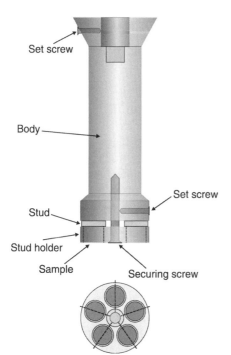

Fig. 4.65. Biofilm sampling device with removable "buttons".

to ensure that subsequent laboratory tests provide representative information. Biofilms in particular are highly sensitive to dehydration, exposure to air, temperature, mechanical damage, and other gross environmental changes that can occur during removal and transport of the coupon (12).

Examination of coupons for microbial populations can be performed either directly or indirectly, using histological embedding techniques to preserve and remove the biofilm. Although fairly involved, the embedding technique offers several advantages over direct observation in that the biofilm and corrosion products are preserved for future analysis. Environmental scanning electron microscopy (ESEM) may also be utilized to examine biofilms on test coupons; however, exopolymers and corrosion products often obscure the cells, making quantification and identification difficult with this method.

4.5.5. Monitoring MIC Effects

The presence of a biofilm on a metallic surface can greatly alter the local corrosion processes. In addition to the electrochemical changes that affect corrosion, biofilms can also modify other readily measured characteristics, such as pressure drop or heat-transfer resistance. Monitoring such microbiological influences can provide a useful indicator that a biofilm is present and that action should be taken to mitigate potential MIC.

4.5.5.1. Deposition Accumulation Monitors

Methods for monitoring deposits can provide an indication of the accumulation of biofilm and other solids on surfaces or in orifices. For example, monitoring pressure drop across an orifice provides a simple method for continuous monitoring of deposit accumulation and biofilm accumulation. The main disadvantage of the pressure drop technique is that it is not specific to the biofilm build-up since it detects the total scaling and deposition effects in a line (12).

These measurements can be made on actual operating units on-line, but they may also be done using model heat exchanger units or instrumented pipe loops run in parallel to system flow. Figure 4.17 shows such an instrumented pipe loop test unit with five parallel, instrumented pipe runs. Water flow from the target system is diverted through this unit, so that conditions are representative of the actual operating system (67).

Measurements of friction factors and heat exchange efficiencies can indicate fouling. While deposits of any sort can affect flow and heat transfer in an operating industrial system, biofilms are especially effective. A 165-pm thick biofilm shows 100 times the relative roughness of a calcite scale and a thermal conductivity close to that of water, that is, almost 100 times less than carbon steel (66). The assumption is generally made that a susceptible system showing extensive fouling is prone to MIC. Systems showing the effects of extensive fouling are operating inefficiently and may warrant remedial action, in any case.

4.5. Monitoring Microbiologically Influenced Corrosion   293

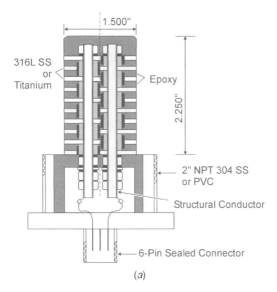

(a)

(b)

Fig. 4.66 (a) Schematic and (b) picture of the BioGEORGE MIC detection probe. (Courtesy of George Licina, Structural Integrity Associates, Inc.)

4.5.5.2. Electrochemical Methods

An electrochemical method for on-line monitoring of biofilm activity has been developed for continuous monitoring of biofilm formation without the need for excessive involvement of plant personnel (Fig. 4.66a and b). A series of stainless steel or titanium disks are exposed to the plant environment. One set of disks is

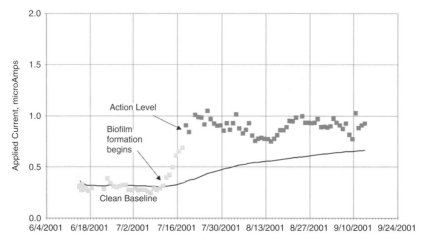

Fig. 4.67 A plot of signals generated by a BioGEORGE MIC detection probe. (Courtesy of George Licina, Structural Integrity Associates, Inc.)

polarized (relative to the other set) for a short period of time each day. The electrodes are connected through a shunt the remainder of the time. Biofilm activity, which is also an electrochemical process, is monitored by tracking changes in the applied current when the external potential is on and the generated current when the potential is off (12).

The onset of biofilm formation on the probe is indicated when either of these independent indicators deviates from the baseline level (Fig. 4.67). Such a departure would then trigger the alarm located in a control box (Fig. 4.68). The level of biofilm activity is also measured by the amount of variation from the baseline. The applied and generated currents from a well-controlled system will be a flat line, devoid of any significant deviations.

Another experimental approach to detect MIC with an electrochemical signals is based on the use of small silver sulfide and silver chloride electrodes capable of detecting sulfides or chlorides by changes in the potential between the Ag/AgCl or Ag/Ag$_2$S electrode pairs (12). The production of sulfides or concentration of chlorides by microbial action was shown to produce a nearly instantaneous change in the potential of the electrode pair, thus providing an indication of specific microbial activity.

4.6. MONITORING PIPELINE CP SYSTEMS

Monitoring pipeline CP systems is a technically complex field. In many cases, condition monitoring requirements are specified by regulatory authorities. Since CP systems are expected to operate in demanding environmental conditions over long time periods, the reliability requirements of the associated hardware are high. Regular monitoring of the equipment is therefore an important aspect of any CP program.

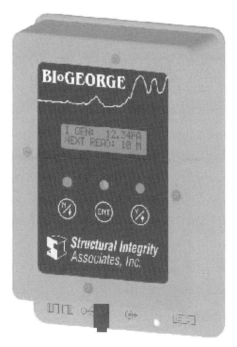

Fig. 4.68 Control box of a BioGEORGE MIC detection probe. (Courtesy of George Licina, Structural Integrity Associates, Inc.)

An increasing trend toward selective remote rectifier monitoring, using modern communication systems and computer networks, is emerging to accomplish these tasks with reduced resources. Wireless cell phone and satellite communication systems are available for interrogating rectifiers in remote locations. The other aspect of monitoring a pipeline CP system is the monitoring of protection afforded to the pipeline. The following sections briefly describe some of the techniques used to carry out this work.

4.6.1. Close Interval Potential Surveys

The principle of a close interval potentials surveys (CIPS) is to record the potential profile of a pipeline over its entire length by taking potential readings at $\sim$1 m intervals. A reference electrode is connected to the pipeline at a test post and positioned in the ground over the pipeline at regular intervals for the measurement of the potential difference between the reference electrode and the pipeline (Fig. 4.69).

A three-person crew is typically required to perform these measurements. One person walking ahead locates the pipeline with a pipe locator to ensure that the potential measurements are performed directly above the pipeline. This person also carries a tape measure and inserts a distance marker (a small flag) at regular intervals over the pipeline. A second person carries a pair of electrodes connected to the test post by means of a trailing thin copper wire and the potential

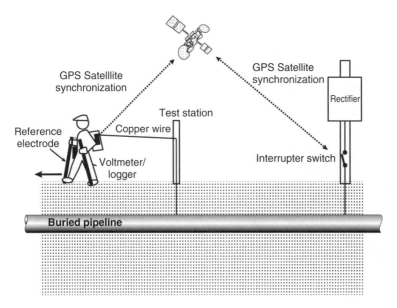

Fig. 4.69. Schematic description of the CIPS methodology.

measuring instrumentation. The second person is also responsible for entering specific features as a function of the measuring distance. The third person collects the trailing wire, after individual survey sections have been completed.

The CIPS measurements are quite demanding on field crews and require extensive logistical support from both the pipeline operator and the CIPS contractor. Field crews are typically required to move over/around fences, roads, highways and other obstacles and difficult terrain. Breakage of the trailing copper wire is not uncommon and special strengthened wire has to be taped down onto road surfaces that are crossed.

An important consideration in the potential readings is the IR or ohmic drop error that is included in the potential measurements when a CP system is operational. A commonly used method to correct for the IR drop, often called making instant-off measurements, can be a very accurate way to make potential measurements. In reality, however, making measurements completely free of voltage drop error is not always possible because some currents cannot be easily interrupted.

Uninterruptible current sources may include sacrificial anodes directly bonded to the structure, foreign rectifiers, stray currents, telluric currents, and long-line cells (76). Modern interrupters are based on solid-state switches and are programmable to perform switching only when the survey is performed during the day time. This feature minimizes the depolarization of the pipeline that may occur gradually due to the cumulative effects of the "off" periods.

When several rectifiers protect a structure, it is necessary that all rectifiers be interrupted at the exact same instant in order to obtain meaningful measurements.

Pipeline operators usually specify that at least two rectifiers ahead of the survey team and two rectifiers behind the survey team have to be interrupted in a fully synchronized manner. The amount of time between current interruption and depolarization can vary from a fraction of a second to several seconds, depending on the details of the structure. In addition, capacitive spikes that occur shortly after current is interrupted may mask the instant-off potential. Measurements made with a recording voltmeter are preferred as they can be subsequently analyzed to determine the real instant-off potential (76).

An example of graphical CIPS data is presented in Fig. 4.70 (77). In the simplest format, the "on" and "off" potentials are plotted as a function of distance. The usual sign convention is for potentials to be plotted as positive values. The difference between the "on" and "off" potential values should be noted. As is usually the case, the "off" potentials are less negative than the "on" values. When the relative position of these two lines is reversed, it indicates that some unusual conditions such as stray current interference may be at play.

The CIPS technique provides a complete pipe-to-soil potential profile and the interpretation of results, including the identification of defects, is relatively straightforward.

4.6.2. Pearson Survey

The Pearson Survey, named after its inventor, is used to locate coating defects in buried pipelines. Once these defects have been identified, the protection levels provided by the CP system can be investigated at these critical locations in more detail.

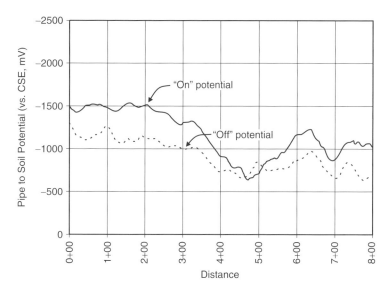

Fig. 4.70. Over-the-line survey with cathodic protection.

During a Pearson survey, an ac signal of ~1000 Hz is imposed onto the pipeline by means of a transmitter (Fig. 4.71), which is connected to the pipeline and an earth spike as illustrated in Fig. 4.72. Two survey operators make earth contact either through aluminum poles or metal studded boots (Fig. 4.73). A distance of several meters typically separates the operators. Essentially, the signal measured by the receiver provides a measure of the potential gradient over the distance between the two operators. Defects are located by a change in the potential gradient, which translates into a change in signal intensity.

As in the CIPS technique, the measurements are usually recorded by walking directly over the pipeline. As the front operator approaches a defect, increasing

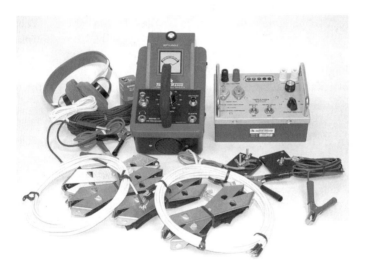

Fig. 4.71 Instrumentation to carry out a Pearson survey. (Photo courtesy of Tinker & Rasor.)

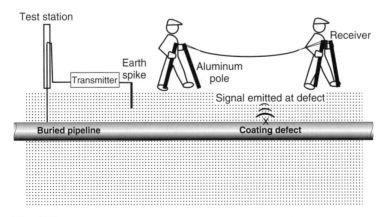

Fig. 4.72. Schematic description of the Pearson survey technique.

Fig. 4.73 Pipeline CP fault testing with Pearson-type detector. (Photo courtesy of Tinker & Rasor.)

signal intensity is recorded. As the front person moves away from the defect, the signal intensity drops and later picks up again as the rear operator approaches the defect. The interpretation of signals may obviously be more difficult when several defects are located between the two operators.

In principle, a Pearson survey can be performed with an impressed CP system still energized. However, sacrificial anodes should be disconnected since the signal from these may otherwise mask actual coating defects. A three-person team is usually required to locate the pipeline, perform the survey measurements, place defect markers into the ground, and move the transmitters periodically.

By walking the entire length of the pipeline, an overall inspection of the right of way can be made together with the measurements. In principle, all significant defects and metallic conductors causing a potential gradient will be detected. There are no trailing wires and the impressed CP current does not have to be turned on and off.

The disadvantages associated with Pearson surveys are similar to those of CIPS, as the entire pipeline has to be walked and contact established with the ground. The technique is therefore unpractical in many areas, such as roads, paved areas, or rivers.

4.6.3. Direct Current Voltage Gradient Surveys

The direct current voltage gradient (DCVG) surveys are a more recent method to locate defects on coated buried pipelines and to assess their severity. The technique again relies on the fundamental effect of a potential gradient being established in the soil at coating defects under the application of CP current. In general, the greater the size of the defect, the greater the potential gradient.

The potential gradient is measured by an operator between two reference electrodes, for example, copper sulfate electrodes, separated by a distance of

300 Chapter 4 Corrosion Monitoring

approximately one- half of a meter. A pulsed dc signal is imposed on the pipeline for DCVG measurements. The pulsed input signal minimizes interference from other current sources (e.g., CP systems, electrified rail transit lines, telluric effects). This signal can be obtained with an interrupter on an existing rectifier or through a secondary current pulse superimposed on the existing "steady" CP current.

The operator walking the pipeline observes voltage deflections on a precision voltmeter to identify defect locations. The presence of a defect is indicated by an increased deflection as the defect is approached, no deflection when the operator is immediately above the defect and a decreasing deflection as the operator walks away from the defect (Fig. 4.74). The high precision in locating defects ($\sim$0.1 to 0.2 m) represents a major advantage in minimizing the work of subsequent digs if corrective action needs to be taken.

The DCVG technique is particularly suited to complex CP systems, for example, areas with a relatively high density of buried structures. These are generally the most difficult survey conditions. The DCVG equipment is relatively simple and involves no trailing wires.

4.6.4. Corrosion Coupons

Cathodic protection (CP) coupons are now being used as an alternative method to make potential measurements that may be substantially free of voltage drop

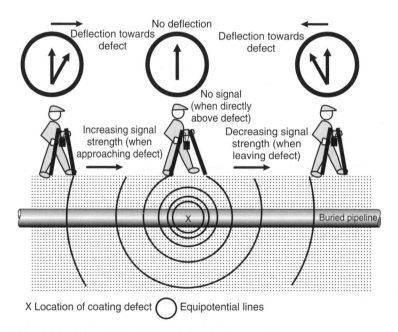

Fig. 4.74. Schematic description of the DCVG methodology.

error. A CP coupon is a small piece of metal that is electrically connected to the structure at a test station. The potential of a coupon will closely approximate the potential of any exposed portion of the structure (holiday) located in the vicinity of the coupon. By disconnecting the coupon from the structure at the test station, an instant-off potential measurement can be made on the coupon without having to interrupt any other current sources. However, these measurements are still not completely free of voltage drop error. Any voltage drop occurring in the electrolyte in the distance between the reference electrode and the coupon surface will still be incorporated into measurements. Placing a reference electrode as close as possible to a coupon can minimize this error. However, the reference electrode must not be placed so close that it shields the coupon (76).

Perhaps the most important consideration in the installation of corrosion coupons is that a coupon must be representative of the actual pipeline surface/defect. The exact metallurgical detail and surface finish as found on the actual pipeline are therefore required on the coupon. The influence of corrosion product build-up may also an important factor. The environmental conditions of the coupon have to be matched with those of the pipe being monitored, for example, temperature, soil conditions, soil compaction, and oxygen concentration.

Measurement of current flow to/from the coupon and its direction can also be determined, for example, by using a shunt resistor in the bond wire. Importantly, it is also possible to determine corrosion rates from the coupon. Electrical resistance sensors provide an option for *in situ* corrosion rate measurements, as an alternative to weight loss coupons.

4.7. ATMOSPHERIC CORROSION MONITORING

The economic losses caused by atmospheric corrosion are tremendous and account for the disappearance of a significant portion of metal produced. Atmospheric corrosion has been reported to account for more failures in terms of cost and tonnage than any other type of material degradation processes. Atmospheres have been classified into four basic types:

Industrial: An industrial atmosphere is characterized by its pollution level, two constituents of which are particularly corrosive (i.e., sulfur compounds and nitrogen oxides). When present, these contaminants may combine with dew or fog to produce a highly corrosive acid film on exposed surfaces.

Marine: Marine atmospheres are often laden with a sea salt mist or aerosol particles that can be carried by the wind to settle on exposed surfaces. The quantity of salt contamination typically decreases with distance from the saline source, and is greatly affected by wind currents.

Rural: A rural environment does not usually contain strong chemical contaminants, but it does contain organic and inorganic dusts. Its principal corrosive constituent is moisture. In tropical environments, in addition to the high average temperature, the daily cycle includes a high relative humidity, intense sunlight, and long periods of condensation during the night.

Indoor: Indoor atmospheres can be generally considered to be quite mild. However, the presence of local contaminants can transform ambient air into a very corrosive environment if it is not properly ventilated or controlled through dehumidification, for example. Additionally, the consequences of indoor corrosion have been drastically amplified by the incredibly small volume of the material that needs to be damaged before causing a fault in modern computer and communication equipment. The microchip in an automobile, for example, is not directly subjected to the same environmental hazards as the car body. However, the tolerance for corrosion loss in electronic devices is many orders of magnitude less, that is, on the order of picograms (10^{-12} g). The minimum line width in the state-of-the-art printed circuit boards (PCBs) in 1997 was already <100 μm. On hybrid integrated circuits (HICs), line spacing may be <5 μm (78).

Various methods have been developed for measuring the factors that influence atmospheric corrosion. Temperature, RH, wind direction and velocity, solar radiation, and amount of rainfall are easily recorded. Not so easily determined are dwelling time of wetness (TOW), and the quantity of sulfur dioxide and chloride contamination. However, methods for these determinations have been developed and are in use at various test stations.

By monitoring these factors and relating them to corrosion rates, a better understanding of atmospheric corrosion can be obtained for planning maintenance procedures and inspection schedules, for the prediction or outdoor performance of materials and systems, and to circumvent the effects of corrosion products on the environment.

4.7.1. Relative Humidity and Time of Wetness

Relative humidity is defined as the ratio of the quantity of water vapor present in the atmosphere to the saturation quantity at a given temperature, and it is expressed as percent (%). A fundamental requirement for atmospheric corrosion processes is the presence of a thin-film electrolyte that can form on metallic surfaces when exposed to a critical level of humidity (Fig. 4.75). While this film is almost invisible, the corrosive contaminants it contains are known to reach relatively high concentrations, especially under conditions of alternate wetting and drying.

The critical humidity level is a variable that depends on the nature of the corroding material, the tendency of corrosion products and surface deposits to absorb moisture, and the presence of atmospheric pollutants (47). It has been shown that, for example, this critical humidity level is 60% for iron if the environment is free of pollutants.

In the atmospheric classification scheme described in ISO 9223 (79). The TOW is an estimated parameter based on the length of time when the relative

Anode Reaction: $2Fe \rightarrow 2Fe^{2+} + 4e^-$

Cathode Reaction: $O_2 + 2H_2O + 4e^- \rightarrow 4OH^-$

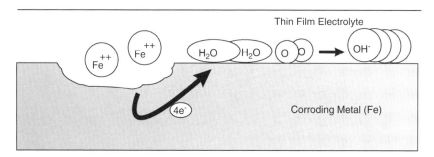

Fig. 4.75. Schematic description of the atmospheric corrosion of iron.

humidity is >80% at a temperature >0°C. It can be expressed as the hours or days per year or the annual percentage of time. Experience from applying the ISO classification system has shown, however, that certain observations need further clarification. Substantial corrosion has also been measured on specimens exposed at temperatures well below 0°C, which is in direct contradiction to the ISO criterion (80). It has been proposed on the basis of these results, that the TOW in cold environments be estimated as the length of time that relative humidity exceeds 50% and the ambient temperature exceeds −10°C.

A method of measuring the TOW has been developed by Sereda and correlated with the corrosion rates encountered in the atmosphere (81). The moisture sensing elements in this sensor are manufactured by plating and selective etching of thin films of appropriate anode (copper) and cathode (gold) materials in an interlaced pattern on a thin nonconductive substrate (Fig. 4.76). When moisture condenses on the sensor it activates the sensor, producing a small voltage (0–100 mV) across a 10^7 Ω resistor.

Thin sensing elements are preferred in order to preclude influencing the surface temperature to any extent. Although a sensor constructed using a 1.5-mm thick glass reinforced polyester base has been found to be satisfactory on plastic surfaces, this will not be the case with the same sensing element on a metal surface with a high-thermal conductivity (82). For metal surfaces, the sensing element should be appreciably thinner. Commercial epoxy sensor backing products of thickness of 1.5 mm, or less, are suitable for this purpose.

4.7.2. Pollutants

Sulfur dioxide (SO_2), which is the gaseous product of the combustion of fuels that contain sulfur such as coal, diesel fuel, gasoline, and natural gas, has been

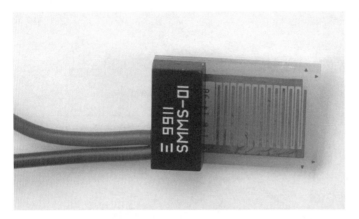

Fig. 4.76 Interlocking combs of gold and copper electrodes in a 'Sereda' humidity sensor. (Courtesy Kingston Technical Software.)

identified as one of the most important air pollutants that contribute to the corrosion of metals.

Less recognized as corrosion promoters, are the nitrogen oxides (NO_x), which are also products of combustion. A major source of NO_x in urban areas is the exhaust fumes from vehicles. Sulfur dioxide, NO_x, and airborne aerosol particles can react with moisture and ultraviolet (UV) light to form new chemicals that can be transported as aerosols. A good example of this is the summertime haze or smog over many large cities. Up to 50% of this haze is a combination of sulfuric and nitric acids.

Sulfur dioxide is usually measured in terms of its concentration in air in units of $\mu g\,m^{-3}$. Precise methods are available to monitor continuously the amount of sulfur dioxide in a given volume of air. However, this is only indirectly related to the effect of sulfur dioxide on corrosion since only the actual amount of hydrated sulfur dioxide or sulfur trioxide deposited on metal surfaces is important.

Since it is the SO_2 deposited on the metal surface that affects the corrosion, deposited SO_2 has been measured as its reaction product in terms of sulfate deposition rate on the surface in units of $mg\,m^{-2}\,day^{-1}$. The pollution levels can also be measured in terms of the concentration of the dissolved sulfate (SO_4^{2-}) in rain water.

4.7.3. Airborne Particles (Chlorides)

The behavior of aerosol particles in outdoor atmospheres can be explained by invoking the laws that govern their formation, movement, and capture. These particles are present throughout the planetary boundary layer and their concentrations depend on a multitude of factors including location, time of day or year, atmospheric conditions, presence of local sources, altitude, and wind velocity. Size is normally used to classify aerosol because it is the most readily measured property and other properties can be inferred from size information (83). The highest mass fraction of particles in an aerosol is characterized by particles having a diameter in the range of 8–80 μm (84). Some studies have also indicated that there is a strong correlation between wind speed and the deposition and capture of aerosols. In such a study of saline winds in Spain, a very good correlation was found between chloride deposition rates and wind speeds above a threshold of 3 m s^{-1} or 11 km h^{-1} (85).

The lifetime of any particular particle depends on its size and location. Studies of the migration of aerosols inland of a sea coast have shown that typically the majority of the aerosol particles are deposited close to the shoreline (typically 400–600 m) and consist of large particles (>10-μm diameter), which have a short residence time and are controlled primarily by gravitational forces (84, 85). The aerosols that form also have mass and are subject to the influence of gravity, wind resistance, droplet dry-out, and possibilities of impingement on a solid surface, as they progress inland.

Airborne salinity refers to the content of gaseous and suspended salt in the atmosphere. It is measured by its concentration in air in units of μg m^3. Since it is the salt that is deposited on the metal surface that affects the corrosion, it is also often measured in terms of deposition rate in units of mg m^{-2} day^{-1}. The chloride levels can also be measured in terms of the concentration of the dissolved salt in rain water.

A number of methods have been employed for determining the contamination of the atmosphere by aerosol transported chlorides (e.g., sea salt and road deicing salts). The "wet candle method", for example, is relatively simple, but has the disadvantage that it also collects particles of dry salt that might not be deposited otherwise (86). This technique uses a wet wick of a known diameter and surface area to measure aerosol deposition (Fig. 4.77). The wick is maintained wet using a reservoir of water or 40% glycol–water solution. Particles of salt or spray are trapped by the wet wick and retained. At intervals, a quantitative determination of the chloride collected by the wick is made and a new wick is exposed.

In reality, the wet candle method gives an indication of the salinity of the atmosphere rather than the contamination of exposed metal surfaces. The technique is considered to measure the total amount of chloride arriving to a vertical surface and its results may not be truly significant for corrosivity estimates.

In order to evaluate the chloride deposition rates in a confined space, such as a ventilated subfloor where corrosion of fasteners can be quite severe, a special collecting box shown in Fig. 4.78 was used in which the airborne chlorides were collected on horizontal and vertical filter papers positioned at different locations from the box openings (Fig. 4.79).

For obvious reasons, the test chambers used in this study were thoroughly cleaned of chlorides before setting them outdoor. Rigid plastic material panels stuck at predetermined positions with self-adhesive pads were used to hold the filter papers with chloride-free plastic paper clips. The roof protected the surface from direct rain, but the filter papers were still exposed to deposition of the airborne chlorides and possibly some rain. The paper was removed every month

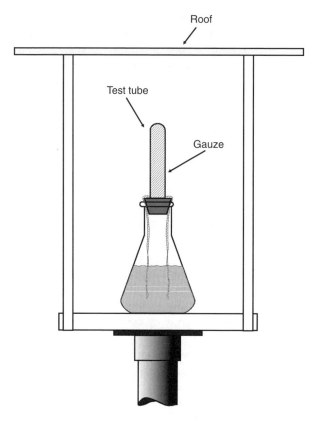

Fig. 4.77. Schematic of a wet candle chloride apparatus.

for chemical analysis. A wet candle with its own roof cover was installed on each box (Fig. 4.78) to provide a measure of the level of chlorides outside the boxes.

4.7.4. Atmospheric Corrosivity

The simplest form of direct atmospheric corrosion measurement is by coupon exposures. Subsequent to their exposure, the coupons can be subjected to weight loss measurements, pit density and depth measurements, and to other types of examination. Flat panels exposed on exposure racks are a common coupon-type device for atmospheric corrosivity measurements. Various other specimen configurations have been used, including stressed U-bend or C-ring specimens for SCC studies. The main drawback associated with conventional coupon measurements is that extremely long exposure times are usually required to obtain meaningful data, even on a relative scale. It is not uncommon for such programs to run for 20 years or longer (87).

Some variations of the basic coupon specimens can provide rapid material–corrosivity evaluations. The helical coil adopted in the ISO 9226 methodology is a high surface area–weight ratio coupon that gives a higher sensitivity

4.7. Atmospheric Corrosion Monitoring 307

Fig. 4.78 A typical subfloor simulation box surmounted by a wet candle chloride measuring device that has been installed at several sites in Australia and New Zealand. (Courtesy of Branz, New Zealand.)

Fig. 4.79 Horizontal and vertical filter papers installed inside a subfloor simulation box. (Courtesy of Branz, New Zealand.)

than panel coupons of the same material (88). The use of bimetallic specimens in which a helical wire is wrapped around a coarsely threaded bolt can provide additional sensitivity and forms the basis of the CLIMAT coupon that was developed to classify industrial and marine atmospheres (89).

An ASTM standard describes the construction of CLIMAT coupons. According to this standard, CLIMAT coupons can be made from 1100 aluminum (UNS A91100) wire wrapped around threaded rods of nylon, 1010 mild steel (UNS G10100 or G10080), and CA110 copper (UNS C11000) (90). The mass loss of aluminum wire after 90 days of exposure is considered to be an indication of atmospheric corrosivity. However, the relative corrosivity of atmospheres could be quite different for the various combinations of materials.

Table 4.9. Mass Loss of Aluminum Wire Exposed in Different Environments

Environment	Helical coil	Mass loss (%)		
		Al–Nylon	Al–Cu	Al–Fe
Rural	0.02		0.88	0.19
Urban	0.03	0.060	0.77	0.25
Marine	0.03	0.066	5.48	3.9
Marine–urban	0.03		4.29	2.83

Fig. 4.80 Aerial view of the NASA Kennedy Space Center beach corrosion test site where atmospheric corrosivity is the highest corrosivity of any test site in the continental United States. (Courtesy of NASA.)

4.7. Atmospheric Corrosion Monitoring

(a)

(b)

Fig. 4.81 A CLIMAT coupon with three copper rods immediately after it was installed at the Kennedy Space Center beach corrosion test site (a), after 30 days (b), and after 60 days (c). (Courtesy of NASA.)

310   Chapter 4   Corrosion Monitoring

(c)

Fig. 4.81. (continued).

The aluminum wire on copper bolts has been found by many to be the most sensitive of the three proposed arrangements in the ASTM standard. This was corroborated in a recent paper that clearly demonstrated the superior sensitivity of the Al–Cu arrangement for all types of simulated environments, as is shown in Table 4.9 (91). While this arrangement is the most sensitive, the use of triplicate coupons on a single holder additionally provides an indication of the reproducibility of the measurements.

Such a CLIMAT coupon with three copper rods installed at the NASA Kennedy Space Center (KSC) beach corrosion test site (Fig. 4.80) is shown immediately after it had been installed (Fig. 4.81a), after 30 days (Fig. 4.81b), and after 60 days (Fig. 4.81c). The KSC has the highest corrosivity of any test site in the continental United States (92). The mass loss recorded after a shorter exposure than usual can be very high. In the present example, it was already 16% of the original aluminum wire after only 60 days.

The CLIMAT coupons sensitivity to atmospheric corrosivity can be used to study fluctuations on a microenvironmental scale (93). In the following example, a supporting panel (Fig. 4.82) was placed ∼2 m above the ground on a pedestrian bridge concrete support ∼4 m from a moderately trafficked highway during the winter months when deicing salts are used. Six sets of CLIMAT units were deployed on the panel with each set in a different microenvironment produced by various baffle geometries. The overall atmospheric conditions, for example, temperature and relative humidity, were therefore the same for each set except for the differences in the rate of aerosol deposition.

The average mass loss experienced by the CLIMATs exposed in different baffling geometries is also shown in Fig. 4.82. One obvious conclusion from

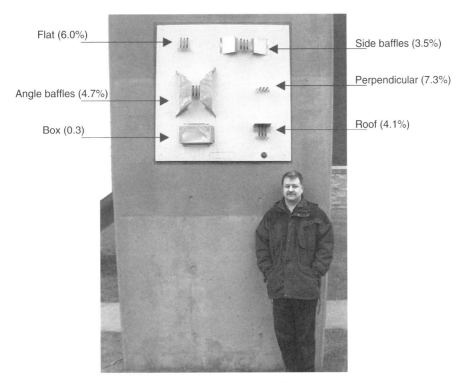

Fig. 4.82. Test panel and CLIMAT coupons exposed in different microenvironments adjacent to a moderately trafficked highway during the winter months when deicing salts are used.

these measurements is that shielding, whether from wind or direct precipitation, can dramatically reduce the corrosion rate of components exposed to the same time-of-wetness factor. In fact, there was a 24-fold difference between the average mass loss in the boxed-in and boldly exposed coupons.

REFERENCES

1. ZOLLARS B, SALAZAR N, GILBERT J, and SANDERS M. *Remote Datalogger for Thin Film Sensors.* Houston, TX: NACE International, 1997.
2. KELLY RG, et al. Embeddable Microinstruments for Corrosion Monitoring. 1–12. 1997. Houston, TX, NACE International. Corrosion 97. 1997.
3. DAVIS GD, RAGHU S, CARKHUFF BG, GARRA F, SRINIVASAN R, and PHILLIPS TE. Corrosion Health Monitor for Ground Vehicles. Paper No. 103. 2005. Tri-Service Corrosion Conference, Orlando, FL, November 14–18, p. 2005.
4. THOMAS MJJS and TERPSTA S. Corrosion Monitoring in Oil and Gas Production. CORROSION 2003, Paper No. 431. 2003. Houston, TX, NACE International.
5. EDEN DC, CAYARD MS, KINTZ JD, SCHRECENGOST RA, BREEN BP, and KRAMER E. Making Credible Corrosion Measurements—Real Corrosion, Real Time. CORROSION 2003, Paper No. 376. 2003. Houston, TX, NACE International.

6. DEAN SW. Overview of Corrosion Monitoring in Modern Industrial Plants. In: Moran GC and Labine P, eds. *Corrosion Monitoring in Industrial Plants Using nondestructive testing and Electrochemical methods, ASTM STP 908.* Philadelphia: American Society for testing and Materials, 1986; pp. 197–220.
7. POWELL DE, MA'RUF DI, and RAHMAN IY. Practical Considerations in Establishing Corrosion Monitoring for Upstream Oil and Gas Gathering Systems. *Materials Performance* 2001; **4**: 50–54.
8. EDGEMON GL. Electrochemical Noise Based Corrosion Monitoring: Hanford Site Program Status. CORROSION 2005, Paper No. 584. 2005. Houston, TX, NACE International.
9. YANG L, SRIDHAR N, PENSADO O, and DUNN DS. An *In-Situ* Galvanically Coupled Multielectrode Array Sensor for Localized Corrosion. *Corrosion* 2002; **58**: 1004–1014.
10. ROBERGE PR, TULLMIN MAA, GRENIER L, and RINGAS C. Corrosion Surveillance for Aircraft. *Materials Performance* 1996; **35**: 50–54.
11. ROBERGE PR and SASTRI VS. On-Line Corrosion Monitoring with Electrochemical Impedance Spectroscopy. *Corrosion* 1994; **50**: 744–754.
12. ZINTEL TP, LICINA GJ, and JACK TR. Techniques for MIC Monitoring. In: Stoecker II JG, ed. *A Practical Manual on Microbiologically Influenced Corrosion.* Houston, TX: NACE international, 2001.
13. Techniques for Monitoring Corrosion and Related Parameters in Field Applications. NACE 3T199. 1999. Houston, TX, NACE International.
14. Standard Guide for Conducting Corrosion Tests in Field Applications. G4-01. 2001. West Conshohocken, PA, American Society for Testing of Materials.
15. DEAN SW. Corrosion Monitoring for Industrial Processes. In: Cramer DS and Covino BS, eds. *Volume 13A: Corrosion: Fundamentals, Testing, and Protection.* Metals Park, OH: ASM International, 2003; pp. 533–541.
16. Standard Practice for Laboratory Immersion Corrosion Testing of Metals. ASTM G31-72. 2004. West Conshohocken, PA, American Society for Testing of Materials.
17. Standard Guide for Conducting and Evaluating Galvanic Corrosion Tests in Electrolytes. Annual Book of ASTM Standards. G71-81[Vol 03.02]. 2003. Philadelphia, PA, American Society for Testing of Materials.
18. Standard Guide for Crevice Corrosion Testing of Iron-Base and Nickel-Base Stainless Alloys in Seawater and Other Chloride-Containing Aqueous Environments. G78-01[Vol 03.02]. 2001. West Conshohocken, PA, American Society for Testing of Materials.
19. Standard Test Methods for Pitting and Crevice Corrosion Resistance of Stainless Steels and Related Alloys by Use of Ferric Chloride Solution. Annual Book of ASTM Standards. G 48-03[Vol 03.02]. 2003. Philadelphia, PA, American Society for Testing of Materials.
20. Standard Practice for Making and Using U-Bend Stress-Corrosion Test Specimens. G30-97[Vol 03.02]. 2003. West Conshohocken, PA, American Society for Testing of Materials.
21. Standard Practice for Making and Using C-Ring Stress-Corrosion Test Specimens. G38-01[Vol 03.02]. 2001. West Conshohocken, PA, American Society for Testing of Materials.
22. Standard Practice for Preparation of Stress-Corrosion Test Specimens for Weldments. G58-85[Vol 03.02]. 1999. West Conshohocken, PA, American Society for Testing of Materials.
23. Standard Test Method for Determining the Susceptibility to Intergranular Corrosion of 5XXX Series Aluminum Alloys by Mass Loss After Exposure to Nitric Acid (NAMLT Test). Annual Book of ASTM Standards. G 67-04[Vol 03.02]. 2004. Philadelphia, PA, American Society for Testing of Materials.
24. Standard Practice for Preparing, Cleaning, and Evaluating Corrosion Test Specimens. G1-03[Vol 03.02]. 2003. West Conshohocken, PA, American Society for Testing of Materials.
25. HAUSLER RH. Corrosion Inhibitors. In: BABOIAN R, ed. *Corrosion Tests and Standards:* 2nd Edition. West Conshohocken, PA: American Society for Testing of Materials, 2005; pp. 480–499.

26. FREEMAN RA and SILVERMAN DC. Error Propagation in Coupon Immersion Tests. *Corrosion* 1992; **48**: 463–466.
27. DRAVNIEKS A and CATALDI HA. Industrial Applications of a Method for Measuring Small Amounts of Corrosion without Removal of Corrosion Products. *Corrosion* 1954; **10**: 224–230.
28. FREEDMAN AJ, TROSCINSKI ES, and DRAVNIEKS A. An Electrical Resistance Method of Corrosion Monitoring in Refinery Equipment. *Corrosion* 1958; **14**: 175t–178t.
29. Standard Guide for On-Line Monitoring of Corrosion in Plant Equipment (Electrical and Electrochemical Methods). Annual Book of ASTM Standards. G 96-90[Vol 03.02]. 2001. Philadelphia, PA, American Society for Testing of Materials.
30. DENZINE AF and READING MS. An Improved, Rapid Corrosion Rate Measurement Technique for All Process Environments. *Materials Performance* 1998; **37**: 35–41.
31. VAN ORDEN, AC Applications and Problem Solving Using the Polarization Technique. CORROSION 98, Paper No. 301. 1998. Houston, TX, NACE International.
32. STERN M, GEARY AL. *Journal of the Electrochemical Society* 1957; **104**: 56.
33. STERN M. *Corrosion* 1958; **14**: 440.
34. GRAUER R, MORELAND PJ, and PINI G. *A Literature Review of Polarisation Resistance Constant (B) Values for the Measurement of Corrosion Rate*. Houston, TX: NACE International, 1982.
35. YANG B. Method for On-Line Determination of Underdeposit Corrosion Rates in Cooling Water Systems. *Corrosion* 1995; **51**: 153–165.
36. YANG, B. Real Time Localized Corrosion Monitoring in Refinery Cooling Water Systems. CORROSION 1998, Paper No. 595. 1998. Houston, TX, NACE International.
37. SILVERMAN, DC Tutorial on Cyclic Potentiodynamic Polarization Technique. CORROSION 98, Paper No. 299. 1998. Houston, TX, NACE International.
38. BOUKAMP, BA Equivalent Circuit (Equivcrt.PAS) Users Manual. Report CT89/214/128. 1989. The Netherlands, University of Twente.
39. ROBERGE, PR Analyzing Electrochemical Impedance Corrosion Measurements by the Systematic Permutation of Data Points. Munn, R. S. [STP 1154], 197–211. 1992. Philadelphia, PA, American Society for Testing and Materials. Computer Modeling in Corrosion.
40. HARUYAMA, S. and TSURU, T. A Corrosion Monitor Based on Impedance Method Electrochemical Corrosion Testing. Mansfeld, F. and Bertocci, U. [STP 727], 167–186. 1981. Philadelphia, PA, American Society for Testing and Materials. Computer Modeling in Corrosion.
41. BOSCH RW, HUBRECHT J, BOGAERTS WF, and SYRETT BC. Electrochemical Frequency Modulation: A New Electrochemical Technique for Online Monitoring. *Corrosion* 2001; **57**: 60–70.
42. JANKOWSKI J. Harmonic Synthesis: A Novel Electrochemical Method for Corrosion Rate Monitoring. *Journal of the Electrochemical Society* 2003; **150**: B181–B191.
43. IVERSON WP. Transient Voltage Changes Produced in Corroding Metals and Alloys. *Journal of the Electrochemical Society* 1968; **115**: 617–618.
44. EDEN DA, HLADKY K, JOHN DG, and DAWSON JL. Simultaneous Monitoring of Potential and Current Noise Signals from Corroding Electrodes. CORROSION 1986, Paper No. 274. 1986. Houston, TX, NACE International.
45. HUET F, BAUTISTA A, and BERTOCCI U. Listening to Corrosion. *The Electrochemical Society Interface* 2001; **10**: 40–43.
46. COTTIS RA, AL-AWADHI MAA, AL-MAZEEDI H, and TURGOOSE S. Measures for the detection of localized corrosion with electrochemical noise. *Electrochimica Acta* 2001; **46**: 3665–3674.
47. ROBERGE PR. *Handbook of Corrosion Engineering*. New York: McGraw-Hill, 2000.
48. COTTIS RA and TURGOOSE S. *Corrosion Testing Made Easy: Electrochemical Impedance and Noise*. Houston, TX: NACE International, 1999.
49. KLASSEN RD and ROBERGE PR Self Linear Polarization Resistance. CORROSION 2002, Paper No. 330. 2002. Houston, TX, NACE International.

50. EDEN DA. Electrochemical Noise. In: REVIE RW, ed. *Uhlig Corrosion Handbook*. New York: John Wiley & Sons, 2000.
51. EDEN DA. Electrochemical Noise—The Third Octave. CORROSION 2005, Paper No. 351. 2005. Houston, TX, NACE International.
52. YANG L and SRIDHAR N. Coupled Multielectrode Online Corrosion Sensor. *Materials Performance* 2003; **42**: 48–52.
53. KLASSEN RD, ROBERGE PR, and HYATT CV. A Novel Approach to Characterizing Localized Corrosion Within a Crevice. *Electrochimica Acta* 2001; **46**: 3705–3713.
54. SCHMITT G, BUSCHMANN R, FOEHN, K-H, and THEUNISSEN H. Simultaneous Monitoring of Potential and Current Noise Signals from Corroding Electrodes. CORROSION 2005, Paper No. 382. 2005. Houston, TX, NACE International.
55. SCANLAN RJ, BOOTHMAN RM, and CLARIDA DR. Corrosion Monitoring Experience in the Refinery Industry Using the FSM Technique. CORROSION 2003, Paper No. 655. 2003. Houston, TX, NACE International.
56. KANE RD and CAYARD MS. Use of Corrosion Monitoring to Minimize Downtime and Equipment Failures. *Chemical Engineering Progress* 1998; **94**[10]: 49–57.
57. VERA JR, MÉNDEZ C, HERNÁNDEZ S, and CERPA S. Field Results of the Hydrogen Permeation Sensor Based on Fuel Cell Technology. CORROSION 2002, Paper No. 346. 2002. Houston, TX, NACE International.
58. Standard Test Methods for Electrical Conductivity and Resistivity of Water. Annual Book of ASTM Standards. D 1125-95. 2005. West Conshohocken, PA, American Society for Testing and Materials.
59. Standard Test Methods for Dissolved Oxygen in Water. Annual Book of ASTM Standards. D 888-03. 2003. West Conshohocken, PA, American Society for Testing and Materials.
60. DEXTER SC. Microbiologically Influenced Corrosion. In: CRAMER DS and COVINO BS, eds. *Volume 13A: Corrosion: Fundamentals, Testing, and Protection*. Metals Park, OH: ASM International, 2003; pp. 398–416.
61. SCOTT PJB. Expert Consensus on MIC: Prevention and Monitoring. *Materials Performance* 2004; **43**: 50–54.
62. GEESEY GG. Introduction Part II—Biofilm Formation. In: Kobrin G, ed. *Microbiologically Influenced Corrosion*. Houston, TX: NACE International, 1993.
63. SANDERS PF. Monitoring and Control of Sessile Microbes: Cost Effective Ways to Reduce Microbial Corrosion. In: Sequeira CAC, Tiller AK, eds. *Microbial Corrosion-1*. New York: Elsevier Applied Science, 1988; pp. 191–223.
64. GILBERT PD and HERBERT BN. Monitoring Microbial Fouling in Flowing Systems Using Coupons. In: Hopton JW and Hill EC, eds. *Industrial Microbiological Testing*. London, UK: Blackwell Scientific Publications, 1987; pp. 79–98.
65. GERHARDT P, et al. *Manual of Methods for General Bacteriology*. Washington, DC: American Society of Microbiology, 1981.
66. JACK TR. Monitoring Microbial Fouling And Corrosion Problems In Industrial Systems. *Corrosion Reviews* 1999; **17**: 1–31.
67. JACK TR. Biological Corrosion Failures. In: Shipley RJ and Becker WT, eds. *ASM Handbook Volume 11: Failure Analysis and Prevention*. Materials Park, OH: ASM International, 2002.
68. HUNIK JH, VAN DEN HOOGEN MP, DE BOER W, SMIT M, TRAMPER J. Quantitative Determination of the Spatial Distribution of *Nitrosomonas europaea* and *Nitrobacter agilis* Cells Immobilized in k-Carrageenan Gel Beads by a Specific Fluorescent-Antibody Labelling Technique. *Applied and Environmental Microbiology* 1993; **59**: 1951–1954.
69. POPE, DH State of the Art Report on Monitoring, Prevention and Mitigation of Microbiologically Influenced Corrosion in the Natural Gas Industry. GRI-92/0382. 1992. Chicago, IL, Gas Research Institute.

70. COSTERTON, JW and COLWELL, RR Native Aquatic Bacteria: Enumeration, Activity and Ecology. [STP 695]. 1977. Philadelphia, PA, American Society for Testing and Materials.
71. JORGENSON BB. A Comparison of Methods for the Quantification of Bacterial Sulfate Reduction in Coastal Marine Sediments. *Geomicrobiology Journal* 1978; **1**: 49–64.
72. ODOM JM, JESSIE K, KNODEL E, and EMPTAGE M. Immunological Cross-Reactivities of Adenosine-5′-Phosphosulfate Reductases from Sulfate-Reducing and Sulfide Oxidizing Bacteria. *Applied and Environmental Microbiology* 1991; **57**: 727–733.
73. BRYANT RD, JANSEN W, BOIVIN J, LAISHLEY EJ, COSTERTON JW. Effect of Hydrogenase and Mixed Sulfate-Reducing Bacterial Populations on the Corrosion of Steel. *Applied and Environmental Microbiology* 1991; **57**: 2804–2809.
74. SCOTT PJB and DAVIES M. Survey of Field Kits for Sulfate Reducing Bacteria. *Materials Performance* 1992; **31**: 64–68.
75. VOORDOUW G, TELANG AJ, JACK TR, FOGHT J, FEDORAK PM, and WESTLAKE DWS. Identification of Sulfate-Reducing Bacteria by Hydrogenase Gene Probes and Reverse Sample Genome Probing. In: Minear RA, Ford AM, Needham LL, and Karch MJ, eds. *Applications of Molecular Biology in Environmental Chemistry*. Boca Raton: Lewis Publishers, 1995.
76. ANSUINI FJ and DIMOND JR. Field Tests on an Advanced Cathodic Protection Coupon. CORROSION 2005, Paper No. 39. 2005. Houston, TX, NACE International.
77. BIANCHETTI RL. Survey Methods and Evaluation Techniques. In: *Peabody's Control of Pipeline Corrosion*. Houston, TX: NACE International, 2001; pp. 65–100.
78. FRANKENTHAL RP. Electronic Materials, Components, and Devices. In: Revie RW, ed. *Uhlig's Corrosion Handbook*. New York: Wiley-Interscience, 2000; pp. 941–947.
79. Corrosion of metals and alloys—Corrosivity of atmospheres—Classification. ISO 9223. 1992. Geneva, Switzerland, International Standard Organization (ISO).
80. KING G, GANTHER W, HUGHES J, GRIGIONI P, and PELLEGRINI A. Studies in Antarctica Help to Better Define the Temperature Criterion for Atmospheric Corrosion. 2001. NACE International Northern Area Region Conference, February 26–28 2001, Anchorage, Alaska.
81. SEREDA PJ, CROLL SG, and SLADE HF. Measurement of the Time-of-Wetness by Moisture Sensors and their Calibration. In: Dean SW and Rhea EC, eds. *Atmospheric Corrosion of Metals*. Philadelphia, PA: ASTM, 1982; p. 48.
82. Standard Practice for Measurement of Time-of-Wetness on Surfaces Exposed to Wetting Conditions as in Atmospheric Corrosion Testing. ASTM G84-89. [Annual Book of ASTM Standards, Vol 03.02]. 1999. West Conshohocken, PA, American Society for Testing of Materials.
83. HIDY GM. *Aerosols: An Industrial and Environmental Science*. Orlando, FL: Academic Press, 1984.
84. FELIU S, MORCILLO M, and CHICO B. Effect of Distance from Sea on Atmospheric Corrosion Rate. *Corrosion* 1999; **55**: 883–891.
85. MORCILLO M, CHICO B, MARIACA L, and OTERO E. Salinity in Marine Atmospheric Corrosion: Its Dependence on the Wind Regime Existing in the Site. *Corrosion Science* 2000; **42**: 91–104.
86. Standard Test Method for Determining Atmospheric Chloride Deposition Rate by Wet Candle Method. ASTM G140-02. [Annual Book of ASTM Standards, Vol 03.02]. 2002. Philadelphia, PA, American Society for Testing of Materials.
87. ROBERGE PR. *Corrosion Basics—An Introduction*. 2nd edn. Houston, TX: NACE International, 2006.
88. Corrosion of metals and alloys—Corrosivity of atmospheres—Determination of corrosion rate of standard specimens for the evaluation of corrosivity. ISO 9226. 1992. Geneva, Switzerland, International Standard Organization (ISO).
89. DOYLE DP and WRIGHT TE. Rapid Method for Determining Atmospheric Corrosivity and Corrosion Resistance. In: Ailor WH, ed. *Atmospheric Corrosion*. New York: John Wiley and Sons, 1982; pp. 227–243.

90. *Standard Practice for Conducting Wire-on-Bolt Test for Atmospheric Galvanic Corrosion*. ASTM G116-99 edn. West Conshohocken, PA: American Society for Testing of Materials, 1999.
91. CALDERON JA, ARROYAVE CE. A Laboratory Approach to the Mechanism of Attack on the Wire-on-Bolt Device Used for Atmospheric Corrosion Studies. *Corrosion* 2005; **61**: 99–110.
92. COBURN S. Atmospheric Corrosion. In: *Metals Handbook*, *9th ed*, *Vol. 1, Properties and Selection, Carbon Steels*. Metals Park, Ohio: American Society for Metals, 1978; p. 720.
93. KLASSEN RD, ROBERGE PR, LENARD DR, and BLENKINSOP GN. Corrosivity Patterns Near Sources of Salt Aerosols. Townsend HE, ed. [ASTM STP 1421], pp. 19–33. 2002. West Conshohocken, PA, American Society for Testing and Materials. Outdoor and Indoor Atmospheric Corrosion.

Chapter 5

Nondestructive Evaluation

5.1. Purpose of Nondestructive Evaluation
5.2. Data Analysis and Representation
    5.2.1. Defect Response Variance
    5.2.2. Validation of Inspection Tools
    5.2.3. Data Presentation
        5.2.3.1. A-Scan Presentation
        5.2.3.2. B-Scan Presentation
        5.2.3.3. C-Scan Presentation
5.3. Visual and Enhanced Visual Inspection
    5.3.1. Borescopes
    5.3.2. Fiberscopes
    5.3.3. Video Imaging Systems
    5.3.4. Liquid Penetrant Inspection
    5.3.5. Magnetic Particle Inspection
    5.3.6. D-Sight
    5.3.7. Edge of Light
    5.3.8. Moiré Interferometry
5.4. Ultrasonic Inspection
    5.4.1. Thickness Measurements
    5.4.2. Defect Sizing
    5.4.3. Time-of-Flight Diffraction
5.5. Radiographic Inspection
    5.5.1. Computed Tomography
    5.5.2. Tangential Radiography
5.6. Electromagnetic Inspection
    5.6.1. Magnetic Flux Leakage
    5.6.2. Eddy Current
        5.6.2.1. Remote-Field Eddy Current
        5.6.2.2. Pulsed Eddy Current
        5.6.2.3. Magneto-Optic Imaging
5.7. Thermographic Inspection
    References

Corrosion Inspection and Monitoring, by Pierre R. Roberge
Copyright © 2007 John Wiley & Sons, Inc.

5.1. PURPOSE OF NONDESTRUCTIVE EVALUATION

Recent advances in nondestructive evaluation (NDE) or nondestructive testing (NDT) technologies have led to improved methods for quality control, in-service inspection, and the development of new options for material diagnostics. Detailed defect sizing and characterization has become an important objective of much NDE work underway today. To address this challenge, the NDE community has turned to a combination of multiple mode inspections and computer-aided data analysis. Success in this activity has generated quantitative NDE capabilities that may be used both as improved quality assurance tools and as new options for material diagnostics.

The NDE techniques provide ways to assess the integrity of a system or component without compromising its performance. These sensors are used to acquire information about the objects being tested and the information is converted into materials and defect parameters for performance and in-service life prediction.

When it is possible to use several inspection techniques, the choice of a specific schedule will depend on the accuracy and cost of the inspection, balancing the money spent on safety measures with the business return of the system being maintained. The accuracy of a given technique must also be sufficient to detect defects considerably smaller than those that could result in failure because these defects can grow in size between inspections.

The diverse nature of different NDE processes results in different sources of variance and possible impact on detection output capabilities. For example, a manually applied liquid penetrant process is dominated by the skill of the operator in process application and interpretation. An automated eddy current process is dominated by calibration, instrument, and procedure variances. It is important to recognize the source and nature of variance in each NDE process and to take such information into consideration when applying margins to the NDE processes (Table 5.1) (1). The NDE methods and procedures are selected

Table 5.1. Dominant Sources of Variance in NDE Procedure Application

	Materials	Equipment	Procedure	Calibration	Criteria	Human factors
Liquid penetrant	X		X			X
Magnetic particle	X	X	X			X
Radiography	X	X	X			X
Manual eddy current		X	X	X	X	X
Automatic eddy current		X	X	X	X	
Manual ultrasonic		X	X	X	X	X
Automatic ultrasonic		X	X	X	X	
Manual thermo—		X	X	X		X
Automatic thermo		X	X	X	X	

Table 5.2. Relative Cost and Requirement Ratings for the Main NDE Techniques

	Cost		Requirement		
	Inspection	Equipment	Skill	Process control	Process variance
Liquid penetrant	Low	Low	High	High	High
Magnetic particle	Low	Moderate	High	High	High
Radiography	Moderate	High	High	High	High
Manual eddy current	Low	Moderate	High	Moderate	Moderate
Automatic eddy current	Moderate	High	Moderate	High	Low
Manual ultrasonic	Low	Moderate	High	Moderate	Moderate
Automatic ultrasonic	Moderate	High	Moderate	High	Low
Manual thermo	Low	High	High	High	Moderate
Automatic thermo	Low	High	Moderate	High	Low

using a variety of practical implementation criteria, such as the relative ratings presented in Table 5.2.

The lowest cost method that produces the required result is usually the method of choice. Table 5.3 presents a general overview of the procedural steps required for the main NDE techniques considered here. However, regardless of which technique is chosen, the critical decision remains to decide on the frequency of application. This decision depends on three factors: (1) the extent of damage that might remain invisible to the technique; (2) the rate of damage occurrence with time; (3) the extent of damage the structure can tolerate.

The measured response is also quite specific to test object attributes. A summary of the technologies that could be capable of detecting, locating, and monitoring of corrosion damage in risers, for example, is shown in Table 3.8 (2). Advantages, disadvantages, and primary corrosion damage detected by each technique are described shortly in the table. In addition, a brief summary of the limitations of each technique is shown in Table 5.4.

Although each method is dependent on different basic principles both in application and output, repeatable and reproducible NDE results depend on specific understanding and control of a series of parameters, such as

- Material composition (magnetic, nonmagnetic, metallic, nonmetallic).
- Part thickness, size, and geometry.
- Material condition (heat treatment, grain size).
- Inspection scanning rate.
- Fabrication method (casting, forging, weldment, adhesive, or brazing bonded).
- Surface condition (rough, plated, bright, scaled).
- Nature or use of the part (critical, noncritical, high or low stress).
- Human factors.

Table 5.3. General Process Steps for the Main NDE Techniques

Liquid penetrant inspection
Test object cleaning to remove both surface and materials in the capillary opening
Application of a penetrant fluid and allowing it to penetrate into the capillary opening
Removal of surface penetrant fluid without removing fluid from the capillary
Application of a developer to provide a visible contrast to the penetrant fluid material)
Visually inspecting the test object to detect, classify and interpret the presence, type and size of the penetrant indication

Magnetic particle inspection
Test object cleaning to remove surface contaminants
Inducing a magnetic field in the object
Applying a fluid or powder containing finely divided particles that are attracted by the presence of a discontinuity in a magnetic field
Visually inspecting the test object to detect, classify and interpret the presence, type and size of indication

X-Radiographic inspection
Locating a sheet of X-ray sensitive film on one side of the test object
Locating an X-ray source on the opposite side of the test object
Activating the X-ray source to "expose the film" in a through transmission mode
Developing the film
Visually inspecting the resultant film image to detect, classify and interpret the presence, type and size (magnitude) of included indications

Eddy current inspection
The eddy current probe is placed in contact with the test object
An alternating magnetic field is induced in the probe by an alternating current in the probe coil
Eddy current flow is induced in the test object
The magnitude and phase of the induced current flow is sensed by a secondary coil in the probe or by change of inductance in the probe
A localized change in induced current flow indicates the presence of a discontinuity in the test object
The size of the discontinuity is indicated by the extent of the response change as the probe is scanned along the test object

Ultrasonic inspection
An ultrasonic transducer is located in contact or in close proximity to the test object
The transducer is energized in a pulsed mode to direct and propagate acoustic energy into the test object
Acoustic energy is transmitted, reflected and scattered within the test object
Energy within the test object is transmitted or redirected by internal interfaces caused by test object geometry features or internal anomalies
Transmitted or redirected energy from the test object is detected by a transducer located on or near the test object
The transmitted or redirected energy is analyzed in the time or frequency domains and interpretation of the internal condition of the test object is made by the pattern and amplitude features

Table 5.3. (*continued*)

Thermographic inspection
A pulse of thermal energy is introduced into the test object
Energy is diffused within the test object according to the thermal conductivity, the thermal mass, inherent temperature differentials and the time of observation
The temperature of the test object surface is monitored by a thermographic camera with capability for detection in the IR energy spectrum
Interpretation is completed by visually monitoring the relative surface temperature as a function of time and relating temperature differences in the time domain to the internal condition and/or structure of the test object
A relative change in surface temperature is indicative of a change in continuity or disbondment in a bonded structure
The size of an unbond is indicated by the location of the temperature gradient on the surface at a specific time and is modified by comparison with responses from similar test objects with similar geometry and thermal mass

No NDE technique produces absolute discrimination of defects, but the end output of a procedure may be quantified and the anomaly or defect detection capability may be measured, analyzed, quantified and documented.

5.2. DATA ANALYSIS AND REPRESENTATION

An NDE output response to a defect within the test object will depend on the form of detection, the magnitude of the feature that is used in detection, and the relative response magnitude of the material surrounding the defect. In an ultrasonic inspection procedure, for example, the amplitude of the response from a defect within a structure may be used to discriminate the response from the background noise due to the grain structure adjacent to a defect (Fig. 5.1) (1). If the NDE measurement is made, repetitively, on the same defect, a distribution of responses to both the defect and the surrounding material will be obtained as shown schematically in Fig. 5.2.

The measured response distribution reflects the variance in the NDE measurement process and is typical of that obtained for any measurement process. The response or noise from the surrounding material constitutes the baseline level for use in discrimination of responses from internal defects. Both the discrimination capability and defect sizing capabilities for the NDE procedure are dependent on the relative amplitudes and the rate of change of the defect response with increasing defect size.

The considerable defect to defect variance and variance in signal response to defects of equal size causes increased spread in the probability density distribution of the signal response. If a threshold decision level is applied to the responses shown in Fig. 5.2, clear defect discrimination can be achieved (Fig. 5.3). If the same threshold decision level is applied to a set of smaller size defects, clear discrimination cannot be accomplished (Fig. 5.4).

Table 5.4. Summary of Methods and Techniques (Technologies) Limitations

Technology can use on steel?	Can use on titanium?	Can use on composites?	Can see through coatings?	Can see through insulation?	Pipe wall thickness range?	Max length of inspection?	Technology can use on steel?
Visual conventional	Yes	Yes	Yes	No	No	N/A	N/A
Visual enhanced	Yes	Yes	Yes	No	No	N/A	N/A
Short-range ultrasonics (manual point-to-point measurements, single echo, or echo to echo)	Yes	Yes	No	Yes, <6 mm	No	1–40 mm	N/A
Short-range ultrasonics (bonded array, single echo, or echo to echo)	Yes	Yes	No	No	No	1-40 mm	N/A
Short-range ultrasonics (semiAUT—TOFD)	Yes	Yes	No	No	No	6 mm+	N/A
Short-range ultrasonics (AUT mapping with single/multiple focused probes or PA)	Yes	Yes	No	No	No	1 mm+	N/A
Short-range ultrasonics (AUT pigging with single/multiple L- or SV-waves probes or PA)	Yes	Yes	No	No	No	6 mm+	N/A
Long-range ultrasonics (guided waves and drycouplant transducers)	Yes	Yes	No	Yes	Yes	1 mm+	<30 m

Table 5.4. (*continued*)

Technology can use on steel?	Can use on titanium?	Can use on composites?	Can see through coatings?	Can see through insulation?	Pipe wall thickness range?	Max length of inspection?	Technology can use on steel?
Long-range ultrasonics (guided waves and MsS)	Yes	Yes	No	Yes	Yes	1 mm+	<30 m
Long-range ultrasonics (guided/SH-waves and EMATs)	Yes	Yes	No	Yes	Yes	1 mm+	<10 m
ET conventional	Yes	Yes	Yes-R	Yes	No	1 mm+	N/A
RFEC	Yes	Yes	No	Yes	No	1 mm+	N/A
Pulsed eddy current	Yes	Yes	No	Yes	Yes	6 mm+	N/A
MFL	Yes	No	Yes-R	Yes	No	12 mm	N/A
ACFM	Yes	Yes	No	Yes	No	1 mm+	N/A
FSM	Yes	Yes	No	No	Yes	1 mm+	N/A
Digital radiography	Yes	Yes	Yes-R	Yes	Yes	1 mm+	N/A
Tangentional radiography	Yes	Yes	No	Yes	Yes	1 mm+	N/A
AE	Yes	Yes	Yes-R	Yes	Yes	1 mm+	N/A
Infrared thermography	Yes	Yes	No	Yes	No	1 mm+	N/A
Magnetic particles	Yes	No	No	No	No	1 mm+	N/A

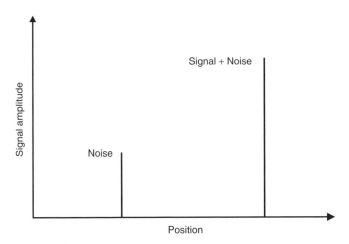

Fig. 5.1 Signal responses for a single defect measurement. (Adapted from Ref. 1.)

In this example, the threshold decision level could be adjusted to a lower signal magnitude to produce detection. As the signal magnitude is adjusted downward to achieve detection, a slight increase in the noise level will result in a ''false call'' since the noise and signal plus noise responses will now overlap. In such cases, a downward adjustment in the threshold decision level to detect all defects will result in an increase in "false calls".

Figure 5.5 shows an example where the threshold decision level has been adjusted to a level where a significant number of false calls may occur. In this example, a slight change in defect signal distribution will also result in failure to detect a defect, meaning that the NDE technique is not appropriate for purposes of primary discrimination. However, the procedure may be useful as a prescreening tool, if it is followed by another procedure that provides discrimination of the residuals. For example, a neural network detection process could be structured to provide discrimination at a high false call rate, but may be a useful in-line tool if other features are used for purposes of discrimination after the defect or variance is identified.

Accept–reject decisions may thus result in missing a defect or in false detection when the NDE procedure is operated near the limit of discrimination (Fig. 5.5). In these conditions, the NDE discrimination/detection process can be improved by using a conditional probability mapping based on decision theory principles. From decision theory, if two types of responses are assumed, that is background noise and a signal plus noise distribution, then the output of a threshold decision level such as shown in Fig. 5.6 will be a combination of accept, reject, misses, and false calls as illustrated in Fig. 5.7. The four possible outcomes are therefore:

1. **True Positive (TP)**,
    where M(Aa) is the total number of TP calls;

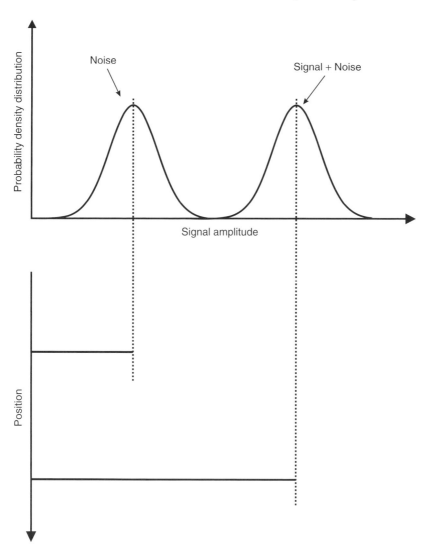

Fig. 5.2 Signal (plus noise) and noise response distributions. (Adapted from Ref. 1.)

and P(A,a) or P(TP) is the probability of TP calls. The defect signaled is truly there.

2. **False Positive (FP)**,
   where M(An) is the total number of FP calls;
   and P(A,n) or P(FP) is the probability of FP calls.
   The defect signaled is not there (false call).

3. **False Negative (FN)**,
   where M(Na) is the total number of FN calls;

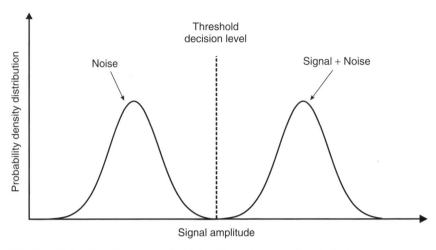

Fig. 5.3 Defect detection at a threshold signal level. (Adapted from Ref. 1.)

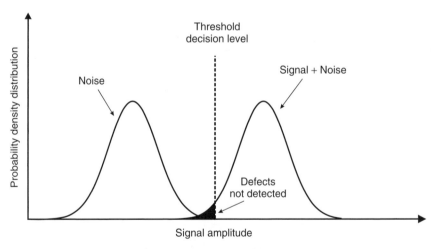

Fig. 5.4 Failure to detect smaller defects at the same threshold signal level. (Adapted from Ref. 1.)

and P(N,a) or P(FN) is the probability of FN calls.
The presence of a defect does not trigger a response (miss)

4. **True Negative (TN)**,
where M(Nn) is the total number of TN calls;
and P(N,n) or P(TN) is the probability of TN calls.
There is no defect and nothing trigger a response.

The interdependence of the matrix quantities is denoted by

$$TP + FN = \text{Total opportunities for positive calls} \tag{5.1}$$

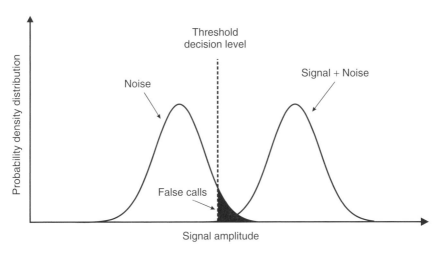

Fig. 5.5 Threshold decision level results in false calls. (Adapted from Ref. 1.)

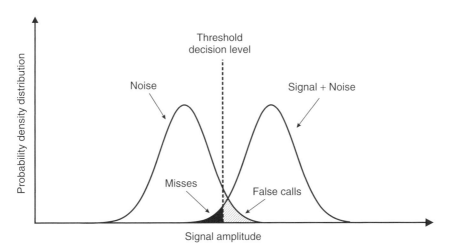

Fig. 5.6 Decisions from signal responses at low discrimination levels. (Adapted from Ref. 1.)

$$\text{FP} + \text{TN} = \text{Total opportunities for negative calls} \qquad (5.2)$$

Therefore, only two independent probabilities need be considered in alternate inspection–decision tasks. The specificity of an NDE procedure or the probability of detection (POD) of defects may be expressed as:

$$\text{POD} = \frac{\text{TP}}{\text{TP} + \text{FN}} = \frac{\text{Number of positive calls}}{\text{Number of opportunities for positive calls}} \qquad (5.3)$$

	Flaw presence	
	Positive (a)	Negative (n)
NDT signal — Positive (a)	M(Aa) True positive (TP) P(A,a) No error	M(An) False positive (FP) P(A,n) Type II error
NDT signal — Negative (n)	M(Na) False negative (FN) P(N,a) Type I error	M(Nn) True negative (TN) P(N,n) No error

Fig. 5.7 Conditional probability in defect or flaw detection (Adapted from Ref. 1.)

Similarly, the nonspecifity of an NDE procedure or the probability of false calls (POFC) may be expressed as:

$$\text{POFC} = \frac{\text{FP}}{\text{TN} + \text{FP}} = \frac{\text{Number of false positive calls}}{\text{Number of opportunities for negative calls}} \quad (5.4)$$

Application of the method requires the use of test objects with known defects. The method also assumes that all defects are of equal size and that the variance in defect response distribution is due solely to the measurement process. Confidence limits for the respective probabilities may be calculated from the data sample size used in the test case. The analysis, model, and plotting procedure differ with type of data produced. A typical hit/miss POD curve is shown in NTIAC-Fig. 5.8, where the 95% statistical confidence is indicated for a 90% demonstrated POD, a commonly used specification for aircraft inspection.

In practical applications, the NDE process characteristic of primary interest is the POD because the POFC includes a dependence on the consequence of a false call for a specific application. If false calls require significant efforts for resolution, a low level of false calls will be required. If an economical and efficient secondary method is used to resolve false calls, acceptance may become part of end to end production process requirements (1).

5.2.1. Defect Response Variance

All defects of equal size do not respond equally when an NDE procedure is applied. The physical nature of defect initiation and growth vary considerably with the origin of the defect, with the type of material involved, with the load history on the component, with the environmental exposure, and with the load levels immediately prior to inspection. Crack opening and crack closure effects have been studied extensively in materials fatigue and fracture properties measurements. Crack closure has a dominant effect on crack detectability by X-ray radiography, liquid penetrant, and by ultrasonic inspection methods. Crack closure has lesser effects on magnetic and eddy current methods. Additionally, defect closure effects on NDE processes are not linear and vary with defect size (1).

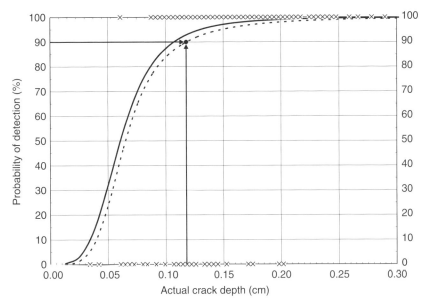

Fig. 5.8 The POD curve (solid line) and 90% probability with 95% confidence point obtained with hand scan ultrasonic on A92219 aluminum stringer stiffened panels after etch with 417 defects and 327 detected. (Adapted from Ref. 1.)

Most POD studies in this area are therefore conducted with defects that are thought to be representative of a "worst case" condition for defect detection and therefore provide a margin for detection of defects in real systems of interest. Defect-to-defect variance in the qualification of test objects is inherent in the variance of an NDE procedure, and therefore documentation of the initiation, growth, and load history of defects used in POD demonstrations is a requirement when using POD data.

Margins in use of NDE data are commonly used in life cycle management of engineering systems. For example, in life, cycle analysis in aircraft maintenance, it is common practice to provide for two or more NDE inspection opportunities for detection of "critical cracks". This does not reduce the requirement for rigor in demonstrating NDE capabilities for detection at the "critical crack length".

The common experimental approach for determining POD is to use specimens simulating the actual parts. However, this approach can be very expensive. In addition, it is extremely difficult to produce representative fatigue cracks in complex components. A more realistic approach is to inspect actual components with service-induced flaws and verify the inspection results and crack sizes by destructive tests. In the aerospace industry, for example, this approach is particularly attractive for engine components where service-retired parts are available but may not be quite practical for airframe components.

An alternative method, proposed in the early 1980s, was to use inspection data obtained from field inspections or from full-scale fatigue tests along with crack sizes to estimate POD. In such case, crack growth curves are needed to estimate crack sizes missed during previous inspections using back-extrapolation. This approach is economical and best suited for airframe structures, but often is very difficult to implement due to the small number of useful NDI data and lack of proper crack growth information. In current practice, if a crack is detected, it is either removed or repaired, and therefore the chance for the generation of realistic crack growth data is eliminated. To properly implement this process, procedures for capturing damage metrics before repair action would need to be defined (3).

5.2.2. Validation of Inspection Tools

Industry and regulators are becoming increasingly aware of the need to assess the performance of pipeline in-line inspection (ILI) tools. This performance consists of a number of separate measures that must be assessed and used to evaluate a particular tool based on the requirements of the inspection on a given pipeline. The ILI tool performance is done by four main measures, that is, detection, identification, accuracy, and locating (4).

In this context, detection is the measure of the tool ability to find and report pipeline features and anomalies that exist on the pipeline. A closely associated measure is the frequency that the tool falsely reports an anomaly where no anomaly exists. Identification is the measure of the ability of the tool ability to properly identify the type of anomaly and to differentiate it from other types of anomalies, (e.g., discrimination of internal and external corrosion). Accuracy is the measure of the ability of the ILI tool to report the severity of an anomaly, once that anomaly has been detected and identified. For corrosion anomalies, ILI tools typically report the depth, length, and width. These size parameters may differ from the actual depth, length, and width of the defects. The difference between the true and reported size of a corrosion anomaly is the inverse measure of the accuracy of the tool. Finally, location is the ability of the ILI tool to correctly report the location of a reported pipeline feature (4).

These performance measures have been presented as if they are unrelated. However, in many cases these measures may be interrelated. The misidentification of a corrosion anomaly may result in it not being reported. Similarly, poor location accuracy may result in welds and repairs being misidentified. Both of these parameters are defined in terms of the total population of anomalies, which is rarely possible because it would mean that the entire pipeline would need to be exposed and thoroughly examined. Rather, only a selected portion of the pipeline is exposed and assessed. Thus, these parameters can only be estimated from the sample of the total population of pipeline features and anomalies.

For example, suppose that following a crack-detection ILI run, a total of 40 excavations on the pipeline are conducted: 25 of those sites are conducted to investigate corrosion anomalies and the remaining 15 are conducted to investigate reported stress corrosion cracking (SCC). Of the 15 SCC excavations, 6 of those sites have SCC, while the remaining 9 sites show no evidence of SCC. However, two of the corrosion dig sites have SCC that was not detected by the tool. For this example, the POD in Eq. (5.5) is calculated as the ratio of the total number of SCC sites successfully detected by the tool, 6, to the total number of SCC sites found, 8 (6 + 2 found by accident).

$$\text{POD} = \frac{6}{8} = 0.75 \tag{5.5}$$

The POFC, on the other hand is calculated in Eq. (5.6) from the total number of SCC sites erroneously reported as SCC by the tool, 9, to the total number of sited reported as SCC, 15.

$$\text{POFC} = \frac{9}{15} = 0.6 \tag{5.6}$$

It is important to note that POD and POFC are often interrelated and are likely to depend on both the type of pipeline features and their size. If the designers of an ILI tool and the analyzers of the data do not want to miss critical defects on the pipeline, then they are apt to err on the side of caution when faced with the choice of whether or not to report ambiguous or ill-defined anomalies. If, on the other hand, they are afraid of reporting erroneous anomalies, then they may report only anomalies for which they are sure, not a likely attitude in the world of pipeline integrity.

The second measure of tool performance is the probability of identification (POI). An ILI tool can sometimes detect and report a pipeline anomaly, but misidentify its type. This misidentification can cause the operator to fail to take appropriate action. Once a pipeline feature is detected and identified, POI expresses the ratio of anomalies correctly identified to the total number of anomalies identified as that particular type.

By using the previous example, of the 15 SCC excavations, six of those sites were found to have crack-like anomalies, the remaining nine sites showing no evidence of cracking. In reality, of those six sites, only five were actually due to SCC. The sixth non-SCC anomaly was due to an inclusion that appeared to be a crack. For this example, POI for SCC is calculated in Eq. (5.7) as the ratio of the total number of SCC sites successfully identified by the tool, five, to the total number of SCC sites reported, six.

$$\text{POI} = \frac{5}{6} = 0.83 \tag{5.7}$$

Sizing accuracy is the measure of a tools ability to report the correct size or severity of certain types of corrosion anomalies. Accuracy is often quoted as a tolerance and confidence level, for example, $\pm 10\%$ depth, 80% of the time, where 10% is the tolerance and 80% is the confidence level. A standard method of comparing ILI sizing depths to field measured depths is by using what is called a unity graph in which each anomaly is plotted as a point on an $x-y$ chart, with the x-coordinate being the field depth and the y coordinate the ILI depth (5). Figure 5.9 illustrates such a unity graph obtained by comparing 40 field depth measurements with the ILI results obtained for the same locations.

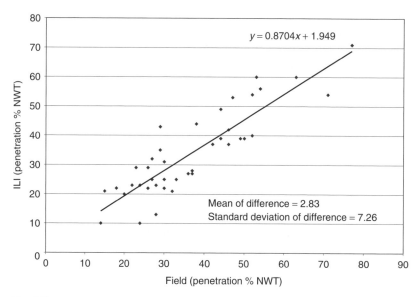

Fig. 5.9 Unity graph comparing the results of 40 field and ILI pit depth measurements. (Adapted from Ref. 4.)

From the comparison of the field excavation results and ILI data, a pipeline operator will want to assess what the accuracy of the ILI tool is and if it meets the required specifications. To answer these questions, the operator needs to estimate the accuracy of the field measurement data and compare it with that of the ILI data. If the best-fit line through the points on the unity graph has a slope that is <1, then a significant error in the field measurements is likely.

The mean of the differences, 2.83 in the data plotted in Fig. 5.9, is a measure of the relative bias between the field and ILI tools. If this mean is significantly different from zero, then either the ILI or the field tool measurements may be biased and systematically reporting depths higher or lower than the true depth. The standard deviation of the difference, 7.26 in this example, is a measure of the random scatter of the data. If the errors in the ILI and field tools are independent, then their expected values are related according to the following Eq. (5.8) rearranged into Eq. (5.9):

$$[S_{tdev}(ILI - field)]^2 = [S_{tdev}(ILI_{error})]^2 + [S_{tdev}(Field_{error})]^2 \quad (5.8)$$

$$[S_{tdev}(ILI_{error})]^2 = [S_{tdev}(ILI - field)]^2 - [S_{tdev}(Field_{error})]^2 \quad (5.9)$$

In this case, the error of the field data is known to have a standard deviation of 4. Thus the standard deviation of the ILI tool is calculated to be 5.98 or 6 according to Eq. (5.10) and (5.11):

$$S_{tdev}(ILI_{error}) = \sqrt{[S_{tdev}(ILI - field)]^2 - [S_{tdev}(Field_{error})]^2} = \sqrt{(7.2^2 - 4^2)} \quad (5.10)$$

$$S_{tdev}(ILI_{error}) = 5.98 \quad (5.11)$$

If the error of the field tool is unknown, then other methods may provide ways of approximating the error of both the ILI and field tool measurements. To convert the standard deviation of the ILI error to an 80% confidence and tolerance, the standard deviation would be multiplied by 1.28, that is, 7.7 or 8 in this example. Thus the ILI accuracy in this example is calculated to be ±8%, 80% of the time. The operator may also wish to know if the ILI tool meets the required or expected performance standards. Typically ILI data is quoted to be accurate within 10%, 80% of the time. According to the calculated accuracy here, the tool would meet this performance standard.

Another method to verify the accuracy of the tool is to simply count the number of times the field tool and ILI tool are within some specific tolerance. To account for the measurement error of the field tool, the tolerance must be calculated using both the field tool and ILI accuracy. For this example, the expected accuracy of the ILI tool is ±10%, 80% of the time, and the field tool accuracy is ±5%, 80% of the time. The standard deviation of the error of the field tool is 4, thus the 80% confidence tolerance is $1.28 \times 4 = 5.12 \sim 5$. The combined 80% confidence tolerance is estimated using Eq. (5.12).

$$\text{Combined 80\% confidence} = \sqrt{(5^2 + 10^2)} = 11.2 \qquad (5.12)$$

In the original data that were used to produce Fig. 5.9, only 5 of the difference values exceeded the value of 11.2, leaving 35 of 40 or 87% to be within tolerance. Again the ILI tool would satisfy the expected performance.

5.2.3. Data Presentation

The NDE data can be collected and displayed in a number of different formats. The three most common graphical methods are called A-scan, B-scan, and C-scan presentations. Each presentation mode provides a different way of visualizing the region of material being inspected. Most modern computerized NDE scanning systems can display data in all three presentation forms simultaneously.

5.2.3.1. A-Scan Presentation

The A-scan presentation of UT results, for example, displays the amount of received ultrasonic energy as a function of time. The relative amount of received energy is plotted along the vertical axis and elapsed time is displayed along the horizontal axis. In the A-scan presentation, relative discontinuity size can be estimated by comparing the signal amplitude obtained from an unknown reflector to that from a known reflector. Reflector depth can be determined by the position of the signal on the horizontal sweep.

For a component being inspected as illustrated in Fig. 5.10, the A-scan presentation shown in Fig. 5.11 has an IP generated by the transducer near time zero. As the transducer is scanned along the surface of the part, four other signals are likely to appear at different times on the screen. When the transducer is in its

334   Chapter 5   Nondestructive Evaluation

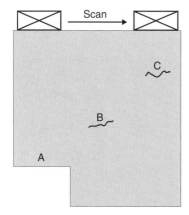

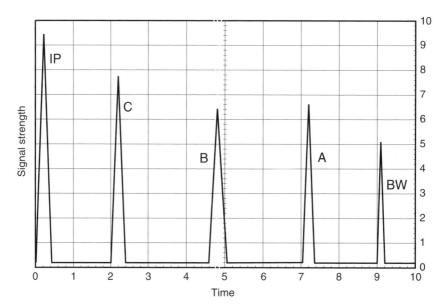

Fig. 5.10. Schematic set-up of a component being inspected with a surface NDT scanning method.

Fig. 5.11 A-scan presentation of UT results obtained during the NDT shown in Fig. 5.10. (IP = initial pulse, BW = back wall.)

far left position, only the IP signal and signal A, the ultrasonic energy reflecting from surface A, will be seen on the trace.

As the transducer is scanned to the right, a signal from the BW will appear latter showing that the ultrasonic wave has traveled farther to reach this surface. When the transducer is over flaw B, signal B will appear at a point on the time scale that is approximately halfway between the IP signal and the BW signal. Since the IP signal corresponds to the front surface of the material, this indicates that flaw B is about halfway between the front and back surfaces of the sample.

When the transducer is moved over flaw C, signal C will appear earlier in time since the ultrasonic wave travel path is shorter and signal B will disappear since the ultrasonic wave will no longer be reflecting from it.

5.2.3.2. B-Scan Presentation

The B-scan presentation provides a cross-sectional view of the test specimen. In the B-scan, the time-of-flight (TOF) of the ultrasonic energy, in the same example, would be displayed along the vertical and the linear position of the transducer is displayed along the horizontal axis. From the B-scan, the depth of the reflector and its approximate linear dimensions in the scan direction can be determined. The B-scan is typically produced by establishing a trigger gate or threshold on the A-scan. Whenever the signal intensity is great enough to trigger the gate, a point is produced on the B-scan. The gate is triggered by the ultrasonic reflecting from the backwall of the specimen and by smaller reflectors within the material.

In the B-scan two dimensional (2D) image shown in Fig. 5.12, line A is produced as the transducer is scanned over the reduced thickness portion of the specimen. When the transducer moves to the right of this section, the backwall line BW is produced. When the transducer is scanned over flaws B and C, lines are drawn on the B-scan to represent the length of these flaws with their position representing their depth within the material. Note that a limitation to this display technique is that reflectors may be masked by larger reflectors near the surface.

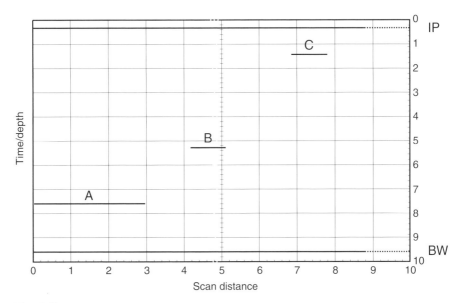

Fig. 5.12 B-scan presentation of UT results obtained during the NDT shown in Fig. 5.10. (IP = initial pulse, BW = back wall.)

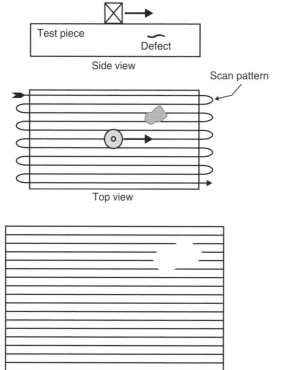

Fig. 5.13. Schematic of scanning pattern with UT and following C-Scan representation of a defect.

5.2.3.3. C-Scan Presentation

The C-scan presentation provides a plan-type view of the location and size of the test specimen anomalies or features. The plane of the image is parallel to the scan pattern of the transducer. C-scan presentations are produced with an automated data acquisition system, such as a computer controlled scanning system. For UT results, a data collection gate is established on the A-scan and the amplitude or the TOF of the signal is recorded at regular intervals as the transducer is scanned over the test piece. The relative signal amplitude or the TOF is displayed as a shade of gray or a color for each of the positions where data was recorded. The C-scan presentation provides an image of the features that reflect and scatter the ultrasonic wave within and on the surfaces of the test piece (Fig. 5.13).

5.3. VISUAL AND ENHANCED VISUAL INSPECTION

Visual inspection is the oldest and most common NDE technique used to inspect for corrosion. It may be a quick and economical way of detecting various types of defects before they degenerate into failures. The physical principle behind visual inspection is that visible light is reflected from a surface, revealing some of its features. By observing the appearance of a component, an inspector can infer its condition. Surface corrosion, exfoliation, pitting, and intergranular corrosion may be detected visually when proper access to the inspection area is possible. Obviously, visual inspection can only detect surface defects. However, some internal corrosion processes do exhibit surface indications, such as pillowing or flaking (6).

Visual inspection and measurement of corrosion damage is desirable, but most frequently not possible without a system shutdown. However, in some instances it may be possible to use visual aids without unnecessarily shutting down the system, for example, by using borescopes or video cameras as described in the following sections. In a shutdown condition, access to vessels can be sufficient to make direct measurements of localized pitting depths with micrometers. In less accessible areas, such as tubes, a calibrated borescope may be used to make similar pit depth measurements (7).

The reliability of visual inspection depends on the ability and experience of the inspector. The inspector must know how to search for critical defects and how to recognize areas where failure may occur. The human eye is usually very discerning and, with training, an operator can interpret images much better than any automated device. When necessary, permanent records can be obtained by photography or digital imaging. Visual inspection is often conducted using a strong flashlight, a mirror mounted on a ball joint, and a magnifying aid. Magnifying aids range in power from $1.5\times$ to $2000\times$. Fields of view typically range from 90 to 0.2 mm with resolutions ranging from 50 to 0.2 μm. A $10\times$ magnifying glass is recommended for positive identification of suspected cracks or corrosion damage.

The disadvantage of visual inspection is that the surface to be inspected must be relatively clean and accessible to either the naked eye or to an optical aid, such as a borescope. Typically, visual inspection lacks the sensitivity of other surface NDE methods. Visual inspection techniques can additionally be labor intensive and monotonous, leading to errors. However, compared to traditional NDE methods, such as eddy current and ultrasonics, visual inspections provide great advantage in speed of inspection and ease of interpretation. This is a significant economic advantage, as the downtime of the system being inspected is usually the greatest cost of performing NDE. Enhanced visual NDE methods attempt to provide similarly fast inspection while providing increased reliability over unassisted visual inspections.

5.3.1. Borescopes

A borescope is a long, thin, rigid rod-like optical device that allows an inspector to see into inaccessible areas by transmitting an image from one end of the scope to the other. Borescopes can get into tight places and provide reliable visual indications of internal surface conditions. Borescopes typically range from 6 to 13 mm in diameter and can be as long as 2 m. In general, the diameter of the borescope determines the size of the minimum opening into which it can be inserted. Certain structures, such as engines, are designed to accept the insertion of borescopes for the inspection of critical areas.

A borescope works by forming an image of the viewing area with an objective lens. That image is transferred along the rod by a system of intermediate lenses. The image arrives at the ocular lens, which creates a viewable virtual image that can be focused for comfortable viewing. Borescopes often incorporate a light near the objective lens to illuminate the viewing area. Different borescopes are designed to provide direct, forward oblique, right angle and retrospective viewing of the area in question (8).

The three main optical components of a borescope are the objective lens system, the relay lens system, and the eyepiece (Fig. 5.14). The objective lens is located at the end of the borescope and it acts similar to a camera lens. It forms the primary image of an object on the back of the lens. Relay lenses reform or relay the primary image along the length of the borescope. In long borescopes, several relay lenses are used. The last set of relay lenses produce the final image at the eyepiece of the borescope. Each of these three components may produce a magnification and the total magnification is the product of each magnification in accordance with Eq. (5.13) (9):

$$M_b = (M_o)(M_r)(M_e) \qquad (5.13)$$

Where M_b is the total borescope magnification, M_o the magnification of objective lens, M_r is the magnification of the relay lens, (usually 1), and $M_e =$ magnification of eyepiece

The manufacturer determines relay lens and eyepiece magnification. The magnification of the objective lens is complicated by the fact that it varies with

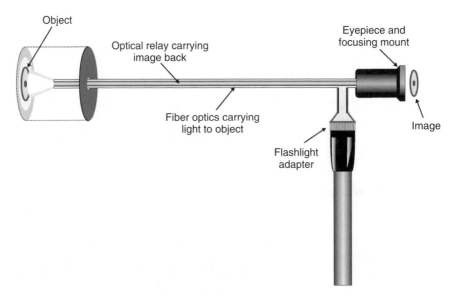

Fig. 5.14. Schematic of a borescope.

distance from the object to the objective lens and needs to be calibrated against a small object, such as a wire of known length positioned near the unknown defect. In specifying the magnification of a borescope, it is necessary to specify the distance of the object being observed since visual magnification is a logarithmic function of that distance (Fig. 5.15). Magnification is highest at very close distances, decreases with increasing object distances, changes quickly at short distances, and then changes only slightly with increasing object distance.

The accuracy of defect image measurement depends on the probe-to-object distance and the operator ability to focus the instrument. Under ideal conditions, the measurement of image defects can be made to accuracy within 25 μm by using a special filar eyepiece.[1] A number of light sources, UV, Vis, high intensity, and light-emitting diode (LED), are also available to aid viewing.

5.3.2. Fiberscopes

Fiberscopes are bundles of fiber optic cables that transmit light from end to end. They are similar to borescopes with the main difference that they are flexible and can curl into otherwise inaccessible areas. They also incorporate light sources for illumination of the subject area and devices for bending the tip in the desired direction (8). For fiberscopes, the image is carried from the objective lens to

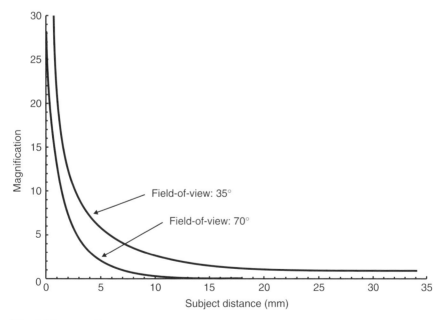

Fig. 5.15. Magnification with a borescope as a function of subject distance for two field-of-views.

[1] A filar eyepiece is graduated reticule mounted on a transverse slide attached to a micrometer.

the eyepiece by a bundle of fiber optic cables, instead of by a rigid system of lenses. Light cannot escape through the side once it enters a fiber optic cable, so it follows the cable around twists and bends.

The ends of the fibers must occupy precisely the same respective positions at both the lens and eyepiece ends otherwise the image would be "scrambled". The resolution of a fiberscope depends on the number of fibers and their diameter. More fibers of smaller diameter provide higher resolution, but they are also more expensive.

5.3.3. Video Imaging Systems

Video imaging systems (or "videoscopes") consist of tiny charge coupled device (CCD) cameras at the end of a flexible probe. Borescopes, fiberscopes and even microscopes can be attached to video imaging systems. These systems consist of a camera to receive the image, processors, and a monitor to view the image. The image on the monitor can be enlarged or overlaid with measurement scales. Images can also be printed on paper or stored digitally to obtain a permanent record. Video images can be processed for enhancing and analyzing video images for defect detection. Specialized processing algorithms may be applied which could identify, measure, and classify defects or objects of interest (8).

Many companies now offer a complete, standalone video information processing system. Features of these closed-circuit television (CCTV) systems include defect image measuring capability, computerized video enhancement, alphanumeric keyboard for data identification, high-resolution display and recording, hardcopy documentation, and audio recording. Video imaging systems are ideal for applications involving long inspection periods, multiple operator viewing, or cases where complete documentation is required (9).

5.3.4. Liquid Penetrant Inspection

The liquid penetrant NDE method is applied to detect faults that have a capillary opening to the test object surface. The nature of this NDE method demands attention to material type, surface condition and rigor of cleaning. Liquid penetrant inspection can be performed with little capital expenditure and materials used are low in cost per use. This technique is applicable to complex shapes and is widely used for general product assurance.

The basic stages of liquid penetrant inspection are illustrated in Fig. 5.16. The surface to be inspected is first cleaned thoroughly to remove all traces of dirt and grease. A brightly colored or fluorescent liquid is then applied liberally usually by soaking the component during 20–30 min to allow the fluid to penetrate any surface-breaking cracks or cavities. After soaking, the excess liquid penetrant is wiped from the surface and a developer applied. The developer is usually a dry white powder, which draws penetrant out of any crack by reverse capillary action

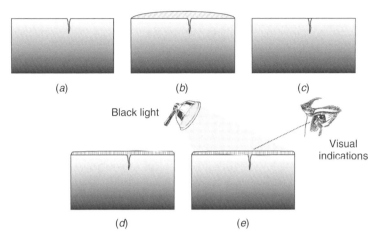

Fig. 5.16. The basic stages of liquid penetrant inspection: (*a*) sample before testing; (*b*) liquid penetrant applied; (*c*) surplus wiped off leaving penetrant in crack; (*d*) developer powder applied, dye soaks into powder; and (*e*) view colored indications, or UV lamp shows up fluorescent indications.

to produce indications on the surface. These colored indications are broader than the actual flaw and are therefore more easily visible.

A number of different liquid penetrant systems are used in industry. Fluorescent penetrants are normally used when the maximum flaw sensitivity is required. However, these penetrants must be viewed under darkened conditions with a UV lamp or black light, which may not always be practical. The most commonly used systems are solvent removable, or water washable, red dye systems.

5.3.5. Magnetic Particle Inspection

Magnetic particle inspection (MPI) is applied to the detection of defects at or near the surface in test objects made of ferromagnetic materials when a magnetic field is applied to the specimen, either locally or overall, using a permanent magnet, or an electromagnet with flexible cables or hand-held prods. It cannot, however, be used to detect deeply embedded flaws, nor can it be used on non-ferromagnetic materials, such as aluminum, copper, or austenitic stainless steel.

The MPI is often used to look for cracking at welded joints and in areas identified as being susceptible to environmental cracking, such as SCC or hydrogen induced cracking, and other cracking mechanisms such as fatigue cracking or creep cracking. Wet fluorescent MPI finds widespread use in looking for environmental damage on the inside of vessels.

Magnetic particle inspection can be performed with little capital expenditure and, as with the liquid penetrant technique, materials used are low in cost per use, the technique is applicable to complex shapes, and is widely used for general product quality assurance. Magnetic inspection can be portable. It only requires a magnetization power source, such as is provided by an electrical outlet. It is most frequently used in evaluating the quality of the weld deposit and subsurface weld indications, such as cracks.

If the material is sound, most of the magnetic flux is concentrated below the surface of the component being examined. However, if a flaw is present, such that it interacts with the magnetic field, the flux is distorted locally and 'leaks' from the surface of the specimen in the region of the flaw. Material characteristics or surface treatments that result in variable magnetic properties may therefore decrease the detectability of defects.

Once a component is magnetized, fine magnetic particles applied to its surface are attracted to the area of flux leakage, creating a visible indication of the defect. The materials commonly used for this purpose are black iron particles and red or yellow iron oxides. In some cases, iron particles are coated with a fluorescent material enabling them to be viewed under a UV lamp in darkened conditions.

Magnetic particles are usually applied as a suspension in water or paraffin. This enables the particles to flow over the surface and to migrate to any flaws. On hot surfaces, or where contamination is a concern, dry powders may be used as an alternative to wet inks. On dark surfaces, a thin layer of white paint may be applied to increase the contrast between the background and the black magnetic particles. The most sensitive technique, however, is to use fluorescent particles viewed under a UV black light.

Magnetic particle inspections is particularly sensitive to surface-breaking or near-surface cracks, even if the crack opening is very narrow. However, if the crack runs parallel to the magnetic field, there is little disturbance to the magnetic field and it is unlikely that the crack will be detected. For this reason, it is recommended that the inspection surface is magnetized in two directions at 90° to each other. Alternatively, techniques using swinging or rotating magnetic fields can be used to ensure that all orientations of crack are detectable.

Permanent magnets are attractive for on-site inspection, as they do not need a power supply. However, they tend to be used only to examine relatively small areas and have to be pulled from the test surface. Despite the need of a power supply, electromagnets find widespread application. Their main attraction is that they are easy to remove once the current has been switched off and that the strength of the magnetic field can be varied. For example, an ac electromagnet can be used to concentrate the field at the surface where it is needed.

Handheld electrical prods are useful in confined spaces. However, they suffer two major disadvantages that can rule out their use altogether. First, arc strikes can occur at the prod contact points and these can damage the specimen surface. Second, because the particles must be applied when the current is turned on, the inspection becomes a two-person operation.

In some cases, MPI can leave residual fields that subsequently interfere with welding repairs. These can be removed by slowly wiping the surface with an energized ac yoke.

5.3.6. D-Sight

The concept of D-sight is related to the schlieren method[2] for visualizing index of refraction gradients or slopes in an optical system (Fig. 5.17). D-Sight has the potential to map areas of surface waviness, as well as identify cracks, depressions, evidence of corrosion, and other surface defects. One possible problem with D-Sight is that the technique shows virtually every deviation on a surface, regardless of whether it is a defect or a normal result of manufacture.

D-Sight requires two operators. The first operator is responsible for positioning the inspection head on the surface being inspected. The second, the pendant controller, uses preplanned placements shown in the pendant to direct the first operator (10). It can be used in direct visual inspection or combined with photographic or video cameras and computer-aided image processing. D-Sight systems usually consist of an inspection head containing the optics, video camera and light source, a computer with inspection planning software, and remote pendant with touch sensitive screen for controlling the acquisition process.

The acquired images are shown on the pendant touch screen. It is possible for the pendant controller to make an immediate assessment, but the recommended procedure is to postpone the image analysis and complete the acquisition process for the whole inspected surface. This reduces the time during which the system has to be available to the inspectors. The results of inspection may then be reported on wire-frame diagrams of the aircraft type, for example, and thus easily be referenced back to locate the detected damage.

The D-Sight image shown in Fig. 5.18 provides a direct comparison to the as-received and paint stripped views of the outer surface of the intact lap joint. As labeled this is a "transformed" compilation image made from four original D-sight. These images were collected after temporarily increasing the surface reflectivity of the lap joint by wetting with a fluid or "highlighting".

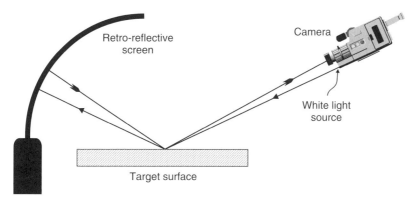

Fig. 5.17 Principle of the D-sight NDE inspection system. (Adapted from Ref. 10.)

---

[2]Schlieren methods evaluate the refractive index distribution inside a test section from the light deflection angle distribution.

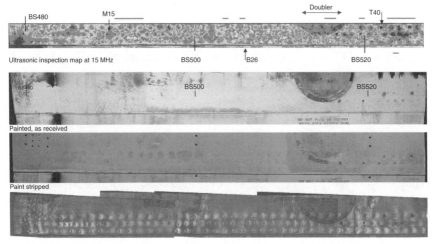

DAIS 250C on painted surface (Transformed images, L to R overlay)

Fig. 5.18 The D-sight image shown in Fig.-DAIS provides a direct comparison to the as-received and paint stripped views of the outer surface of the intact lap joint. As labeled this is a "transformed" compilation image made from four original D-sight. (Photo courtesy of the National Research Council Canada.)

When viewing the D-sight inspection image of the corroded lap joint shown in Fig. 5.18 the overall deformation pattern is readily visible but the location that is different because it is smoother reveals the presence of a broken fastener with only the paint holding it in place. There are few inspection techniques that can find failed rivets before visual inspection notices them by their absence. Rivet M15 in Fig. 5.18 fell out when the paint was subsequently stripped.

5.3.7. Edge of Light

The edge of light (EOL) enhanced optical technique is a relatively new approach to nondestructive inspection. It works by converting changes in surface slope into light intensity variation in an image. The optical path in the EOL scanner converts surface slope changes into light intensity variation. A light source behind a slit is used to produce a rectangular band of light on the surface of the object to be scanned. In the main zone of the band, the intensity is constant. At the edge, the intensity does not fall to zero immediately, due to diffraction effects, although it does drop off rapidly with distance (Fig. 5.19).

The detector of the EOL scanner is set up to operate in the middle of this "edge-of-light" zone of rapidly changing intensity. The detector is held at a constant position with respect to the light source, and records the light intensity reflected from the surface. The intensity of light is constant if the surface is smooth. A change in slope appearing in the edge zone changes the angle at which the light is reflected from the surface, which changes the location of the

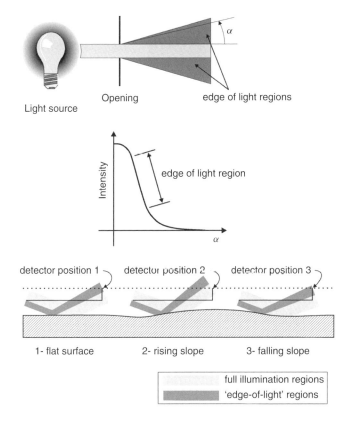

Fig. 5.19 Principle of the edge-of-light technique used to reveal surface inequalities. (Adapted from Ref. 11.)

detector in the edge-of-light band, and therefore the intensity of light appearing at the detector.

The EOL has unique applications in the inspection of small-scale surface phenomena, and has been used for the detection of surface breaking fatigue cracks in turbine engine components, as well as for forensics applications. For many applications, EOL performs better than inspections using liquid penetrant, magnetic particle, ultrasonic inspections, optical microscopy, or even scanning electron microscopy (Fig. 5.20a and b).

The technique is relatively quick, with scanning speeds in the order of 2–20 linear cm s^{-1} and line widths of 10 cm or more. Additionally, EOL inspection results are easily interpreted since they closely resemble the actual subject (11).

5.3.8. Moiré Interferometry

Moiré interferometry is a family of techniques that visualize surface irregularities. Many variations are possible, but the technique most applicable to corrosion

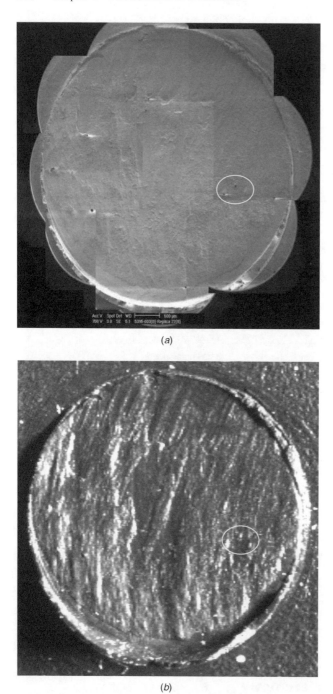

Fig. 5.20. (*a*) Assemblage of 12 SEM images of fractured shank of 5-mm diameter rivet. (*b*) Single EOL scanned image of fractured shank of 5-mm diameter rivet. *Note*: The white oval locates the same three points in each image. (Photo courtesy of the National Research Council Canada.).

inspection is shadow or projection moiré. The structured light technique is geometrically similar to projected or shadow moiré methods, and can be thought of as an optical straight edge. Instead of fringe contours being the resultant observation, the departure from straightness of a projected line is what is observed and analyzed. Using image processing techniques, the surface profile can be calculated.

5.4. ULTRASONIC INSPECTION

Ultrasonic inspection or ultrasonic testing (UT), one of the most widely used NDE techniques, is applied to measure a variety of material characteristics and conditions. Ultrasonic examination is performed using a device that generates an ultrasonic wave with a piezoelectric crystal at a frequency between 0.1 and 25 MHz into the piece being examined and analyzes the return signal. The method consists in measuring the time it takes for the signal to return and the amount and shape of that signal (6).

The accuracy of ultrasonic monitoring is limited by variations of ultrasonic wave velocity in different metals, by temperature variations in the substrate, and discrimination of the acoustic reflections. Test objects must support the propagation of acoustic energy and have a geometric configuration that allows the introduction and detection of acoustic energy in the reflection, transmission or, scattered energy configurations. The frequencies of the transducer and the probe diameter have a direct effect on what is detected. Lowering the testing wave frequency increases depth of penetration while increasing the probe diameter reduces the beam spread. Increasing the frequency also increases the beam spread for a given diameter (6).

Manual scanning is performed using instruments that have an oscilloscope-type read-out. Operator interpretation is made by pattern recognition, signal magnitude, timing, and respective hand-scan position. Variations in instrument read-out and variations in scanning can be significant. Automated scanning is performed using an instrumented scanner that keeps track of probe position and automated signal detection such that a response map of the internal structure of the component can be generated.

Techniques have been developed to employ different types of waves, depending on the type of inspection desired. Compression or longitudinal waves are the type most widely used. They occur when the beam enters the surface at an angle close to 90°. These waves travel through materials as a series of alternate compressions and dilations in which the vibrations are parallel to the direction of the wave travel (Fig. 5.21). This wave is easily generated, easily detected, and has a high velocity of travel in most materials. Compression waves are typically used for thickness measurements and for the detection and location of defects that present a reasonably large frontal area parallel to the surface from which the test is being made, such as for corrosion loss and delaminations. They are not very effective, however, for the detection of cracks that are perpendicular to the surface.

Wave length

Compression or longitudinal wave

Shear or transverse wave

Fig. 5.21. Schematic description of longitudinal and shear waves used for UT testing.

Shear or transverse waves are also used extensively in ultrasonic inspection and are generated when the beam enters the surface at moderate angles (Fig. 5.21). Shear wave motion is similar to the vibrations of a rope that is being shaken rhythmically. Shear waves have a velocity that is ~50% of longitudinal waves in the same material. They also have a shorter wavelength than longitudinal waves, which makes them more sensitive to small inclusions. This also makes them more easily scattered and reduces penetration.

Surface waves or Rayleigh waves occur when the beam enters the material at a shallow angle. These waves travel with little attenuation in the direction of the propagation, but their energy decreases rapidly as the wave penetrates below the surface. They are affected by variations in hardness, plated coatings, shot peening, and surface cracks, and are easily dampened by dirt or grease on the specimen.

Lamb waves, also known as plate waves and guided waves, occur when ultrasonic vibrations are introduced at an angle into a relatively thin sheet. A lamb wave consists of a complex vibration that occurs throughout the thickness of the material, somewhat like the motion of surface waves. The propagation characteristics of lamb waves depend on the density, elastic properties, and structure of the material, as well as the thickness of the test piece and the frequency of vibrations. Lamb waves can be used for detecting voids in laminated structures, such as sandwich panels and other thin, bonded, laminated structures.

5.4.1. Thickness Measurements

Ultrasonic inspection has been used for decades to measure the thickness of solid objects. A piezoelectric crystal serves as a transducer to oscillate at high frequencies, coupled directly or indirectly to one surface of the object whose thickness is to be measured. The time the wave of known velocity takes to travel through the material is used to determine its thickness. Since the late 1970s, ultrasonic equipment has been enhanced greatly by combining the basic electronics with computers. However, many instruments still in use today are for single-point thickness measurements, which do not provide the capability of the more sophisticated systems.

Rugged instruments based on portable computers are now available from many vendors. These systems, complete with motor-driven robotic devices to manipulate the transducer(s), have created the ability to measure wall thickness of corroded components at tens of thousands of points >0.1 m^2, which can be converted into mass loss and pitting rates. This capability, coupled with increased precision of field measurements achievable with computer-controlled systems, has made these automated ultrasonic systems well suited for on-line corrosion monitoring (7). However, the sensitivity of even these sophisticated techniques generally excludes their use for real-time measurements.

Developments are now being made with individual transducers or transducer arrays that are left in place to provide continuous monitoring. Permanently attached transducers improve accuracy by removing errors in relocating a transducer to exactly the same point with exactly the same couplant thickness.

With proper transducer selection, equipment setup, and controlled temperature conditions, the accuracy of controlled ultrasonic inspection can exceed ± 0.025 mm in a laboratory setting. Field inspections are typically to within ± 0.1 mm. Robotic systems have permitted us to increase the precision of thickness reading by achieving more consistent transducer manipulation than can be accomplished with manual ultrasonic inspection.

Uncoated components having a smooth external surface after cleaning off any biomass or debris can be inspected for internal corrosion or erosion wall losses with the traditional single backwall echo approach (Fig. 5.22). Through-coating measurements allow coated systems to be inspected without removal of the coating after applying echo-to-echo technique and A-scan imaging provided the coating is well bonded to the metal surface and its thickness <6 mm (Fig. 5.23).

The thickness of the metal substrate is determined simply by the TOF for the ultrasonic signal to reach the back surface and return to the transducer measured using either signal T_2 or T_3, as shown in Fig. 5.24. The UT digital gauges with 4- to 5-MHz, dual-element transducers are able to inspect carbon steel or austenitic stainless steel for walls thicker than 1 mm (2). Dual-element transducers can focus the ultrasonic beam at a specific depth range thus enabling optimum sensitivity on corroded, eroded, or irregular internal riser surfaces (Fig. 5.23). These transducers are highly sensitive to small pits in their optimum thickness range.

External inspection for internal general loss or localized pitting without surface coatings or with coatings thinner than 6 mm typically involves marking the component surface into a grid pattern, followed by point-by-point ultrasonic thickness measurements of individual grid sections using manually manipulated

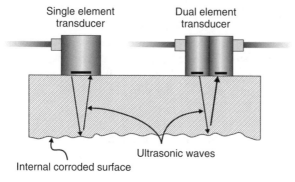

Fig. 5.22 Thickness measurements taken using two types of UT probes. (Adapted from Ref. 2.)

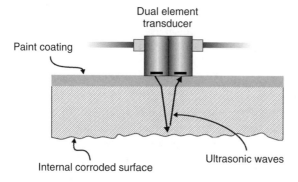

Fig. 5.23 Thickness measurements obtained through a paint coating using a dual UT probe. (Adapted from Ref. 2.)

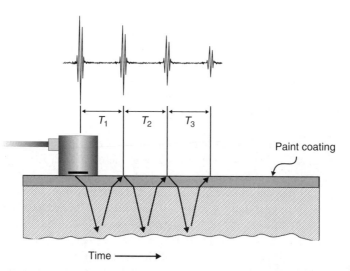

Fig. 5.24 Thickness testing using longitudinal waves (wave propagation is actually perpendicular to the surface, but is spread out in the image to show the source of ultrasonic probing signals). (Adapted from Ref. 2.)

measuring instruments. This tedious task often results in limited measurement accuracy. The costs and difficulties of this process are compounded by the need for inspection personnel.

5.4.2. Defect Sizing

The available combinations of equipment and techniques make UT well-suited for crack depth sizing over a wide range of industry applications. Ultrasonic length and depth sizing examinations are common on a wide variety of aluminum, carbon steel, and stainless steel components in industry. Less common materials, such as plastics and cast stainless steels, can also be examined with special equipment.

In pulse-echo testing, a transducer sends out a pulse of energy and the same or a second transducer listens for reflected energy or an echo (Figs. 5.11 or 5.12). Reflections occur due to the presence of discontinuities and the surfaces of the test article. The amount of reflected sound energy is displayed versus time, which provides the inspector information about the size and the location of features that reflect the ultrasonic wave. Ultrasonic sizing of the length and depth of defects is accomplished using a variety of transducers and techniques. For decades, the amplitude drop method[3] was used for length and depth sizing. However, in the mid-1980s the amplitude drop method was proven to be accurate only for length sizing (7).

There is no one depth-sizing technique that is accurate through the full range of thicknesses encountered in industry. Multiple depth-sizing methods are normally utilized to assess crack depths accurately. Crack length sizing with the amplitude drop method can typically be performed using the same equipment that is used for crack detection. Accurate crack depth sizing often uses special transducers and calibrations. The crack is typically first classified as shallow, mid-wall, or deep, then accurately sized with the appropriate equipment for the applicable range.

Defect sizing with UT depends on specialized training and experience to provide relatively accurate and consistent results. Through-wall depth sizing accuracy of ±1.3 mm is normally achievable in the field.

5.4.3. Time-of-Flight Diffraction

Conventional UT is limited because the pulse-echo signal is strongly influenced by reflector type, shape, and orientation, and is not related just to size. A relatively recent development is what is called the time-of-flight diffraction (TOFD) method.

---

[3]The popular 6-decibel (dB) amplitude drop method, for example, refers to the fact that a 6 dB decrease in gain results in a 50% decrease in screen amplitude. The operator manipulates the transducer to maximize the signal from an indication. When this is achieved, the maximized signal is then set to 80% full screen height (FSH) and the transducer is moved slowly along parallel to the axis of the defect until the screen signal drops to 40% FSH, or 50% of the maximized signal (a 6-dB drop).

With TOFD, one transducer transmits an ultrasonic signal through the material being inspected. This signal is then received by a second transducer (Fig. 5.25). This produces a single waveform or plot of signal amplitude versus time. If there is a defect in the material between the transmitter and the receiver, the ultrasonic signal will be diffracted and this will alter the waveform. The time of arrival of diffracted ultrasonic signals is used to calculate crack location and depth. Since identifying a flaw from a single waveform can be difficult, multiple waveforms or B-scans are "stacked" together to create a D-scan. The advantages of the TOFD method over other ultrasonic methods are that it allows large volumes of material to be inspected efficiently, produces permanent records of data, and reduces operator misjudgment and subjectivity.

The TOFD relies on the detection of forward scattered diffracted signals, as opposed to backward reflected or diffracted signals that are used in conventional pulse-echo ultrasonic inspection. For weld inspection, two transducers are used, one mounted on either side of the weld, with both aimed at the same point in the weld volume. The tests are conducted in a pitch-catch mode, with one transducer being the transmitter and the second the receiver. Broad beam transducers are used so that the entire weld is flooded with ultrasound and, consequently, the entire volume is inspected using a single uniaxial scan pass along the weld length.

Because the technique relies on detecting the forward scattered diffracted signals originating at the flaw edges, precise measurement of flaw size, location, and orientation is possible, and inspection reliability improves. Point reflectors are precisely determined by using geometric considerations, beam angles, and the appropriate sound propagation velocity. The TOFD has been accepted for the examination of critical components and welds in many industries such as fossil and nuclear power, petrochemical, oil and gas refineries, and pipelines.

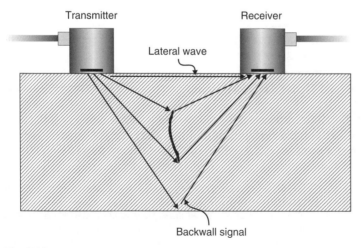

Fig. 5.25. Schematic description of the TOFD method to carry out NDE inspection.

5.5. RADIOGRAPHIC INSPECTION

Radiographic inspection methods utilize radiation in the form of either X-rays or other high energy rays, which can penetrate the material being inspected and be relatively absorbed depending on the thickness or the density of this material. Variations in the material can be detected by recording the differences in absorption of the transmitted waves. The variations in transmitted waves may be recorded by either film (radiography) or electronic devices (radioscopy) thus providing a 2D image that requires interpretation. The method is sensitive to any discontinuities that affect the absorption characteristics of the material (6).

The thickness of corroded piping and other equipment can be deduced from radiographic images in several ways. For example, the difference in optical density of the film in a noncorroded area of the image compared with the optical density in the pitted area can be correlated with the difference in thickness of the two areas, and thereby the depth of corrosion pits can be estimated. With repeated surveys of specific areas at intervals determined by the severity of the corrosion damage, the changing depth and area of corrosion can be resolved and corrosion rates approximated. The method can be used on-line, but is too insensitive to provide real-time measurements (7).

The radiographic technique can be used to determine the integrity of piping and equipment over large areas relatively inexpensively, using either manual techniques or automated, real-time radiographic inspection systems. This technique does not require access to the component being inspected, which means it can be used to inspect insulated, clad, bundled, or otherwise inaccessible components (Fig. 5.26).

However, absolute thickness of the inspected object is not normally discernible from the radiographic image. From the differential optical density of the film, a difference in thickness can be estimated, but the remaining wall thickness is assumed based on other information, such as the nominal wall thickness of the piping.

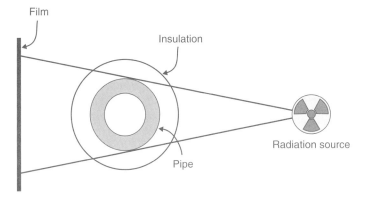

Fig. 5.26 Tangential radiography setup to measure wall thickness. (Adapted from Ref. 2.)

Scale or other debris in the area of corrosion can significantly affect the accuracy of calculated pit depths. The radiographic source-to-detection distance is also limited by the strength of the source. In field radiography, this limits the technique for use on piping and other objects with cross-sections not larger than 1 m.

The design of the X-ray tube to perform X-ray radiography is a fundamental characteristic of the technique. There are many different types of tubes used for special applications. The most common is the directional tube, which emits radiation perpendicular to the long axis of the tube in a cone of ~40°. Another type is the panoramic tube, which emits X-rays in a complete 360° circle. This type of tube would be used, for example, to examine girth welds in a jet engine with a single exposure (6).

Radioisotopes or neutron sources may be used in place of X-ray tubes. Radioisotope equipment has inherent hazards, and great care must be taken with its use. Only fully trained and licensed personnel should work with this equipment. As with X-rays, the most common method of measuring γ-ray transmission is with film.

In X-ray and γ radiography, attenuation increases uniformly with mass number and density of the elements in the material being examined, whereas, with neutrons, attenuation is random with a tendency for certain light elements, such as hydrogen to absorb and scatter neutrons rather well. Thus, neutron radiography is particularly well suited to detect corrosion and moisture entrapment, especially in aircraft composite components. For example, neutron radiography has been used successfully to indicate areas of moisture ingress, honeycomb cell corrosion, and adhesive–composite hydration in F-18 Hornet flight control surfaces. Figure 5.27 illustrates the results obtained with three NDE techniques on an F-18 rudder suspected of having been penetrated with water, a potentially very damaging situation.

5.5.1. Computed Tomography

Advanced uses of radiography are being made with the aid of computers and complex algorithms to manipulate the data. This is termed computed tomography, or CT scanning. By scanning a part from many directions in the same plane, a cross-sectional view of the part can be generated, and the internal structure may be displayed in a 2D view. The tremendous advantage of this method is that internal dimensions can be measured very accurately to determine such conditions as wall thinning in tubes, size of internal discontinuities, relative shapes, and contours. More advanced systems can generate three-dimensional (3D) scans when more than one plane is scanned. Computed tomography scanning is costly and time consuming.

5.5.2. Tangential Radiography

Tangential radiography, also known as profile radiography, is used for detailed inspection of small pipe sections under insulation. Figure 5.26 illustrates a scheme

Fig. 5.27 Results obtained with three NDE techniques on an F-18 rudder suspected of containing water: UT C-scan (*a*), X-ray radiography (*b*), and neutron radiography (*c*). (Courtesy of Dr. LGI Bennett, Royal Military College of Canada.)

that can be used to inspect for corrosion in pipe walls. From the contrast in the radiograph image, the thickness of the pipe wall can be measured for both edges of the pipe. In order to ensure complete inspection, successive measurements must be made while rotating the source and film around the circumference of the pipe.

Among the drawbacks of tangential radiography is the difficulty of inspecting pipes with diameters greater than ~25 cm, radiation concerns that require care to ensure nearby workers are not exposed to unhealthy levels of radiation, and the expense of film. In addition, film-based tangential radiography is a slow process, requiring images to be acquired from many angles since only a small quadrant of the pipe can be inspected at one time.

5.6. ELECTROMAGNETIC INSPECTION

5.6.1. Magnetic Flux Leakage

Magnetic flux leakage (MFL) is the oldest and most commonly used in-line inspection method for finding metal-loss regions in gas-transmission pipelines. The MFL is typically used to detect metal loss due to corrosion or gouging, but can sometimes find other metallurgical conditions, such as inclusions or weld porosity. This technique is not well suited for detection of cracks or other long, narrow, or shallow defects. However, there may be some limited success finding deep circumferential cracks (2). It is useful for locating defects where volumetric material loss, such as pitting, occurs and it is a relatively fast technique for inspecting equipment with large surface areas, such as tanks, or long distances in the case of pipelines.

As illustrated in Fig. 5.28, the MFL technique uses a combination of permanent magnets and sensor coils to identify corrosion in steel tubes, pipes, and plates. Magnetic flux is channeled into the component being inspected between the opposite poles of two magnets. If there is a defect in the metal, some flux that leaks out may be detected and measured by the sensors positioned near the metal surface and between the magnets. Corrosion causes a decrease in the thickness of the wall, leading to flux leakage at both sides of the wall. At metal loss regions, a higher flux density is recorded. These changes in the magnetic field depend on the radial depth, axial length, circumferential width, and shape of the defects. The amplitude of the leaking flux is compared with results from known defects in similar materials after the inspection and the analyst then makes wall-loss estimates and predictions.

The MFL has applications in tanks, heat exchangers, pipelines, and tubular products. Different types of MFL systems are used for these different applications. Some systems, such as in-line pipeline inspection tools, are very sophisticated self-contained systems that gather and store various data remotely. Signal analysis can be either real time or done at a later date, depending on the application and the inspection system (7).

Magnetic flux leakage can work through coatings of up to 10 mm thickness and is generally limited to pipe thicknesses of 12–15 mm. Above this thickness, it is difficult to obtain magnetic saturation. Typical wall thickness sizing accuracy is on the order of $\pm 10\%$ and length accuracy of ± 1 cm with a 80% confidence level. It depends on surface cleanliness so that sensors have good contact with the part being inspected and sizing is impaired by noise signals produced by oxides and scale on the metal surface. In some applications, strong magnets can make moving the sensors difficult.

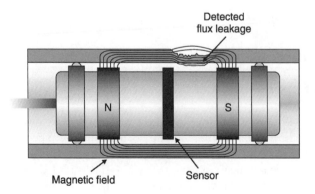

Fig. 5.28 Pipe inspection using MFL. (Adapted from Ref. 2.)

5.6.2. Eddy Current

Eddy current (EC) testing uses the principles of electromagnetic induction to identify specific material characteristics and conditions. When an electrically conductive, but nonmagnetic material is exposed to an alternating magnetic field that is generated by a coil of wire carrying an alternating current, eddy currents are induced on and below the surface of the material. These eddy currents, in turn, generate their own magnetic field that opposes the magnetic field of the test coil. This magnetic field interaction causes a resistance of current flow, or impedance, in the test coil. Slightly magnetic materials, such as S41000 stainless steels, can still be examined using a magnetically biased probe that is used to neutralize the permeability effect on the eddy currents.

By measuring changes in impedance, the test coil or a separate sensing coil can be used to detect any condition that would affect the current carrying properties of the test material. Eddy currents are sensitive to changes in electrical conductivity, changes in magnetic permeability[4], geometry or shape of the part being analyzed, and defects. Among the latter defects are cracks, inclusions, porosity, and corrosion (6). The size of the defects and the detection limits are important. The EC is extremely sensitive to very small changes in the material structure, hence is capable of finding small discontinuities of <100 μm in highly conductive materials (2). Depending on the orientation of the defect being measured, an accuracy of ±5% of wall thickness can be obtained.

An EC system incorporates the electronic signal generator–processor and a probe containing at least one coil. The fields in the generator and sensor coils are balanced by adjustments of frequency, amplitude, and distance. A change in the balance of the two fields indicates various defects or material thinning.

Eddy current has several distinct advantages. First, the process does not require direct contact of the test probe to the specimen. However, the probe must be in close proximity to the test specimen in order to induce eddy currents in the specimen. The technique is fast when used for tube bundles inspection for which external access to tubes is restricted, a serious limiting factor for most other NDE techniques. In industrial and aerospace settings, the same system may be used to examine many different pieces of equipment. Another important advantage of eddy current NDE is that it requires only minimum part preparation.

Reliable inspections can be performed through a nonconductive coating up to thickness of ∼0.4 mm. Inspection can also be performed very rapidly in production applications, as the technique is conducive to automation. The process requires a skilled operator to calibrate and interpret indications. Once the range of the inspected component conditions is known the system can be set for automated interpretation provided a reliable set of reference standards, preferably of production conditions, has been used to calibrate the system.

Eddy currents are fairly localized in the immediate region of the coil. The exact depth of penetration by eddy currents is dependent on the magnetic permeability and conductivity of a specimen, as well as the frequency of the ac

---

[4]Magnetic permeability provides a measure of the ability of a material to be magnetized.

probing signal. The standard depth of penetration (δ) expressed in Eq. (5.14) is the distance below the surface of a flat specimen in which eddy currents are at 37% of the density at the surface.

$$\partial = \frac{1}{\sqrt{\pi M T f}} \qquad (5.14)$$

Where, M is the relative permeability, T is its electrical conductivity, and f is the frequency used.

Sophisticated, digitized, multichannel eddy current instruments are used extensively throughout industry. Computer-controlled robotic delivery systems to manipulate probes are used in areas such as manufacturing and nuclear power generation. Currently, phase analysis instruments provide both impedance and phase information. This information is displayed on an oscilloscope or an integrated LCD display on the instrument (7).

Another type of EC instrument displays its results on planar form on a screen. This format allows both coil impedance components to be viewed. One component consists of the electrical resistance due to the metal path of the coil wire and the conductive test part. The other component consists of the resistance developed by the inducted magnetic field on the coil magnetic field. The combination of these two components on a single display is known as an impedance plane.

Automated scanning is performed using an instrumented scanner that keeps track of probe position and automated signal detection such that a response map of the test object surface can be generated. The results of eddy inspection are extremely accurate if the instrument is properly calibrated. Most modern EC instruments in use today are relatively small and battery powered. In general, surface detection is accomplished with probes containing small coils (3-mm diameter) operating at a high frequency, generally 100 kHz and above. Low frequency eddy current (LFEC)[5] is used to penetrate deeper into a part to detect subsurface defects or cracks in underlying structure. The lower the frequency, the deeper is the penetration as indicated in Eq. (5.14).

5.6.2.1. Remote-Field Eddy Current

Remote-field eddy current (RFEC) was developed in the 1950s and is widely used for inspection of metallic pipes and tubing. Its inspection is conducted by exciting a relatively large, low-frequency ac coil inside the pipe. A pick-up coil, offset by approximately two pipe diameters, can then be used to detect changes in the flux field due to the tube wall condition, thickness, permeability, and conductivity. Figure 5.29 shows a schematic of RFEC testing (2).

In operation, an electromagnetic field is transmitted into the pipe thickness. This direct path is attenuated rapidly to create circumferential eddy currents, which diffuse radially toward the outer wall. Once reaching the outer wall, the field spreads rapidly with little attenuation. This field then rediffuses back to the inner wall, interacting with defects in the remote field region, before being

---
[5] The LFEC is generally considered to be between 100 Hz and 50 kHz.

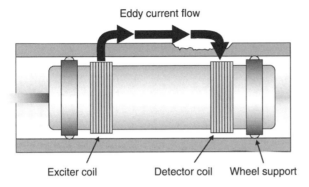

Fig. 5.29 Risers remote field ET of a pipe. (Adapted from Ref. 2.)

detected by the pick-up coil. Because this is a through-wall technique, defects at any depth in the pipe wall can be detected with similar sensitivities instead of just surface defects that can be detected by conventional EC.

In general, the wall thickness affects the phase between the excitation and detector signals, and the extent of the defect affects the amplitude. Although magnetic flux can be used to measure changes in thickness of both magnetic and nonmagnetic materials, it is generally used to identify defects in magnetic material for which eddy currents are not effective. Sophisticated multichannel digital instruments are now available to measure magnetic flux. The system is composed of electronics to generate, manipulate, and display the magnetic flux signals, and a probe to introduce the magnetic field into the component being inspected. Originally used to inspect oilfield tubulars for corrosion, it now has applications in small-bore tubes and in plate. Typical applications include inspecting heat exchanger and boiler tubes for corrosion or erosion (7).

Advantages over conventional EC include equal sensitivity to defects on either the inner or outer surface, insensitivity to probe wobble or lift-off[6], and not being limited by the penetration depth. However, RFEC is usually limited to the detection of wall loss. Attempts have been made to detect pits using differential receivers and multiple receiver coils, but the sensitivity is usually insufficient due to the interference of corrosion byproducts with signals from pits. Additionally, the speed of inspection using RFEC is significantly slower than with conventional EC testing.

5.6.2.2. Pulsed Eddy Current

Pulsed eddy current (PEC), another EC variant technology, employs the response to an input pulse instead of a sinusoidal wave to characterize the loss of material. The advantages of a PEC technique over conventional EC are its larger

---

[6]Life-off refers to the distance between the coil of a surface probe and sample. It is a measure of coupling between probe and sample.

penetration depth, relative insensibility to lift-off, and the possibility to obtain a quantitative measurement result for wall thickness. This technology has been shown to measure material loss on the bottom of a top layer, the top of a bottom layer, and the bottom of a bottom layer in two-layer samples with an accuracy of ~5%. A mechanical bond is not necessary as it is with ultrasonic testing. The instrument and scanner are rugged, portable, and can use conventional coils and commercial probes.

Additionally, pulsed EC can detect of wall-thinning areas without the need to remove the outside protection coatings. The pulsed magnetic field sent by the probe coil penetrates through any nonmagnetic material between the probe and the object under inspection, for example, insulation material. The detected EC signal is processed and compared to a reference signal in order to eliminate the normal effects due to the material properties and a reading for the average wall thickness within magnetic field area results. This technique can be used for nominal wall thickness between 6 and 65 mm and insulation thickness <150 mm using transducers with a diameter between 50 and 100 mm (2).

5.6.2.3. Magneto-Optic Imaging

Magneto-optic imaging (MOI) is provided by the response of a magneto-optic sensor to weak magnetic fields that are generated when eddy currents induced by the MOI interact with defects in the inspected material. Since its introduction, the use of MOI has steadily increased particularly in areas where it outperforms traditional methods, such as EC spot probes and sliding probes. Multiple tests have shown MOI to be a fast and reliable method of finding surface and subsurface cracks and sometimes corrosion in metal aircraft skins (12).

The magneto-optic/EC nondestructive testing instrument is based in part on the principles of Faraday magneto-optic rotation. In 1845, Michael Faraday first observed the effect when linearly polarized light was transmitted through a piece of glass placed in an external magnetic field. It was observed that magnetic fields affect optical properties of certain materials so that when linearly polarized light is transmitted through the material in the direction of an applied magnetic field, the plane of polarization is rotated.

A schematic of a MOI set-up is shown in Fig. 5.30. A foil carrying alternating current serves as the excitation source and induces eddy currents in a conducting test specimen. Under normal conditions, the associated magnetic flux is tangential to the specimen surface. Defects in the specimen generate a normal component of the magnetic flux density that is detected by the magneto-optic sensor. The usual sensor consists of a thin film of bismuth-doped iron garnet grown on a substrate of gadolinium gallium garnet.

Images appear directly at the sensor and can be viewed directly or imaged by a small video camera located inside the imaging unit. The operator can view these images while moving the imaging head continuously along the area to be inspected. In contrast to conventional EC methods, the MOI images resemble the defects that produce them, making the interpretation of the results more intuitive

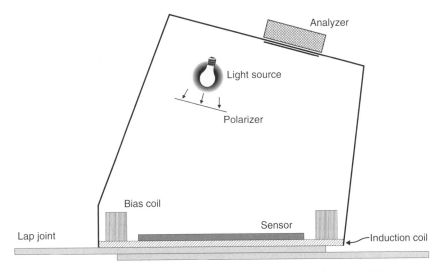

Fig. 5.30 Schematic of a magneto-optic imaging system. (Adapted from Ref. 12.)

than the interpretation of traces on a screen. Rivet holes, cracks, and subsurface corrosion are readily visible. The image is in video format, and therefore easily recorded for documentation.

5.7. THERMOGRAPHIC INSPECTION

Thermography is an emerging inspection technique that monitors the changes in the thermal patterns of an object as it is heated, cooled, or kept in a steady-state ambient condition. Thermography is becoming an increasingly popular NDE tool as equipment costs and computer technologies become more affordable and able to handle the large quantities of data generated by the thermographic equipment. Commercial thermographic equipment is available that can detect a 0.2°C temperature difference in materials up to 500°C. Beyond that temperature, a 2°C difference is detectable for temperatures up to 3500°C (2).

Thermographic inspection methods may be used to measure a variety of material characteristics and conditions. They are generally applied in the defect detection mode for the detection of interfaces and variation of the properties on interfaces within layered test objects. Inspected components must be thermally conductive and the test object surface must be reasonably uniform in color and texture. The method is a noninvasive area–volume inspection process and therefore loses resolution near edges and at locations of non-uniform geometry changes (6). Because it is an area-type technique, it is most useful for identifying areas that should be inspected more carefully with other NDE techniques offering a greater precision, such as EC and UT methods.

With a passive thermographic measurement system, the thermal camera images the IR energy radiated, scattered, and reflected from the surface of the material undergoing inspection. An embedded defect (corrosion pit or void) has a thermal emissivity that is different from the substrate material. Differences in thermal emissivity and thermal diffusion will be imaged as variations in image brightness, allowing damage to be detected and characterized (13).

In certain instances, the thermographic measurements can be made in a spectral bandpass window that allows the IR energy to propagate efficiently through a coating layer and probe the material substrate underneath, such being the case for typical US Air Force primer–topcoat paint combinations (14). Figure 5.31 depicts the spectral transmission window for a 75-μm thick paint layer in the 2–12-μm wavelength range, which occurs between 3.5 and 5.8 μm, peaking at 5.2 μm. The corrosion regions can therefore be imaged directly through the paint by using a mid-wave IR camera sensitive to 3–5-μm thermal energies, and/or using bandpass filters in that wavelength range.

Thermal wave imaging overcomes some of the limitations of a passive system by measuring the time response of a thermal pulse rather than the temperature response. The thermal pulse penetrates multiple layers, where there is a good mechanical bond between the layers.

The raw image displayed by an IR camera only conveys information about the temperature and emissivity of the surface of the target it views. To gain information about the internal structure of the target, it is necessary to observe the target as it is either being heated or as it cools. Since it takes heat from the surface longer to reach a deeper obstruction than a shallow one, the effect of the

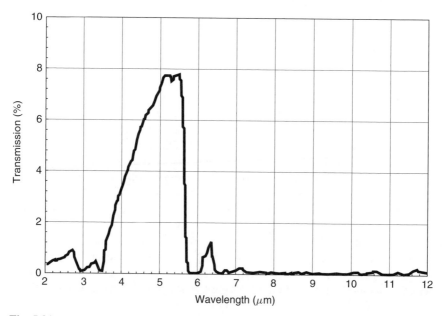

Fig. 5.31 Mid-wave IR spectral transmission window for standard Air Force and primer. (Adapted from Ref. 13.)

shallow obstruction appears at the surface earlier than that of a deep one. The thermal response to a pulse over time, color coded by time of arrival, may be displayed as a two dimensional C-scan image for interpretation by the operator.

Recently, a process has been developed for IR imaging of subsurface defects by ultrasonically stimulating the component. In this technique, an ultrasonic transducer was coupled to the part and the ultrasonic energy from the transducer caused the defect areas to heat up. Consequently, these heated areas were detectable by the thermal imaging system (15).

REFERENCES

1. RUMMEL WD and Matzkanin GA. *Nondestructive Evaluation (NDE) Capabilities Data Book.* 3rd edition edn. Austin, TX: Nondestructive Testing Information Analysis Center (NTIAC), 1997.
2. LOZEV M, GRIMMETT B, SHELL E, and SPENCER R. Evaluation of Methods for Detecting and Monitoring of Corrosion and Fatigue Damage in Risers. Project No. 45891GTH. 11-4-2003. Washington, D.C., Minerals Management Service, U.S. Department of the Interior.
3. FAHR A and FORSYTH DS. A Perspective on Inspection Reliability. *Canadian Aeronautics and Space Journal* 2001; **47**: 253–258.
4. DESJARDINS G. Assessment of ILI Tool Performance. CORROSION 2005, Paper No. 164. 2005. Houston, TX, NACE International.
5. In-line Inspection Systems Qualification Standard. [API 1163]. 2005. Washington, DC, American Petroleum Institute (API).
6. ROBERGE PR. *Handbook of Corrosion Engineering.* New York: McGraw-Hill, 2000.
7. Techniques for Monitoring Corrosion and Related Parameters in Field Applications. NACE 3T199. 1999. Houston, TX, NACE International.
8. BDM Federal. Corrosion Detection Technologies: Sector Study. 1998. North American Technology and Industrial Base Organization (NATIBO).
9. MIX PE. *Introduction to Nondestructive Testing: A Training Guide.* 2nd ed. New York: John Wiley & Sons, 2005.
10. KOMOROWSKI JP, SIMPSON DL, and GOULD RW. A Technique for Rapid Impact Damage Detection with Implication for Composite Aircraft Structures. *Composites* 1990; 169–173.
11. FORSYTH DS, KOMOROWSKI JP, MARINCAK A, and GOULD RW. The Edge Of Light Enhanced Optical NDI Technique. *Canadian Aeronautics and Space Journal* 1997; **43**: 231–235.
12. SHIH WCL and FITZPATRICK GL. Magneto-Optic Imaging Technology: A New Tool for Aircraft Inspection. *AMPTIAC Quarterly* 2002; **6**[3]: 17–22.
13. BLACKSHIRE JL, BUYNAK C, STEFFES G, and MARSHALL R. Nondestructive Evaluation through Aircraft Coatings: A State-of-the-Art Assessment. The 9th Joint FAA/DoD/NASA Aging Aircraft Conference, March 6–9, Atlanta. 2006. Federal Aviation Administration (FAA).
14. PEREZ I. and KULOWITCH P. Thermography for Characterization of Corrosion Damage. CORROSION 2000, Paper No. 290. 2000. Houston, TX, NACE International.
15. THOMAS RL, FAVRO LD, HAN X, OUYANG Z, SUI H, and SUNG G. Infrared Imaging of Ultrasonically Excited Subsurface Defects In Materials. [US Patent 6,236,049]. 2001.

Appendix A

SI Units Conversion Table

HOW TO READ THIS TABLE

This table provides conversion factors to Systeme International (SI) units. These factors can be considered as unity multipliers. For example:
Length: m/X

$$0.0254 \text{ in.} \quad 0.3048 \text{ ft}$$

Means that

$$1 = 0.0254 \text{(m/in.)} \text{ or } 1 \text{ in.} = 0.0254 \text{ m}$$
$$1 = 0.3048 \text{ m/ft}^{-1} \text{ or } 1 \text{ ft} = 0.3048 \text{ m}$$

The SI units are listed immediately after the quantity; in this case: Length: m/X. The m stands for meter, and the "X" designates the non-SI units for the same quantity. These non-SI units follow the numerical conversion factors.

Note: In the following table at all locations, ton refers to United States rather than metric ton.

Area: m^2/X	1.0×10^{-4}	cm^2
	1.0×10^{-12}	μm^2
	0.0929	ft^2
	6.452×10^{-4}	$in.^2$
	0.8361	yd^2
	4,047	acre
	$2.59 \times 10^{+6}$	mi^2/e
Density: (kg/m^{-3})/X	1000.0	g/cm^3
	16.02	lbm/ft^3
	119.8	lbm/gal
	27,700	$lbm/in.^{-3}$
	2.289×10^{-3}	$grain/ft^3$

Corrosion Inspection and Monitoring, by Pierre R. Roberge
Copyright © 2007 John Wiley & Sons, Inc.

Appendix A SI Units Conversion Table

Diffusion coefficient: $(m^2/s)/X$	1.0×10^{-4}	cm^2/s
	2.78×10^{-4}	m^2/h
	0.0929	ft^2/s
	2.58×10^{-5}	ft^2/h^{-1}
Electrical charge: C/X	1	A-s
	10	abcoulomb
	3.336×10^{-10}	statcoulomb
Electrical conductance: (Siemens)/X	1	Ω^{-1}
Electrical field strength: $(V/m)/X$	1	$kg\text{-}m/A\text{-}s^3$
	100	V/cm
	39.4	V/in
Electrical resistivity: $(\Omega\text{-}m)/X$	1	$kg\text{-}m^5/A^2\text{-}s^3$
	1.0×10^{-9}	abohm-m
	$8.988 \times 10^{+11}$	statohm-m
Energy: J/X	$3.6 \times 10^{+6}$	kWh
	4.187	cal
	4187	kcal
	1.0×10^{-7}	erg
	1.356	ft-lbf
	1055	Btu
	0.04214	ft-pdl
	$2.685 \times 10^{+6}$	hp-h
	$1.055 \times 10^{+8}$	therm
	0.113	in-lbf
	$4.48 \times 10^{+4}$	hp-min
	745.8	hp-s
Energy density: $(J/m^{-3})/X$	$3.6 \times 10^{+6}$	kWh/m^{13}
	$4.187 \times 10^{+6}$	cal/cm^{-3}
	$4.187 \times 10^{+9}$	$kcal/cm^{-3}$
	0.1	erg/cm^{-3}
	47.9	$ft\text{-}lbf/ft^3$
	$3.73 \times 10^{+4}$	Btu/ft^3
	$1.271 \times 10^{+8}$	kWh/ft^3
	$9.48 \times 10^{+7}$	$hp\text{-}h/ft^3$
Energy, linear: $(J/m)/X$	418.7	cal/cm
	$4.187 \times 10^{+5}$	kcal/cm
	1.0×10^{-5}	erg/cm
	4.449	ft-lbf/ft

	3461	Btu/ft
	$8.81 \times 10^{+6}$	hp-h/ft
	$1.18 \times 10^{+7}$	kWh/ft
Energy per area: $(J/m^{-2})/X$	41,868	cal/cm^2
	$4.187 \times 10^{+7}$	kcal/cm^2
	0.001	erg/cm^2
	14.60	ft-lbf/ft^2
	11,360	Btu/ft^2
	$2.89 \times 10^{+7}$	hp-h/ft^2
	$3.87 \times 10^{+7}$	kWh/ft^2
Flow rate, mass: (kg/s)/X	1.0×10^{-3}	g/s
	2.78×10^{-4}	kg/h
	0.4536	lbm/s
	7.56×10^{-3}	lbm/min
	1.26×10^{-4}	lbm/h
Flow rate, volume: $(m^{-3}/s)/X$	1.0×10^{-6}	cm^3/s
	0.02832	cfs
	1.639×10^{-5}	in.^{3}/s
	4.72×10^{-4}	cfm
	7.87×10^{-6}	cfh
	3.785×10^{-3}	gal/s
	6.308×10^{-5}	gpm
	1.051×10^{-6}	gph
Force: N/X	1.0×10^{-5}	dyn
	1	kg-m/s
	9.8067	kg(force)
	9.807×10^{-3}	g(force)
	0.1383	pdl
	4.448	lbf
	4,448	kip
	8,896	ton(force)
Heat-transfer coefficient: (W/m-K)/X	41,868	cal/s-cm^2-°C
	1.163	kcal/h-m^2-°C
	1.0×10^{-3}	erg/s-cm^2-°C
	5.679	Btu/h-ft^2-°F
	12.52	kcal/h-ft^2-°C
Henry's constant: (N/m2)/X	$1.01326 \times 10^{+5}$	atm
	133.3	mmHg

Appendix A SI Units Conversion Table

	6893	lbf/in²
	47.89	lbf/ft²
Length: m/X	0.01	cm
	1.0×10^{-6}	μm
	1.0×10^{-10}	Å
	0.3048	ft
	0.0254	in.
	0.9144	yd
	1609.3	mi/e
Mass: kg/X	1.0×10^{-3}	g
	0.4536	lbm
	6.48×10^{-5}	grain
	0.2835	oz(avdp)
	907.2	ton (U.S.)
	14.59	slug
Mass per area: (kg/m^{-2})/X	10	g/cm^{-2}
	4.883	lbm/ft^{-2}
	703.0	lbm/in².
	3.5×10^{-4}	ton/mi²
Power: W/X	4.187	cal/s
	4187	kcal/s
	1.0×10^{-7}	erg/s
	1.356	ft-lbf/s
	0.293	Btu/h
	1055	Btu/s
	745.8	hp
	0.04214	ft-pdl/s
	0.1130	in-lbf/s
	3517	ton refrigeration
	17.6	Btu/min
Power density: (W/m³)/X	$4.187 \times 10^{+6}$	cal/s-cm³
	$4.187 \times 10^{+9}$	kcal/s-cm³
	0.1	erg/s-cm³
	47.9	ft-lbf/s-ft³
	$3.73 \times 10^{+4}$	Btu/s-ft³
	10.36	Btu/h-ft³
	$3.53 \times 10^{+4}$	kW/ft³
	$2.63 \times 10^{+4}$	hp/ft³

Appendix A SI Units Conversion Table

Pressure, stress: (Pa)/X	0.1	dyn/cm^2
	1	N/m^2
	9.8067	$kg(f)/m^2$
	$1.0 \times 10^{+5}$	bar
	$1.0133 \times 10^{+5}$	std. atm
	1.489	pdl/ft^2
	47.88	lbf/ft^2
	6,894	lbf/in^2. (psi)
	$1.38 \times 10^{+7}$	$ton(f)/in^2$
	249.1	in H_2O
	2,989	ft H_2O
	133.3	torr, mmHg
	3,386	in Hg
Resistance: Ω/X	1	$kg\text{-}m^2/A^2\text{-}s^3$
	1	V/A
	1.0×10^{-9}	abohm
	$8.988 \times 10^{+11}$	statohm
Specific heat, gas constant: (J/kg-K)/X	1	m^2/s^2-K
	4187	cal/g-°C
	1.0×10^{-4}	erg/g-°C
	4187	Btu/lbm-°F
	5.38	ft-lbf/lbm-°F
Surface tension: (N/m)/X	1.0×10^{-3}	dyn/cm
	14.6	lbf/ft
	175.0	lbf/in.
Temperature: K/X (difference)	0.5555	°R
	0.5555	°F
	1.0	°C
Thermal conductivity: (W/m-K)/X	418.7	cal/s-cm-°C
	1.163	kcal/h-m-°C
	1.0×10^{-5}	erg/s-cm-°C
	1.731	Btu/h-ft-°F
	0.1442	Btu-in/h-ft^2-°F
	2.22×10^{-3}	ft-lbf/h-ft-°F
Time: s/X	60.0	min
	3,600.	h

Appendix A SI Units Conversion Table

	86,400	day
	$3.156 \times 10^{+7}$	year
Torque: N-m/X	1.0×10^{-7}	dyn-cm
	1.356	lbf-ft
	0.0421	pdl-ft
	2.989	kg(f)-ft
Velocity: (m/s)/X	0.01	cm/s
	2.78×10^{-4}	m/h
	0.278	km/h
	0.3048	ft/s
	5.08×10^{-3}	ft/min
	0.477	mi/h
Velocity, angular: (rad/s)/X	0.01667	rad/min
	2.78×10^{-4}	rad/h
	0.1047	rev/min
Viscosity, dynamic: (Kg/m-s)/X	1	$N-s/m^2$
	0.1	P
	0.001	cP
	2.78×10^{-4}	kg/m-h
	1.488	lbm/ft-s
	4.134×10^{-4}	lbm/ft-h
	47.91	$lbf-s/ft^2$
(g/cm-s)/X	1	P
Viscosity, kinematic: $(m^2/s)/X$	1.0×10^{-4}	St
	2.778×10^{-4}	m^2/h
	0.0929	ft^2/s
	2.581×10^{-5}	ft^2/h
$(cm^2/s)/X$	1	St
Volume: m^3/X	1.0×10^{-6}	cm^3
	1.0×10^{-3}	l
	1.0×10^{-18}	μm^3
	0.02832	ft^3
	1.639×10^{-5}	$in.^3$
	3.785×10^{-3}	gal (US)
Voltage, electrical potential: V/X	1.0	$kg-m^2/A-s^3$
	1	W/A

Appendix A SI Units Conversion Table

Using the Table: The quantity in braces {} is selected from the table.

Example 1: To calculate how many meters are in 10 ft, the table provides the conversion factor as 0.3048 m/ft. Hence multiply 10 ft × {0.3048 m/ft} = 3.048 m

Example 2: Convert Thermal Conductivity of 10 kcal/h-m-°C to SI units. Select the appropriate conversion factor for these units, 10 (kcal/h-m-°C) × {1.163(w/m-K)/(kcal/h-m-°C)} = 11.63 w/m-K

Appendix B

B-1. LANGELIER SATURATION INDEX

The Langelier saturation index (LSI) is probably the most widely used indicator of cooling water scale potential. This index indicates the driving force for scale formation and growth in terms of pH as a master variable. In order to calculate the LSI, it is necessary to know the alkalinity (mg L^{-1} as $CaCO_3$ or calcite), the calcium hardness (mg L^{-1} Ca^{2+} as $CaCO_3$), the total dissolved solids (mg L^{-1} TDS), the actual pH, and the temperature of the water (°C). If TDS is unknown, but conductivity is, one can estimate mg L^{-1} TDS using a conversion table (Table B-1). The LSI is defined as.

$$LSI = pH - pH_s \qquad (0.1)$$

where pH is the measured water pH
pH_s is the pH at saturation in calcite or calcium carbonate and is defined as

$$pH_s = (9.3 + A + B) - (C + D) \qquad (0.2)$$

where:

$$A = \frac{(Log_{10}[TDS] - 1)}{10} \qquad (0.3)$$
$$B = -13.12 \times Log_{10}(°C + 273) + 34.55 \qquad (0.4)$$
$$C = Log_{10}[Ca^{2+} \text{ as } CaCO_3] - 0.4 \qquad (0.5)$$
$$D = Log_{10}[\text{alkalinity as } CaCO_3] \qquad (0.6)$$

As for the SL reasoning described earlier, the LSI indicates three situations

- If LSI is negative: No potential to scale, the water will dissolve $CaCO_3$.
- If LSI is positive: Scale can form and $CaCO_3$ precipitation may occur.
- If LSI is close to zero: Borderline scale potential. Water quality or changes in temperature, or evaporation could change the index.

As an example, suppose the drinking water supplied to animals has the following analysis:

$$pH = 7.5$$

Corrosion Inspection and Monitoring, by Pierre R. Roberge
Copyright © 2007 John Wiley & Sons, Inc.

Appendix B

Table B-1. Conversion Table between the Conductivity of Natural Water and the TDS It Contains

Conductivity (micro-mho/cm)	TDS (mg/L as $CaCO_3$)
1	0.42
10.6	4.2
21.2	8.5
42.4	17.0
63.7	25.5
84.8	34.0
106.0	42.5
127.3	51.0
148.5	59.5
169.6	68.0
190.8	76.5
212.0	85.0
410.0	170.0
610.0	255.0
812.0	340.0
1008.0	425.0

$$TDS = 320 \text{ mg/L}$$
$$Calcium = 150 \text{ mg L}^{-1} \text{(or ppm) as } CaCO_3$$
$$Alkalinity = 34 \text{ mg L}^{-1} \text{(or ppm) as } CaCO_3$$

The LSI index is calculated at two temperatures, that is, 25°C (room temperature) and 82°C (cage wash cycle). The colder incoming water will warm to room temperature in the manifolds. Residual water in the rack manifold can be heated to 82°C when the rack is in the cage washer.

LSI Formula:

$LSI = pH - pH_s$
$pH_s = (9.3 + A + B) - (C + D)$, where $A = (Log_{10}[TDS] - 1)/10 = 0.15$
$B = -13.12 \times Log_{10}(°C + 273) + 34.55 = 2.09$ at 25°C and 1.09 at 82°C
$C = Log_{10} [Ca^{2+} \text{as } CaCO_3] - 0.4 = 1.78$
$D = Log_{10} [\text{alkalinity as CaCO3}] = 1.53$

Calculation at 25°C:

$$pH_s = (9.3 + 0.15 + 2.09) - (1.78 + 1.53) = 8.2$$
$$LSI = 7.5 - 8.2 = -0.7$$

Hence, no tendency to scale.
Calculation at 82°C:

$$pH_s = (9.3 + 0.15 + 1.09) - (1.78 + 1.53) = 7.2$$
$$LSI = 7.5 - 7.2 = +0.3$$

Hence, slight tendency to scale.

B-2. RYZNAR STABILITY INDEX

The Ryznar stability index (RSI) uses a correlation established between an empirical database of scale thickness observed in municipal water systems and associated water chemistry data. Like the LSI, the RSI has its basis in the concept of saturation level. The Ryznar index takes the form:

$$RSI = 2(pH_s) - pH$$

The empirical correlation of the Ryznar stability index can be summarized as follows:

- RSI < 6 the scale tendency increases as the index decreases.
- RSI > 7 the calcium carbonate formation probably does not lead to a protective corrosion inhibitor film.
- RSI > 8 mild steel corrosion becomes an increasing problem.

B-3. PUCKORIUS SCALING INDEX

The Puckorius scaling index (PSI) is based on the buffering capacity of the water, and the maximum quantity of precipitate that can form in bringing water to equilibrium. Water high in calcium, but low in alkalinity and buffering capacity, can have a high calcite saturation level. The high calcium level increases the ion activity product. Such water might have a high tendency to form scale due to the driving force, but scale formed might be of such a small quantity as to be unobservable. The water has the driving force, but not the capacity and ability to maintain pH as precipitate matter forms.

The PSI index is calculated in a manner similar to the Ryznar stability index. Puckorius uses an equilibrium pH rather than the actual system pH to account for the buffering effects:

$$PSI = 2(pH_s) - pH_{eq}$$

where pH_s is still the pH at saturation in calcite or calcium carbonate

$$pH_{eq} = 1.465 \times \log_{10} [\text{Alkalinity}] + 4.54$$
$$[\text{Alkalinity}] = [HCO_3^-] + 2[CO_3^{2-}] + [OH^-]$$

B-4. LARSON–SKOLD INDEX

The Larson–Skold index is based upon evaluation of *in situ* corrosion of mild steel lines transporting Great Lakes waters. The index is the ratio of equivalents per million (epm) of sulfate (SO_4^{2-}) and chloride (Cl^-) to the epm of alkalinity in the form bicarbonate plus carbonate:

$$\text{Larson–Skold index} = (\text{epm } Cl^- + \text{epm } SO_4^{2-})/(\text{epm } HCO_3^- + \text{epm } CO_3^{2-}) \tag{0.7}$$

Extrapolation to other waters than the Great Lakes, such as those of low alkalinity or extreme alkalinity, goes beyond the range of the original data. The index has proven to be a useful tool in predicting the aggressiveness of once through cooling waters. The Larson–Skold index might be interpreted by the following guidelines:

- Index < 0.8 chlorides and sulfate probably will not interfere with natural film formation.
- 0.8 < index < 1.2 chlorides and sulfates may interfere with natural film formation. Higher than desired corrosion rates might be anticipated.
- Index > 1.2 the tendency toward high corrosion rates of a local type should be expected as the index increases.

B-5. ODDO–TOMSON INDEX

The Oddo–Tomson index accounts for the impact of pressure and partial pressure of carbon dioxide on the pH of water, and on the solubility of calcium carbonate. This empirical model also incorporates corrections for the presence of two or three phases (water, gas, and oil). Interpretation of the index is by the same scale as for the LSI and Stiff–Davis indices (1).

REFERENCES

1. Oddo JE and Tomson MB. Scale Control, Prediction and Treatment or How Companies Evaluate a Scaling Problem and What They Do Wrong. [Corrosion 92, paper 34]. 1992. Houston, TX, NACE International.

Index

A
Acoustic emission, (AE), 266–267
Age-reliability characteristics, 42
Aircraft maintenance, 161–166
   corrosion assessment, 163
   corrosion definition, 161–162
   environmental deterioration analysis, 165–166
   hot spots, 92–93
   maintenance schedule, 162–163
   maintenance steering group, 163–166
ALARP, 111
Alkalinity monitoring, 277
Aluminum alloys
   corrosion products, 16
   stress corrosion cracking, 21
   three dimensional (3D) grain structure, 22
A-scan, 333–335
Atmospheric corrosion
   airborne particles, 304–307
   chloride, 304–307
   CLIMATs, 308–311
   corrosion probe, 210–212
   coupons, 308–311
   humidity sensor, 304
   Kennedy Space Center beach corrosion test site, 308–310
   model, 303
   monitoring, 301–311
   pollutants, 303–304
   relative humidity, 302–303
   time of wetness (TOW), 302–303
   types of atmospheres, 301–302
   wet candle method, 305–306

B
BioGEORGE MIC probe, 293–295

Bhopal accident, 35
Boiler water reactor (BWR), 157–158
Borescopes, 337–339
B-scan, 335–336

C
Carlsbad pipeline explosion, 35
Cathodic protection
   close interval potential survey (CIPS), 295–297
   corrosion coupons, 300–301
   direct current voltage gradient (DCVG) survey, 299–300
   monitoring, 294–301
   Pearson survey, 297–299
   water mains, 169–170
Chemical analysis of process fluids, 279–281
Circuit board corrosion, 31
CLIMATs, 308–311
Close interval potential survey (CIPS), 135, 295–297
Coating health monitoring system, 248–250
Communication technologies, 90–91
Computed tomography, 354
Computer technologies, 90–91
Condition assessment of water mains, 174–180
Conductivity, 272–273
Conductivity cell, 273
Connector corrosion, 30
Consequence of failure (COF), 107
   assessment, 109
Copper nickel alloys
   pitting, 18
   biofouling, 18

Corrosion Inspection and Monitoring, by Pierre R. Roberge
Copyright © 2007 John Wiley & Sons, Inc.

378   Index

Corrosion
 costs, 22–26
 definition, 2
 factors, 3
 failures, examples of, 34–42
 flow effects, 5–7, 128
 forms, 45–77
 importance, 1
 local cells, 4
 material factor, 15–18
 microbes, 5–6, 129
 strategic impact, 22
 stress effect, 12–16
 temperature effect, 6–12, 128
 impurities, 128
 stray currents, 129
Corrosion fatigue, 72–75
Corrosion management, 82
 water utilities, 168–174
Corrosion monitoring
 availability, 217
 considerations, 194
 data communication and analysis, 216
 definition, 191
 hardware location, 211–214
 inspection and, 105–106
 locations, 197–201
 objectives, 197
 oil and gas production, 199–201
 points, 200
 properties, 217
 role, 192–194
 system reliability, 216–217
 techniques, 217–281
Corrosion monitoring techniques, 217–281
 acoustic emission, (AE), 266–267
 direct intrusive, 218, 220–262
 direct non-intrusive, 218, 262–267
 electrical resistance (ER), 229–232
 electrochemical impedance spectroscopy (EIS), 246–250
 electrochemical methods, 232–262
 electrochemical noise, 251–262
 field signature method (FSM), 263–267
 galvanodynamic polarization, 241–246
 gamma radiography, 262–263
 harmonic distortion analysis, 249–251
 indirect on-line, 267–277

 indirect off-line, 277–281
 linear polarization resistance (LPR), 236–240
 mass loss coupons, 220–229
 off-line, 219
 on-line, 219
 potentiodynamic polarization, 241–246
 thin layer activation, 262–263
 zero resistance ammetry (ZRA), 240–241
Corrosion potential monitoring, 270
Corrosion prediction, 17–18
Corrosion probes
 applications, 203–209
 atmospheric corrosion, 210–212
 coupled multi-electrode array system (CMAS), 207–209
 design, 202–212
 electrical resistance (ER) 230,231
 electrochemical noise (EN), 203–204
 flush mounted, 199, 202–203
 heat adjacent zone (HAZ), 205–206
 inductive resistance, 232–234
 linear polarization resistance (LPR), 238–239
 microbial influenced corrosion (MIC), 293–294
 oil and gas industry, 205
 orientation, 201
 protruding, 203
 response time, 214–216
 retrievable, 221–224
 ring pair electrodes, 205–207
 selection, 202–212
 sensitivity, 214–216
 stress corrosion cracking, 203–205
 thin film, 212
Corrosion rates conversion, 235
Corrosion sensors
 electrochemical noise (EN), 253–255
 flush mounted, 199
 microbial influenced corrosion (MIC), 293–294
 thin film, 212
 wireless, 196
Coupled multi-electrode array system (CMAS), 207–209

Crevice corrosion, 51–55
   coupons, 225–227
   three stages of, 54–55
C-scan, 336

D
Davis-Besse nuclear reactor incident, 38
Dealloying, 64–65
Defect
   definition, 28–29
   response variance, 328–330
Dendrites, 31
Dewpoint, 276
Dezincification, 65
Direct current voltage gradient (DCVG)
      survey, 299–300
Dissolved oxygen, 10–12, 273–274
Dissolved solids, 278
D-sight inspection, 343–344

E
Eddy current inspection, 320, 357–361
Edge of light, 344–346
El Al Boeing 747 crash, 37
Electrical resistance (ER), 229–232
   element selection, 231
   probe, 230–231
Electrochemical impedance spectroscopy
      (EIS), 246–250
   constant phase element, 246
   field monitoring, 246–250
   polarization resistance, 246–248
   RC model circuit, 247
   three point method, 247–248
   two point method, 246–247
   wireless system, 248–250
Electrochemical noise (EN), 251–262
   analysis, 252–258, 260–261
   electrode arrangement, 259–261
   noise records, 254–255, 257–258
   pitting activity, 255–258
   probe, 203–204
   probe design, 258–261
Electromagnetic inspection, 355–361
Electrochemical monitoring, 232–262
   categories, 236
   corrosion rates conversion, 235
   fuel cell probe, 269–270
   hydrogen patch probe, 269
   interface properties, 234

microbial influenced corrosion (MIC),
      293–294
Environmental cracking, 65–68
Environmental deterioration analysis
      (EDA), 165–166
Erika tanker, 40
Event-tree analysis, 119–122
External corrosion damage assessment
      (ECDA), 133–137
   direct examination, 136
   indirect inspection, 135–136
   post-assessment, 137
   pre-assessment, 134–135
   process, 133–134
Extreme value statistics, 50

F
F-16 fighter aircraft corrosion, 30
Failure
   analysis, 122–129
   consequences of, 33–34
   corrosion failure statistics, 46
   corrosion failures, 34–42
   definition, 28, 31–33
   patterns of, 42–43
   probability of, 44
   root causes, 123–125
   types of, 32–33
Failure analysis, 122–129
   conducting a failure analysis, 125–129
   decision tree, 123
   failure site, 125
   historical information, 126
   information, 127–129
   operating conditions, 126
   planning, 125
   plant geometry, 124
   root causes, 123–125
   sample evaluation, 127
   sampling, 126
   types of corrosion, 127
Fault-tree analysis, 116–120
Fiberscopes, 339–340
Field signature method (FSM), 263–267
Flixborough chemical plant explosion, 41
Flow-assisted corrosion (FAC), 5–8, 157
Flow influenced corrosion, 57–60
   cavitation, 60, 62
   differential flow cell, 239
   erosion-corrosion, 60, 62

Flow influenced corrosion (*continued*)
  mass transport control, 59
  phase transport control, 60
Flow regime, 275
Fluid catalytic cracking unit (FCCU), 121–122
FMEA and FMECA, 114
Fouling, 276–277
Fretting corrosion, 61,63
Fuel cell hydrogen probe, 269–270

G

Galvanic corrosion, 53–59
  coupons, 225
  effect on rivets, 58–59
  example of, 60
  surface area ratio, 55
Galvanic series
  in seawater, 56
  in soils, 57
Galvanodynamic polarization, 241–246
Gamma radiography, 262–263
Gas analysis, 278
Guadalajara sewer explosion, 36

H

Harmonic distortion analysis (HAD), 249–250
Hanford site, 204–205
HAZOP, 112–114
Humidity sensor, 304
Hydrogen blistering, 76–77
Hydrogen embrittlement, 75–77
Hydrogen induced cracking (HIC), 75–77
Hydrogen monitoring, 267–270
Hydrogen probes, 268–269
Hydrostatic testing, 139–140

I

Inductive resistance, 232–234
In-line inspection (ILI), 140–143
  crack detection, 142
  geometry tools, 142–143
  magnetic flux leakage, 140–141
  metal loss tools, 140–142
  tool validation, 330–333
  ultrasonic wall measurements, 141–142
Inspection strategies, 91–122
  columns, 93–95
  condition-based design analysis, 95–98

corrosion monitoring, 105–106
heat exchangers, 94
hot spots, 92–99
key performance indicators, 99–105
locations for analysis, 97–98
optimization, 110
piping, 95–96
reactors, 92–94
reinforced concrete structures, 182
risk-based inspection (RBI), 106–111
water mains, 174–180
what to inspect, 91–92
when to inspect, 99–105
Isocorrosion charts, 8–9
Intergranular corrosion, 63–64
Internal corrosion damage assessment (ICDA), 137–139
  detailed examination, 139
  indirect inspection, 138–139
  post-assessment, 139
  pre-assessment, 138
  process, 137

K

Key performance indicators (KPIs), 99–105
  asset performance metrics, 102
  completed monitoring, 101
  corrosion inhibition level, 100–101
  cost of corrosion, 99–100
  selection, 102–105
  tactical perspectives, 102–105
Kennedy Space Center beach corrosion test site, 308–310

L

Langelier saturation index, 373–375
Larsen-Skold saturation index, 376
Life cycle asset management, 86–90
  condition assessment, 88–89
  life cycle costing, 86–88
  prioritization, 89–90
Linear polarization resistance (LPR), 236–240
  differential flow cell, 239
  polarization plot, 237
  sensor element, 238–239
  Stern Geary equation, 236

Tafel plot, 238
Tafel slopes, 237
Liquid penetrant inspection, 320, 340–341

M

MV Erika tanker, 40
MV Kirki, 40
Magnetic particle inspection, 320, 341–342
Magnetic flux leakage, 355–356
Magneto optic imaging, 360–361
Maintenance
    condition based, 84
    corrective, 83
    cost of poor maintenance, 81–82
    predictive, 84
    preventive, 84
    reliability centered, 85
    strategies, 83–86
Mass loss coupons, 220–229
    cleaning, 228–229
    crevice corrosion, 225–227
    evaluation, 228–229
    galvanic corrosion, 225
    heat transfer, 226
    length of exposure, 222–224
    limitations, 222–224
    retrievable, 221–224
    sensitized metal, 228
    uniform corrosion, 224–225
    welded, 226–227
Metal ion analysis, 277–278
Mesa corrosion, 145
Microbial influenced corrosion (MIC), 5–6
    activity assay, 287–291
    analysis, 278
    biofilm formation, 284
    biological assessment, 286–287
    cell components, 290
    coupon testing, 291–292
    deposition accumulation monitor, 292–293
    detection probe, 293–295
    direct inspection, 286–288
    electrochemical methods, 293–294
    enzyme-based assay, 289–290
    fatty acid profile, 290
    field testing, 212–213
    growth assay, 289
    metabolites, 290
    monitoring, 281–295
    nucleic acid method, 290
    planktonic organisms, 283
    problems, 282
    sampling, 285
    sessile organisms, 283
    whole cell assay, 289
Mihama nuclear power plant accident, 39
Moiré interferometry, 345, 347
MSG-3, 164–166

N

Neutron radiography, 354–355
Non-destructive evaluation, 317–363
    A-scan, 333–335
    B-scan, 335–336
    cost rating, 319
    C-scan, 336
    data analysis, 321–336
    data representation, 321–336
    defect response, 328–330
    eddy current inspection, 320, 357–361
    electromagnetic inspection, 355–361
    limitations, 322–323
    liquid penetrant inspection, 320, 340–341
    magnetic particle inspection, 320, 341–342
    process steps, 320
    purpose, 318–321
    radiographic inspection, 320, 353–355
    requirement rating, 319
    signal response, 324–328
    thermographic inspection, 320, 361–363
    ultrasonic inspection, 320, 347–352
    validation of inspection tools, 330–333
    variance, 318
    visual inspection, 336–347
Nuclear power industry
    activation of corrosion products, 150–152
    fuel cladding, 156
    PWR steam generator corrosion, 152

O

Oddo-Tomson saturation index, 376
Offshore oil and gas industry, 159–160
    pipelines, 143–145

382 Index

Offshore oil and gas industry (*continued*)
　RIMAP, 155–160
　risers, 143–145, 146–147
Oil and gas gathering lines, 199–201
On-line water chemistry, 271–275
Oxygen sensor, 274

P

Pack rust, 14
Pearson survey, 297–299
pH monitoring, 271–272
Pillowing, 14–16
Pipelines
　accidents, 132
　external corrosion damage assessment
　　(ECDA), 133–137
　fault-tree analysis, 117–120
　hazardous liquids, 132
　incidents, 131–133
　internal corrosion damage assessment
　　(ICDA), 137–139
　offshore, 143–145
　transmission, 130–143
Pitting corrosion, 48–52
　electrochemical noise (EN), 257–258
　field signature method (FSM), 264–266
POD (see Probability of detection)
Potentiodynamic polarization, 241–246
　polarization diagram, 242
　polarization plot, 241
　polarization scan, 244–246
Power industry, 150–158
　corrosion costs, 151–152
　boiler tube corrosion, 152, 154
　steam chemistry, 153
　raw water piping, 157
　turbine corrosion, 156
　electric generators, 156–157
　flow accelerated corrosion, 157
　heat exchangers, 154–156
　RIMAP, 160–161
　field monitoring, 155
　BWR piping, 152–158
Pressure monitoring, 275–276
Pressurized water reactor (PWR)
　location for analysis, 95–98
　measurement points, 154
　steam generator tube corrosion, 152
Probability of failure (POF), 44, 107
　assessment, 108

Probability of detection (POD), 44
　crack detection example, 330–333
　curve, 329
　definition, 327
Probability of false calls (POFC), 328
Probability of identification (POI), 331
Process fluids analysis, 279–281
Process industry, 145–150
　asset loss risk, 148
　corrosion monitoring, 198–199
　leaks, 149
　piping systems, 148–149
　RIMAP, 158–159
Process variables monitoring, 275–277
Puckorius saturation index, 373–375
Pulse eddy current, 359–360

Q

Quantitative risk assessment, 111

R

Radiographic inspection, 320, 353–355
Redox potential monitoring, 274–275
Reinforced concrete, 180–185
　chloride content, 184
　conventional methods, 182
　electrochemical measurements, 184
　evaluation survey, 183
　inspection methods, 177
　new methods, 182
　permeability tests, 185
　petrographic examination, 184–185
　SHRP guide, 183
Remote field eddy current, 358–359
Residual oxidant, 278
Residual inhibitor, 278–279
RIMAP, 158–161
　process industry, 158–159
　power industry, 160–161
Risers, offshore pipelines, 143–147
Risk assessment, 111–122
　event-tree analysis, 119–122
　fault-tree analysis, 116–120
　FMEA and FMECA, 114
　HAZOP, 112–114
　quantitative risk assessment, 111
　risk matrix, 115–116
Risk-based inspection (RBI), 106–111
　application, 109
　changes in materials, 110

key process parameters, 111
optimization of inspection, 110
optimization of monitoring, 110
Risk matrix, 115–116
Ryznar saturation index, 375

S
Ship corrosion, 40
SI units, 365–371
Soil corrosivity, 179–181
Splash zone, 145
Stainless steels
iron-chromium phase diagram, 21
sensitization, 18–20
Steam generator, 97–98
Stern Geary equation, 236
Stress corrosion cracking (SCC), 66–72
aluminum, 22
BWR piping, 157–158
columns, 93–95
crack detection, 330–333
fault-tree analysis, 117
heat exchangers, 94
intergranular SCC, 67
pipeline, 117
piping, 95–96
reactors, 92–94
reinforced concrete, 13
stress definition, 12
transgranular SCC, 68
turbines, 156
corrosion probe, 203–205
coupons, 225–227
Structure significant item (SSI), 164
Swimming pool roof collapse, 41

T
Tafel plot, 238
Tafel slopes, 237
Tangential radiography, 354–355
Temperature monitoring, 276

Thermographic inspection, 320, 361–363
Time of wetness (TOW), 302–303
Total acid number (TAN), 280
Thin layer activation, 262–263
Tubercles, 168

U
Uniform corrosion, 48
coupons, 224–225
Ultrasonic inspection, 320, 347–352
defect sizing, 351
thickness measurements, 348–358
time-of-flight diffraction, 351–352

V
Video imaging system, 340
Visual inspection, 336–347

W
Water utilities, 165–181
acoustic leak detection, 178
condition assessment techniques, 174–180
corrosion control implementation, 172
corrosion impact, 167
corrosion management, 168–174
customer perception, 167
deterioration factors, 171
diagnostic techniques, 175–176
environmental concerns, 168
health and regulations, 167
piping deterioration, 167–168
prestressed concrete pipes, 177
soil corrosivity, 179–181
water audits, 176–178
Wet candle method, 305–306
Wireless corrosion sensor, 196
electrochemical impedance spectroscopy (EIS), 248–250

Z
Zero resistance ammetry (ZRA), 240–241